Means Residential **Square Foot** Costs

Contractor's Pricing Guide 2006

Senior Editor
Robert W. Mewis, CCC

Contributing Editors
Christopher Babbitt
Ted Baker
Barbara Balboni
Robert A. Bastoni
John H. Chiang, PE
John Kane
Robert J. Kuchta
Robert C. McNichols
Melville J. Mossman, PE
John J. Moylan
Jeannene D. Murphy
Stephen C. Plotner
Eugene R. Spencer
Marshall J. Stetson
Phillip R. Waier, PE

Senior Engineering Operations Manager
John H. Ferguson, PE

Senior Vice President & General Manager
John Ware

Vice President of Sales
John M. Shea

Production Manager
Michael Kokernak

Technical Support
Wayne D. Anderson
Thomas J. Dion
Jonathan Forgit
Mary Lou Geary
Gary L. Hoitt
Genevieve Medeiros
Paula Reale-Camelio
Kathryn S. Rodriguez
Sheryl A. Rose
Laurie Thom

Book & Cover Design
Norman R. Forgit

RSMeans

Means Residential Square Foot Costs

Contractor's Pricing Guide 2006

- Residential Cost Models for All Standard Building Classes
- Costs for Modifications & Additions
- Costs for Hundreds of Residential Building Systems & Components
- Cost Adjustment Factors for Your Location
- Illustrations

$39.95 per copy (in United States).
Price subject to change without prior notice.

*RS**Means***
Construction Publishers & Consultants
63 Smiths Lane
Kingston, MA 02364-0800
(781) 422-5000

Printed in the United States of America.

10 9 8 7 6 5 4 3 2 1

ISSN 1074-049X

ISBN 0-87629-809-9

Foreword

RSMeans is a product line of Reed Construction Data, Inc., a leading provider of construction information, products, and services in North America and globally. Reed Construction Data's project information products include more than 100 regional editions, national construction data, sales leads, and local plan rooms in major business centers. Reed Construction Data's PlansDirect provides surveys, plans, and specifications. The First Source suite of products consists of *First Source for Products,* SPEC-DATA™, MANU-SPEC™, CADBlocks, Manufacturer Catalogs, and First Source Exchange (www.firstsourceexchange.com) for the selection of nationally available building products. Reed Construction Data also publishes ProFile, a database of more than 20,000 U.S. architectural firms. RSMeans provides construction cost data, training, and consulting services in print, CD-ROM, and online. Reed Construction Data, headquartered in Atlanta, is owned by Reed Business Information (www.reedconstructiondata.com), a leading provider of critical information and marketing solutions to business professionals in the media, manufacturing, electronics, construction, and retail industries. Its market-leading properties include more than 135 business-to-business publications and over 125 Webzines and Web portals, as well as online services, custom publishing, directories, research, and direct-marketing lists. Reed Business Information is a member of the Reed Elsevier plc group (NYSE: RUK and ENL)—a world-leading publisher and information provider operating in the science and medical, legal, education, and business-to-business industry sectors.

Our Mission

Since 1942, RSMeans has been actively engaged in construction cost publishing and consulting throughout North America.

Today, over 60 years after RSMeans began, our primary objective remains the same: to provide you, the construction and facilities professional, with the most current and comprehensive construction cost data possible.

Whether you are a contractor, an owner, an architect, an engineer, a facilities manager, or anyone else who needs a fast and reliable construction cost estimate, you'll find this publication to be a highly useful and necessary tool.

Today, with the constant flow of new construction methods and materials, it's difficult to find the time to look at and evaluate all the different construction cost possibilities. In addition, because labor and material costs keep changing, last year's cost information is not a reliable basis for today's estimate or budget.

That's why so many construction professionals turn to RSMeans. We keep track of the costs for you, along with a wide range of other key information, from city cost indexes . . . to productivity rates . . . to crew composition . . . to contractor's overhead and profit rates.

RSMeans performs these functions by collecting data from all facets of the industry and organizing it in a format that is instantly accessible to you. From the preliminary budget to the detailed unit price estimate, you'll find the data in this book useful for all phases of construction cost determination.

The Staff, the Organization, and Our Services

When you purchase one of RSMeans' publications, you are, in effect, hiring the services of a full-time staff of construction and engineering professionals.

Our thoroughly experienced and highly qualified staff works daily at collecting, analyzing, and disseminating comprehensive cost information for your needs. These staff members have years of practical construction experience and engineering training prior to joining the firm. As a result, you can count on them not only for the cost figures, but also for additional background reference information that will help you create a realistic estimate.

The RSMeans organization is always prepared to help you solve construction problems through its five major divisions: Construction and Cost Data Publishing, Electronic Products and Services, Consulting Services, Insurance Services, and Educational Services.

Besides a full array of construction cost estimating books, RSMeans also publishes a number of other reference works for the construction industry. Subjects include construction estimating and project and business management; special topics such as HVAC, roofing, plumbing, and hazardous waste remediation; and a library of facility management references.

In addition, you can access all of our construction cost data through your computer with *Means CostWorks 2006* CD-ROM, an electronic tool that offers over 50,000 lines of RSMeans detailed construction cost data, along with assembly and whole building cost data. You can also access RSMeans cost information from our Web site at www.rsmeans.com.

What's more, you can increase your knowledge and improve your construction estimating and management performance with an RSMeans Construction Seminar or In-House Training Program. These two-day seminar programs offer unparalleled opportunities for everyone in your organization to get updated on a wide variety of construction-related issues.

RSMeans also is a worldwide provider of construction cost management and analysis services for commercial and government owners and of claims and valuation services for insurers.

In short, RSMeans can provide you with the tools and expertise for constructing accurate and dependable construction estimates and budgets in a variety of ways.

Robert Snow Means Established a Tradition of Quality That Continues Today

Robert Snow Means spent years building RSMeans, making certain he always delivered a quality product.

Today, at RSMeans, we do more than talk about the quality of our data and the usefulness of our books. We stand behind all of our data, from historical cost indexes to construction materials and techniques to current costs.

If you have any questions about our products or services, please call us toll-free at 1-800-334-3509. Our customer service representatives will be happy to assist you. You can also visit our Web site at www.rsmeans.com.

Table of Contents

How the Book is Built: An Overview

A Powerful Construction Tool

You have in your hands one of the most powerful construction tools available today. A successful project is built on the foundation of an accurate and dependable estimate. This book will enable you to construct just such an estimate.

For the casual user the book is designed to be:

- quickly and easily understood so you can get right to your estimate.
- filled with valuable information so you can understand the necessary factors that go into the cost estimate.

For the regular user, the book is designed to be:

- a handy desk reference that can be quickly referred to for key costs.
- a comprehensive, fully reliable source of current construction costs so you'll be prepared to estimate any project.
- a source book for preliminary project cost, product selections, and alternate materials and methods.

To meet all of these requirements we have organized the book into the following clearly defined sections.

Square Foot Cost Section

This section lists Square Foot costs for typical residential construction projects. The organizational format used divides the projects into basic building classes. These classes are defined at the beginning of the section. The individual projects are further divided into ten common components of construction. A Table of Contents, an explanation of Square Foot prices, and an outline of a typical page layout are located at the beginning of the section.

Assemblies Cost Section

This section uses an Assemblies (sometimes referred to as systems) format grouping all the functional elements of a building into nine construction divisions.

At the top of each Assemblies cost table is an illustration, a brief description, and the design criteria used to develop the cost. Each of the components and its contributing cost to the system is shown.

Material: These cost figures include a standard 10% markup for profit. They are national average material costs as of January of the current year and include delivery to the job site.

Installation: The installation costs include labor and equipment, plus a markup for the installing contractor's overhead and profit.

For a complete breakdown and explanation of a typical Assemblies page, see "How to Use the Assemblies Section at the beginning of the Assemblies Section.

Location Factors: You can adjust total project costs to over 900 locations throughout the U.S. and Canada by using the data in this section.

Abbreviations: A listing of the abbreviations used throughout this book, along with the terms they represent, is included in this section.

Index

A comprehensive listing of all terms and subjects in this book to help you find what you need quickly.

The Scope of This Book

This book is designed to be as comprehensive and as easy to use as possible. To that end we have made certain assumptions and limited its scope in three key ways:

1. We have established material prices based on a national average.
2. We have computed labor costs based on a seven major-region average of residential wage rates.
3. We have targeted the data for projects of a certain size range.

Project Size

This book is intended for use by those involved primarily in Residential construction costing less than $850,000. This includes the construction of homes, row houses, townhouses, condominiums, and apartments.

With reasonable exercise of judgment the figures can be used for any building work. For other types of projects, such as repair and remodeling or commercial buildings, consult the appropriate RSMeans publication for more information.

How to Use the Book: The Details

What's Behind the Numbers? The Development of Cost Data

The staff at RSMeans continuously monitors developments in the construction industry in order to ensure reliable, thorough, and up-to-date cost information.

While **overall** construction costs may vary relative to general economic conditions, price fluctuations within the industry are dependent upon many factors. Individual price variations may, in fact, be opposite to overall economic trends. Therefore, costs are continually monitored and complete updates are published yearly. Also, new items are frequently added in response to changes in materials and methods.

Costs – $ (U.S.)

All costs represent U.S. national averages and are given in U.S. dollars. The RSMeans Location Factors can be used to adjust costs to a particular location. The Location Factors for Canada can be used to adjust U.S. national averages to local costs in Canadian dollars. No exchange rate conversion is necessary.

Material Costs

The RSMeans staff contacts manufacturers, dealers, distributors, and contractors all across the U.S. and Canada to determine national average material costs. If you have access to current material costs for your specific location, you may wish to make adjustments to reflect differences from the national average. Included within material costs are fasteners for a normal installation. RSMeans engineers use manufacturers' recommendations, written specifications and/or standard construction practice for size and spacing of fasteners. Adjustments to material costs may be required for your specific application or location. Material costs do not include sales tax.

Labor Costs

Labor costs are based on the average of residential wages from across the U.S. for the current year. Rates, along with overhead and profit markups, are listed on the inside back cover of this book.

- If wage rates in your area vary from those used in this book, or if rate increases are expected within a given year, labor costs should be adjusted accordingly.

Labor costs reflect productivity based on actual working conditions. These figures include time spent during a normal workday on tasks other than actual installation, such as material receiving and handling, mobilization at site, site movement, breaks, and cleanup.

Productivity data is developed over an extended period so as not to be influenced by abnormal variations and reflects a typical average.

Equipment Costs

Equipment costs include not only rental, but also operating costs for equipment under normal use. Equipment and rental rates are obtained from industry sources throughout North America—contractors, suppliers, dealers, manufacturers, and distributers.

Factors Affecting Costs

Costs can vary depending upon a number of variables. Here's how we have handled the main factors affecting costs.

Quality—The prices for materials and the workmanship upon which productivity is based represent sound construction work. They are also in line with U.S. government specifications.

Overtime—We have made no allowance for overtime. If you anticipate premium time or work beyond normal working hours, be sure to make an appropriate adjustment to your labor costs.

Productivity—The productivity, daily output, and labor-hour figures for each line item are based on working an eight-hour day in daylight hours in moderate temperatures. For work that extends beyond normal work hours or is performed under adverse conditions, productivity may decrease.

Size of Project—The size, scope of work, and type of construction project will have a significant impact on cost. Economies of scale can reduce costs for large projects. Unit costs can often run higher for small projects. Costs in this book are intended for the size and type of project as previously described in "How the Book Is Built: An Overview." Costs for projects of a significantly different size or type should be adjusted accordingly.

Location—Material prices in this book are for metropolitan areas. However, in dense urban areas, traffic and site storage limitations may increase costs. Beyond a 20-mile radius of large cities, extra trucking or transportation charges may also increase the material costs slightly. On the other hand, lower wage rates may be in effect. Be sure to consider both of these factors when preparing an estimate, particularly if the job site is located in a central city or remote rural location.

In addition, highly specialized subcontract items may require travel and per-diem expenses for mechanics.

Other Factors–

- season of year
- contractor management
- weather conditions
- local union restrictions
- building code requirements
- availability of:
    - adequate energy
    - skilled labor
    - building materials
- owner's special requirements/ restrictions
- safety requirements
- environmental considerations

General Conditions—The "Square Foot Cost" and "Assemblies" sections of this book use costs that include the installing contractor's overhead and profit (O&P). An allowance covering the general contractor's markup must be added to these figures. The general contractor can include this price in the bid with a normal markup ranging from 5% to 15%. The markup depends on economic conditions plus the supervision and troubleshooting expected by the general contractor. For purposes of this book, it is best for a general contractor to add an allowance of 10% to the figures in the Assemblies and Square Foot Cost sections.

Overhead & Profit—For systems costs and square foot costs, simply add 10% to the estimate for general contractor's profit.

Unpredictable Factors—General business conditions influence "in-place" costs of all items. Substitute materials and construction methods may have to be employed. These may affect the installed cost and/or life cycle costs. Such factors may be difficult to evaluate and cannot necessarily be predicted on the basis of the job's location in a particular section of the country. Thus, where these factors apply, you may find significant but unavoidable cost variations for which you will have to apply a measure of judgment to your estimate.

Rounding of Costs

In general, all unit prices in excess of $5.00 have been rounded to make them easier to use and still maintain adequate precision of the results. The rounding rules we have chosen are in the following table.

Prices from . . .	Rounded to the nearest . . .
$.01 to $5.00	$.01
$5.01 to $20.00	$.05
$20.01 to $100.00	$.50
$100.01 to $300.00	$1.00
$300.01 to $1,000.00	$5.00
$1,000.01 to $10,000.00	$25.00
$10,000.01 to $50,000.00	$100.00
$50,000.01 and above	$500.00

Final Checklist

Estimating can be a straightforward process provided you remember the basics. Here's a checklist of some of the steps you should remember to complete before finalizing your estimate.

Did you remember to . . .

- factor in the Location Factor for your locale?
- take into consideration which items have been marked up and by how much?
- mark up the entire estimate sufficiently for your purposes?
- include all components of your project in the final estimate?
- double check your figures for accuracy?
- call RSMeans if you have any questions about your estimate or the data you've found in our publications?

Remember, RSMeans stands behind its publications. If you have any questions about your estimate . . . about the costs you've used from our books . . . or even about the technical aspects of the job that may affect your estimate, feel free to call the RSMeans editors at 1-800-334-3509.

Square Foot Cost Section

Table of Contents

How to Use the Square Foot Cost Pages

Introduction: This section contains costs per square foot for four classes of construction in seven building types. Costs are listed for various exterior wall materials which are typical of the class and building type. There are cost tables for wings and ells with modification tables to adjust the base cost of each class of building. Non-standard items can easily be added to the standard structures.

Accompanying each building type in each class is a list of components used in a typical residence. The components are divided into ten primary estimating divisions. The divisions correspond with the "Assemblies" section of this manual.

Cost estimating for a residence is a three-step process:

(1) Identification
(2) Listing dimensions
(3) Calculations

Guidelines and a sample cost estimating form are shown on the following pages.

Identification: To properly identify a residential building, the class of construction, type, and exterior wall material must be determined. Located at the beginning of this section are drawings and guidelines for determining the class of construction. There are also detailed specifications accompanying each type of building, along with additional drawings at the beginning of each set of tables, to further aid in proper building class and type identification.

Sketches for seven types of residential buildings and their configurations are shown along with definitions of living area next to each sketch. Sketches and definitions of garage types follow the residential buildings.

Living Area: Base cost tables are prepared as costs per square foot of living area. The living area of a residence is that area which is suitable and normally designed for full-time living. It does not include basement recreation rooms or finished attics, although these areas are often considered full-time living areas by owners.

Living area is calculated from the exterior dimensions without the need to adjust for exterior wall thickness. When calculating the living area of a 1-1/2-story, two-story, three-story or tri-level residence, overhangs and other differences in size and shape between floors must be considered.

Only the floor area with a ceiling height of six feet or more in a 1-1/2 story-residence is considered living area. In bi-levels and tri-levels, the areas that are below grade are considered living area, even when these areas may not be completely finished.

Base Tables and Modifications: Base cost tables show the base cost per square foot without a basement, with one full bath and one full kitchen. Adjustments for finished and unfinished basements are part of the base cost tables. Adjustments for multi-family residences, additional bathrooms, townhouses, alternative roofs, and air conditioning and heating systems are listed in Modifications, adjustments, and Alternatives tables below the base cost tables.

The component list for each residence type should also be consulted when preparing an estimate. If the components listed are not appropriate, modifications can be made by consulting the "Assemblies" section of this manual.

Costs for other modifications, adjustments, and alternatives, including garages, breezeways, and site improvements, are in the modification tables at the end of this section. For additional information on contractor overhead and architectural fees, consult the "Reference" section of this manual.

Listing of Dimensions: To use this section of the manual, only the dimensions used to calculate the horizontal area of the building and additions and modifications are needed. The dimensions, normally the length and width, can come from drawings or field measurements. For ease in calculation, consider measuring in tenths of feet, i.e., 9 ft. 6 in. = 9.5 ft., 9 ft. 4 in. = 9.3 ft.

In all cases, make a sketch of the building. Any protrusions or other variations in shape should be noted on the sketch with dimensions.

Calculations: The calculations portion of the estimate is a two-step activity:

(1) The selection of appropriate costs from the tables
(2) Computations

Selection of Appropriate Costs: To select the appropriate cost from the base tables the following information is needed:

(1) Class of construction
(2) Type of residence
(3) Occupancy
(4) Building configuration
(5) Exterior wall construction
(6) Living area

Consult the tables and accompanying information to make the appropriate selections. Modifications, adjustments, and alternatives are classified by class, type and size. Further modifications can be made using the "Assemblies" Section.

Computations: The computation process should take the following sequence:

(1) Multiply the base cost by the area
(2) Add or subtract the modifications
(3) Apply the location modifier

When selecting costs, interpolate or use the cost that most nearly matches the structure under study. This applies to size, exterior wall construction, and class.

How to Use the Square Foot Cost Pages

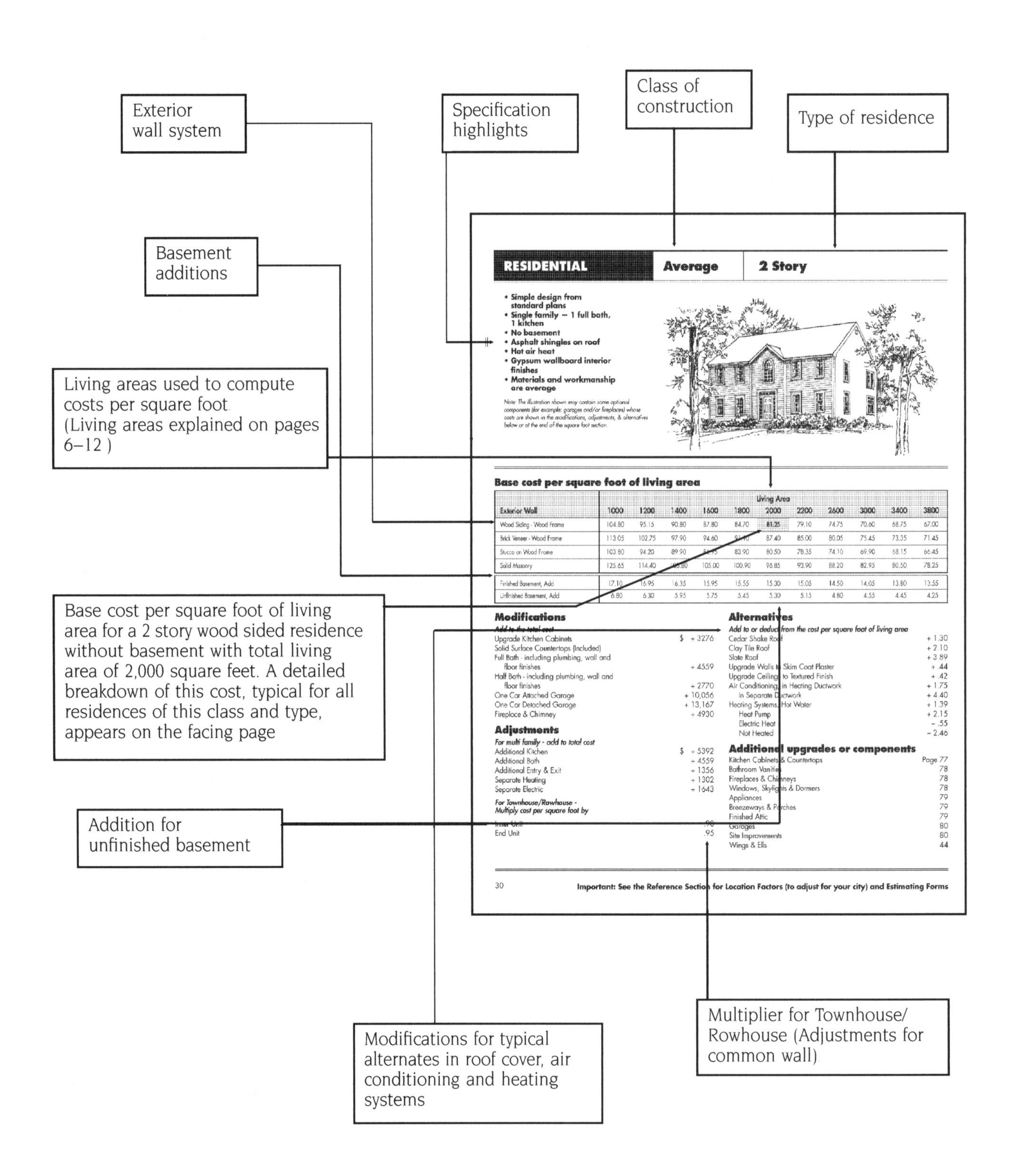

RESIDENTIAL | **Average** | **2 Story**

- **Simple design from standard plans**
- **Single family – 1 full bath, 1 kitchen**
- **No basement**
- **Asphalt shingles on roof**
- **Hot air heat**
- **Gypsum wallboard interior finishes**
- **Materials and workmanship are average**

Note: The illustration shown may contain some optional components (for example: garages and/or fireplaces) whose costs are shown in the modifications, adjustments, & alternatives below or at the end of the square foot section.

Base cost per square foot of living area

	Living Area										
Exterior Wall	1000	1200	1400	1600	1800	2000	2200	2600	3000	3400	3800
Wood Siding - Wood Frame	104.80	95.15	90.80	87.80	84.70	81.25	79.10	74.75	70.60	68.75	67.00
Brick Veneer - Wood Frame	113.05	102.75	97.90	94.60	91.10	87.40	85.00	80.05	75.45	73.35	71.45
Stucco on Wood Frame	103.80	94.20	89.90	[illegible]	83.90	80.50	78.35	74.10	69.90	68.15	66.45
Solid Masonry	125.65	114.40	[illegible]	105.00	100.90	96.85	93.90	88.20	82.95	80.50	78.25
Finished Basement, Add	17.10	16.95	16.35	15.95	15.55	15.30	15.05	14.50	14.05	13.80	13.55
Unfinished Basement, Add	6.80	6.30	5.95	5.75	5.45	5.30	5.15	4.80	4.55	4.45	4.25

Modifications

Add to the total cost

Upgrade Kitchen Cabinets	$ + 3276
Solid Surface Countertops (Included)	
Full Bath - including plumbing, wall and floor finishes	+ 4559
Half Bath - including plumbing, wall and floor finishes	+ 2770
One Car Attached Garage	+ 10,056
One Car Detached Garage	+ 13,167
Fireplace & Chimney	+ 4930

Adjustments

For multi family - add to total cost

Additional Kitchen	$ + 5392
Additional Bath	+ 4559
Additional Entry & Exit	+ 1356
Separate Heating	+ 1302
Separate Electric	+ 1643

For Townhouse/Rowhouse - Multiply cost per square foot by

Inner Unit	.90
End Unit	.95

Alternatives

Add to or deduct from the cost per square foot of living area

Cedar Shake Roof	+ 1.30
Clay Tile Roof	+ 2.10
Slate Roof	+ 3.89
Upgrade Walls to Skim Coat Plaster	+ .44
Upgrade Ceilings to Textured Finish	+ .42
Air Conditioning, in Heating Ductwork	+ 1.75
In Separate Ductwork	+ 4.40
Heating Systems, Hot Water	+ 1.39
Heat Pump	+ 2.15
Electric Heat	– .55
Not Heated	– 2.46

Additional upgrades or components

Kitchen Cabinets & Countertops	Page 77
Bathroom Vanities	78
Fireplaces & Chimneys	78
Windows, Skylights & Dormers	78
Appliances	79
Breezeways & Porches	79
Finished Attic	79
Garages	80
Site Improvements	80
Wings & Ells	44

30 **Important: See the Reference Section for Location Factors (to adjust for your city) and Estimating Forms**

Components
This page contains the ten components needed to develop the complete square foot cost of the typical dwelling specified. All components are defined with a description of the materials and/or task involved. Use cost figures from each component to estimate the cost per square foot of that section of the project.

Specifications
The parameters for an example dwelling from the facing page are listed here. Included are the square foot dimensions of the proposed building. LIVING AREA takes into account the number of floors and other factors needed to define a building's TOTAL SQUARE FOOTAGE. Perimeter and partition dimensions are defined in terms of linear feet.

Line Totals
The extreme right-hand column lists the sum of two figures. Use this total to determine the sum of MATERIAL COST plus INSTALLATION COST. The result is a convenient total cost for each of the ten components.

Average 2 Story

Living Area - 2000 S.F.
Perimeter - 135 L.F.

		Labor-Hours	Cost Per Square Foot Of Living Area		
			Mat.	Labor	Total
1 Site Work	Site preparation for slab; 4' deep trench excavation for foundation wall.	.034		.62	.62
2 Foundation	Continuous reinforced concrete footing 8" deep x 18" wide; dampproofed and insulated reinforced concrete foundation wall, 8" thick, 4' deep, 4" concrete slab on 4" crushed stone base and polyethylene vapor barrier, trowel finish.	.066	2.57	3.22	5.79
3 Framing	Exterior walls - 2" x 4" wood studs, 16" O.C.; 1/2" plywood sheathing; 2" x 6" rafters 16" O.C. with 1/2" plywood sheathing, 4 in 12 pitch; 2" x 6" ceiling joists 16" O.C.; 2" x 8" floor joists 16" O.C. with 5/8" plywood subfloor; 1/2" plywood subfloor on 1" by 2" wood sleepers 16" O.C.	.131	6.46	7.13	13.59
4 Exterior Walls	Beveled wood siding and building paper on insulated wood frame walls; 6" attic insulation; double hung windows; 3 flush solid core wood exterior doors with storms.	.111	9.40	4.78	14.18
5 Roofing	25 year asphalt shingles; #15 felt building paper; aluminum gutters, downspouts, drip edge and flashings.	.024	.61	.98	1.59
6 Interiors	Walls and ceilings, 1/2" taped and finished gypsum wallboard, primed and painted with 2 coats; painted baseboard and trim, finished hardwood floor 40%, carpet with 1/2" underlayment 40%, vinyl tile with 1/2" underlayment 15%, ceramic tile with 1/2" underlayment 5%; hollow core and louvered doors.	.232	12.26	11.81	24.07
7 Specialties	Average grade kitchen cabinets - 14 L.F. wall and base with solid surface counter top and kitchen sink; 40 gallon electric water heater.	.021	1.56	.71	2.27
8 Mechanical	1 lavatory, white, wall hung; 1 water closet, white; 1 bathtub with shower; enameled steel, white; gas fired warm air heat.	.060	2.74	2.33	5.07
9 Electrical	200 Amp. service; romex wiring; incandescent lighting fixtures, switches, receptacles.	.039	.95	1.33	2.28
10 Overhead	Contractor's overhead and profit and plans.		6.20	5.59	11.79
		Total	42.75	38.50	**81.25**

31

Labor-hours
Use this column to determine the unit of measure in LABOR-HOURS needed to perform a task. This figure will give the builder LABOR-HOURS PER SQUARE FOOT of the building. The TOTAL LABOR-HOURS PER COMPONENT are determined by multiplying the LIVING AREA times the LABOR-HOURS listed on that line. (TOTAL LABOR-HOURS PER COMPONENT = LIVING AREA x LABOR-HOURS).

Installation
The labor rates included here incorporate the overhead and profit costs for the installing contractor. The average mark-up used to create these figures is 69.5% over and above BARE LABOR COST including fringe benefits.

Bottom Line Total
This figure is the complete square foot cost for the construction project. To determine TOTAL PROJECT COST, multiply the BOTTOM LINE TOTAL times the LIVING AREA. (TOTAL PROJECT COST = BOTTOM LINE TOTAL x LIVING AREA).

Materials
This column gives the unit needed to develop the COST OF MATERIALS.
Note: The figures given here are not BARE COSTS. Ten percent has been added to BARE MATERIAL COST to cover handling.

***NOTE**
The components listed on this page are typical of all the sizes of residences from the facing page. Specific quantities of components required would vary with the size of the dwelling and the exterior wall system.

Building Classes

Given below are the four general definitions of building classes. Each building — Economy, Average, Custom and Luxury — is common in residential construction. All four are used in this book to determine costs per square foot.

Economy Class

An economy class residence is usually mass-produced from stock plans. The materials and workmanship are sufficient only to satisfy minimum building codes. Low construction cost is more important than distinctive features. Design is seldom other than square or rectangular.

Average Class

An average class residence is simple in design and is built from standard designer plans. Material and workmanship are average, but often exceed the minimum building codes. There are frequently special features that give the residence some distinctive characteristics.

Custom Class

A custom class residence is usually built from a designer's plans which have been modified to give the building a distinction of design. Material and workmanship are generally above average with obvious attention given to construction details. Construction normally exceeds building code requirements.

Luxury Class

A luxury class residence is built from an architect's plan for a specific owner. It is unique in design and workmanship. There are many special features, and construction usually exceeds all building codes. It is obvious that primary attention is placed on the owner's comfort and pleasure. Construction is supervised by an architect.

Residential Building Types

One Story

This is an example of a one-story dwelling. The living area of this type of residence is confined to the ground floor. The headroom in the attic is usually too low for use as a living area.

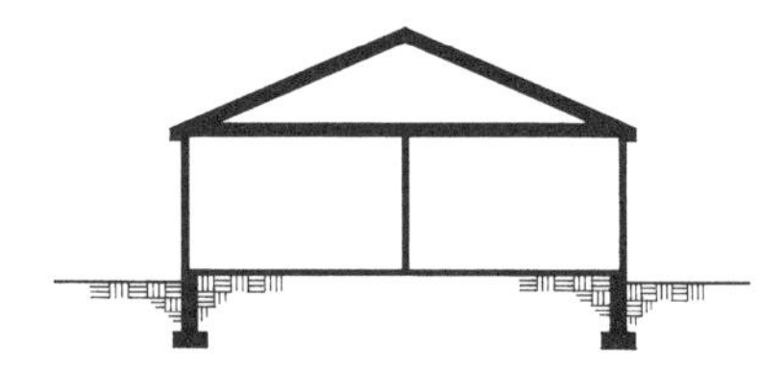

One-and-a-half Story

The living area in the upper level of this type of residence is 50% to 90% of the ground floor. This is made possible by a combination of this design's high-peaked roof and/or dormers. Only the upper level area with a ceiling height of 6′ or more is considered living area. The living area of this residence is the sum of the ground floor area plus the area on the second level with a ceiling height of 6′ or more.

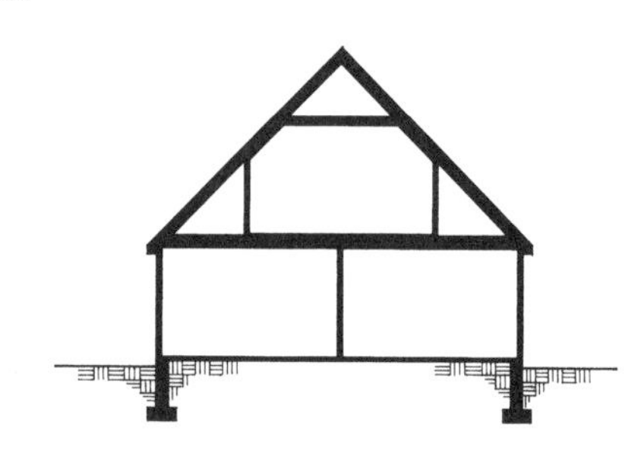

Residential Building Types

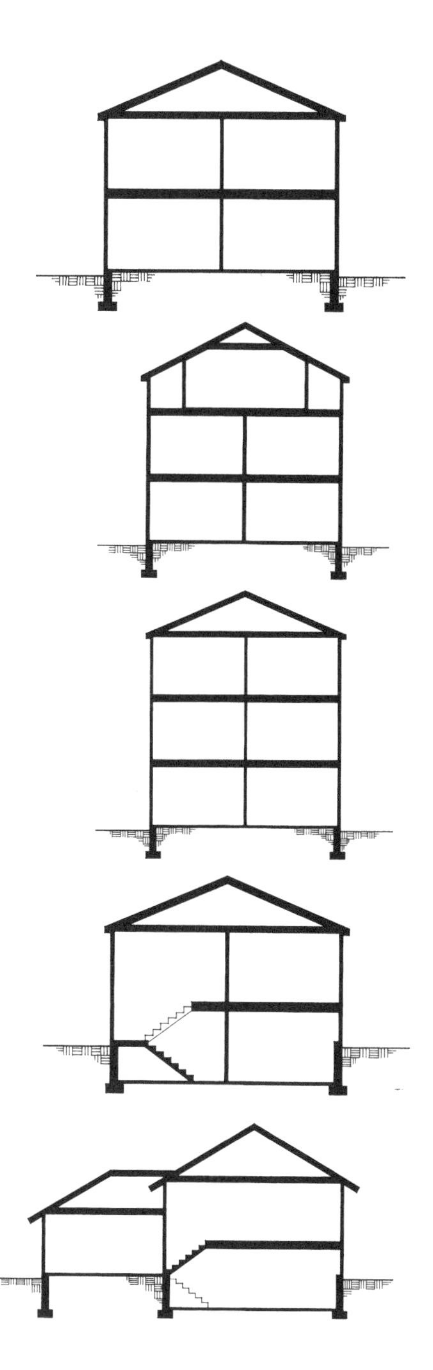

Two Story

This type of residence has a second floor or upper level area which is equal or nearly equal to the ground floor area. The upper level of this type of residence can range from 90% to 110% of the ground floor area, depending on setbacks or overhangs. The living area is the sum of the ground floor area and the upper level floor area.

Two-and-one-half Story

This type of residence has two levels of equal or nearly equal area and a third level which has a living area that is 50% to 90% of the ground floor. This is made possible by a high peaked roof, extended wall heights and/or dormers. Only the upper level area with a ceiling height of 6 feet or more is considered living area. The living area of this residence is the sum of the ground floor area, the second floor area and the area on the third level with a ceiling height of 6 feet or more.

Three Story

This type of residence has three levels which are equal or nearly equal. As in the 2 story residence, the second and third floor areas may vary slightly depending on the setbacks or overhangs. The living area is the sum of the ground floor area and the two upper level floor areas.

Bi-Level

This type of residence has two living areas, one above the other. One area is about 4 feet below grade and the second is about 4 feet above grade. Both are equal in size. The lower level in this type of residence is originally designed and built to serve as a living area and not as a basement. Both levels have full ceiling heights. The living area is the sum of the lower level area and the upper level area.

Tri-Level

This type of residence has three levels of living area. One is at grade level, the second is about 4 feet below grade, and the third is about 4 feet above grade. All levels are originally designed to serve as living areas. All levels have full ceiling heights. The living area is the sum of the areas of each of the three levels.

Exterior Wall Construction

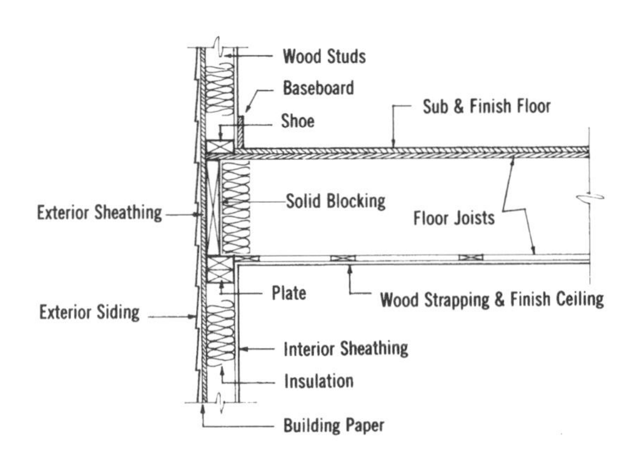

Typical Frame Construction

Typical wood frame construction consists of wood studs with insulation between them. A typical exterior surface is made up of sheathing, building paper and exterior siding consisting of wood, vinyl, aluminum or stucco over the wood sheathing.

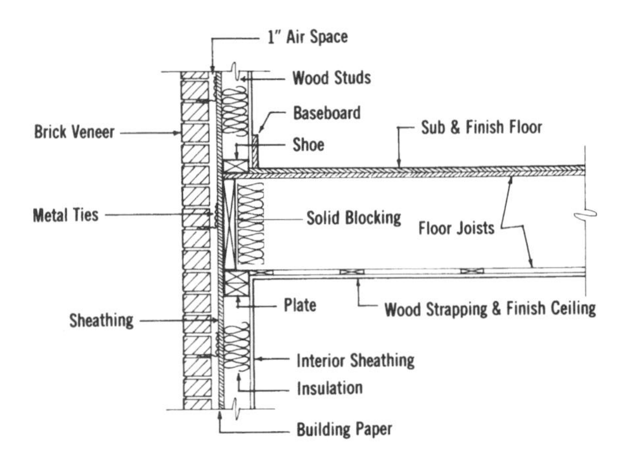

Brick Veneer

Typical brick veneer construction consists of wood studs with insulation between them. A typical exterior surface is sheathing, building paper and an exterior of brick tied to the sheathing with metal strips.

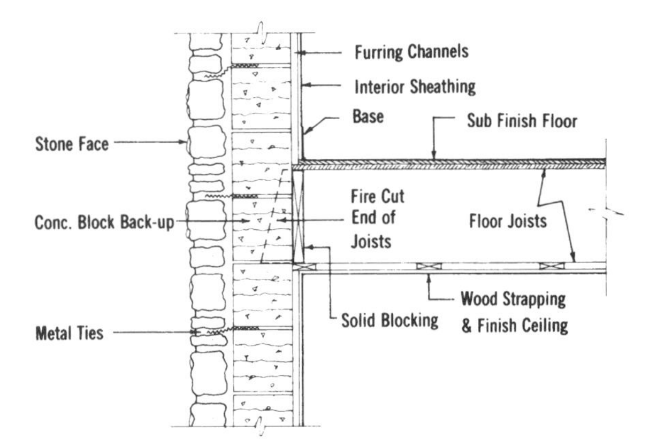

Stone

Typical solid masonry construction consists of a stone or block wall covered on the exterior with brick, stone or other masonry.

Residential Configurations

Detached House

This category of residence is a freestanding separate building with or without an attached garage. It has four complete walls.

Town/Row House

This category of residence has a number of attached units made up of inner units and end units. The units are joined by common walls. The inner units have only two exterior walls. The common walls are fireproof. The end units have three walls and a common wall. Town houses/row houses can be any of the five types.

Semi-Detached House

This category of residence has two living units side-by-side. The common wall is a fireproof wall. Semi-detached residences can be treated as a row house with two end units. Semi-detached residences can be any of the five types.

Residential Garage Types

Attached Garage

Shares a common wall with the dwelling. Access is typically through a door between dwelling and garage.

Built-In Garage

Constructed under the second floor living space and above basement level of dwelling. Reduces gross square feet of living area.

Basement Garage

Constructed under the roof of the dwelling but below the living area.

Detached Garage

Constructed apart from the main dwelling. Shares no common area or wall with the dwelling.

RESIDENTIAL
COST ESTIMATE

OWNER'S NAME: ______ APPRAISER: ______

RESIDENCE ADDRESS: ______ PROJECT: ______

CITY, STATE, ZIP CODE: ______ DATE: ______

CLASS OF CONSTRUCTION	RESIDENCE TYPE	CONFIGURATION	EXTERIOR WALL SYSTEM
☐ ECONOMY	☐ 1 STORY	☐ DETACHED	☐ WOOD SIDING - WOOD FRAME
☐ AVERAGE	☐ 1-1/2 STORY	☐ TOWN/ROW HOUSE	☐ BRICK VENEER - WOOD FRAME
☐ CUSTOM	☐ 2 STORY	☐ SEMI-DETACHED	☐ STUCCO ON WOOD FRAME
☐ LUXURY	☐ 2-1/2 STORY		☐ PAINTED CONCRETE BLOCK
	☐ 3 STORY	OCCUPANCY	☐ SOLID MASONRY (AVERAGE & CUSTOM)
	☐ BI-LEVEL	☐ ONE FAMILY	☐ STONE VENEER - WOOD FRAME
	☐ TRI-LEVEL	☐ TWO FAMILY	☐ SOLID BRICK (LUXURY)
		☐ THREE FAMILY	☐ SOLID STONE (LUXURY)
		☐ OTHER ______	

* LIVING AREA (Main Building)		* LIVING AREA (Wing or Ell) ( )		* LIVING AREA (WING or ELL) ( )	
First Level	S.F.	First Level	S.F.	First Level	S.F.
Second level	S.F.	Second level	S.F.	Second level	S.F.
Third Level	S.F.	Third Level	S.F.	Third Level	S.F.
Total	S.F.	Total	S.F.	Total	S.F.

* Basement Area is not part of living area.

MAIN BUILDING	COSTS PER S.F. LIVING AREA
Cost per Square Foot of Living Area, from Page ______	$
Basement Addition: ______ % Finished, ______ % Unfinished	+
Roof Cover Adjustment: ______ Type, Page ______ (Add or Deduct)	( )
Central Air Conditioning: ☐ Separate Ducts ☐ Heating Ducts, Page ______	+
Heating System Adjustment: ______ Type, Page ______ (Add or Deduct)	( )
Main Building: Adjusted Cost per S.F. of Living Area	$

MAIN BUILDING TOTAL COST: $ ______ /S.F. (Cost per S.F. Living Area) x ______ S.F. (Living Area) x ______ (Town/Row House Multiplier (Use 1 for Detached)) = $ ______ TOTAL COST

WING OR ELL ( ) ______ STORY	COSTS PER S.F. LIVING AREA
Cost per Square Foot of Living Area, from Page ______	$
Basement Addition: ______ % Finished, ______ % Unfinished	+
Roof Cover Adjustment: ______ Type, Page ______ (Add or Deduct)	( )
Central Air Conditioning: ☐ Separate Ducts ☐ Heating Ducts, Page ______	+
Heating System Adjustment: ______ Type, Page ______ (Add or Deduct)	( )
Wing or Ell ( ): Adjusted Cost per S.F. of Living Area	$

WING OR ELL (   ) TOTAL COST: $ ______ /S.F. (Cost per S.F. Living Area) x ______ S.F. (Living Area) x = $ ______ TOTAL COST

WING OR ELL ( ) ______ STORY	COSTS PER S.F. LIVING AREA
Cost per Square Foot of Living Area, from Page ______	$
Basement Addition: ______ % Finished, ______ % Unfinished	+
Roof Cover Adjustment: ______ Type, Page ______ (Add or Deduct)	( )
Central Air Conditioning: ☐ Separate Ducts ☐ Heating Ducts, Page ______	+
Heating System Adjustment: ______ Type, Page ______ (Add or Deduct)	( )
Wing or Ell ( ): Adjusted Cost per S.F. of Living Area	$

WING OR ELL (   ) TOTAL COST: $ ______ /S.F. (Cost per S.F. Living Area) x ______ S.F. (Living Area) x = $ ______ TOTAL COST

TOTAL THIS PAGE ______

Page 1 of 2

RESIDENTIAL
COST ESTIMATE

Total Page 1			$
	QUANTITY	UNIT COST	
Additional Bathrooms: ________ Full ________ Half			
Finished Attic: ________ Ft. x ________ Ft.	S.F.		+
Breezeway: ☐ Open ☐ Enclosed ________ Ft. x ________ Ft.	S.F.		+
Covered Porch: ☐ Open ☐ Enclosed ________ Ft. x ________ Ft.	S.F.		+
Fireplace: ☐ Interior Chimney ☐ Exterior Chimney ☐ No. of Flues ☐ Additional Fireplaces			+
Appliances:			+
Kitchen Cabinets Adjustments: (±)			
☐ Garage ☐ Carport: ________ Car(s) Description ________ (±)			
Miscellaneous:			+

ADJUSTED TOTAL BUILDING COST $

REPLACEMENT COST		
ADJUSTED TOTAL BUILDING COST	$	________
Site Improvements		
(A) Paving & Sidewalks	$	________
(B) Landscaping	$	________
(C) Fences	$	________
(D) Swimming Pools	$	________
(E) Miscellaneous	$	________
TOTAL	$	________
Location Factor	x	________
Location Replacement Cost	$	________
Depreciation	- $	________
LOCAL DEPRECIATED COST	$	________

INSURANCE COST		
ADJUSTED TOTAL BUILDING COST	$	________
Insurance Exclusions		
(A) Footings, Site work, Underground Piping	- $	________
(B) Architects Fees	- $	________
Total Building Cost Less Exclusion	$	________
Location Factor	x	________
LOCAL INSURABLE REPLACEMENT COST	$	________

SKETCH AND ADDITIONAL CALCULATIONS

1 Story

1-1/2 Story

2 Story

Bi-Level

Tri-Level

©Design Basics, Inc.

- **Mass produced from stock plans**
- **Single family — 1 full bath, 1 kitchen**
- **No basement**
- **Asphalt shingles on roof**
- **Hot air heat**
- **Gypsum wallboard interior finishes**
- **Materials and workmanship are sufficient to meet codes**

Note: The illustration shown may contain some optional components (for example: garages and/or fireplaces) whose costs are shown in the modifications, adjustments, & alternatives below or at the end of the square foot section.

Base cost per square foot of living area

Exterior Wall	Living Area 600	800	1000	1200	1400	1600	1800	2000	2400	2800	3200
Wood Siding - Wood Frame	95.60	86.80	80.10	74.70	69.85	66.80	65.35	63.35	59.05	56.05	54.00
Brick Veneer - Wood Frame	102.45	92.95	85.65	79.70	74.40	71.05	69.50	67.20	62.60	59.35	57.05
Stucco on Wood Frame	92.05	83.70	77.25	**72.05**	67.45	64.60	63.20	61.35	57.25	54.40	52.45
Painted Concrete Block	95.95	87.10	80.45	74.95	70.05	67.05	65.60	63.50	59.30	56.25	54.15
Finished Basement, Add	24.15	22.75	21.70	20.80	20.05	19.60	19.30	18.95	18.35	18.00	17.55
Unfinished Basement, Add	10.90	9.80	9.00	8.30	7.65	7.30	7.10	6.80	6.35	6.00	5.75

Modifications

Add to the total cost

Upgrade Kitchen Cabinets	$ + 714
Solid Surface Countertops	+ 696
Full Bath - including plumbing, wall and floor finishes	+ 3633
Half Bath - including plumbing, wall and floor finishes	+ 2208
One Car Attached Garage	+ 9356
One Car Detached Garage	+ 12,048
Fireplace & Chimney	+ 4435

Adjustments

For multi family - add to total cost

Additional Kitchen	$ + 2755
Additional Bath	+ 3633
Additional Entry & Exit	+ 1356
Separate Heating	+ 1302
Separate Electric	+ 908

For Townhouse/Rowhouse - Multiply cost per square foot by

Inner Unit	.95
End Unit	.97

Alternatives

Add to or deduct from the cost per square foot of living area

Composition Roll Roofing	– .75
Cedar Shake Roof	+ 2.90
Upgrade Walls and Ceilings to Skim Coat Plaster	+ .70
Upgrade Ceilings to Textured Finish	+ .42
Air Conditioning, in Heating Ductwork	+ 2.80
In Separate Ductwork	+ 5.35
Heating Systems, Hot Water	+ 1.46
Heat Pump	+ 1.77
Electric Heat	– 1.25
Not Heated	– 3.16

Additional upgrades or components

Kitchen Cabinets & Countertops	Page 77
Bathroom Vanities	78
Fireplaces & Chimneys	78
Windows, Skylights & Dormers	78
Appliances	79
Breezeways & Porches	79
Finished Attic	79
Garages	80
Site Improvements	80
Wings & Ells	24

Economy 1 Story

Living Area - 1200 S.F.
Perimeter - 146 L.F.

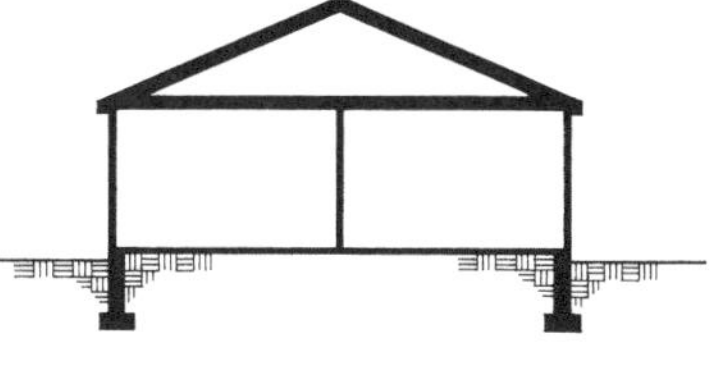

		Labor-Hours	Cost Per Square Foot Of Living Area		
			Mat.	Labor	Total
1 Site Work	Site preparation for slab; 4′ deep trench excavation for foundation wall.	.060		1.00	1.00
2 Foundation	Continuous reinforced concrete footing, 8″ deep x 18″ wide; dampproofed and insulated 8″ thick reinforced concrete block foundation wall, 4′ deep; 4″ concrete slab on 4″ crushed stone base and polyethylene vapor barrier, trowel finish.	.131	4.68	5.77	10.45
3 Framing	Exterior walls - 2″ x 4″ wood studs, 16″ O.C.; 1/2″ insulation board sheathing; wood truss roof frame, 24″ O.C. with 1/2″ plywood sheathing, 4 in 12 pitch.	.098	4.78	4.96	9.74
4 Exterior Walls	Metal lath reinforced stucco exterior on insulated wood frame walls; 6″ attic insulation; sliding sash wood windows; 2 flush solid core wood exterior doors with storms.	.110	5.26	5.19	10.45
5 Roofing	20 year asphalt shingles; #15 felt building paper; aluminum gutters, downspouts, drip edge and flashings.	.047	1.17	1.90	3.07
6 Interiors	Walls and ceilings, 1/2″ taped and finished gypsum wallboard, primed and painted with 2 coats; painted baseboard and trim; rubber backed carpeting 80%, asphalt tile 20%; hollow core wood interior doors.	.243	7.79	9.79	17.58
7 Specialties	Economy grade kitchen cabinets - 6 L.F. wall and base with plastic laminate counter top and kitchen sink; 30 gallon electric water heater.	.004	1.48	.74	2.22
8 Mechanical	1 lavatory, white, wall hung; 1 water closet, white; 1 bathtub, enameled steel, white; gas fired warm air heat.	.086	3.50	2.74	6.24
9 Electrical	100 Amp. service; romex wiring; incandescent lighting fixtures, switches, receptacles.	.036	.74	1.18	1.92
10 Overhead	Contractor's overhead and profit		4.40	4.98	9.38
		Total	33.80	38.25	**72.05**

- **Mass produced from stock plans**
- **Single family — 1 full bath, 1 kitchen**
- **No basement**
- **Asphalt shingles on roof**
- **Hot air heat**
- **Gypsum wallboard interior finishes**
- **Materials and workmanship are sufficient to meet codes**

Note: The illustration shown may contain some optional components (for example: garages and/or fireplaces) whose costs are shown in the modifications, adjustments, & alternatives below or at the end of the square foot section.

Base cost per square foot of living area

Exterior Wall	Living Area 600	800	1000	1200	1400	1600	1800	2000	2400	2800	3200
Wood Siding - Wood Frame	108.80	90.95	81.35	77.00	73.90	69.00	66.60	64.15	58.95	57.05	55.00
Brick Veneer - Wood Frame	118.35	97.90	87.70	83.05	79.60	74.20	71.55	68.85	63.10	60.95	58.65
Stucco on Wood Frame	103.85	87.35	78.00	73.90	70.95	66.30	64.10	61.75	56.80	55.00	53.15
Painted Concrete Block	109.30	91.30	81.65	77.30	74.15	69.30	66.90	64.45	59.15	57.25	55.20
Finished Basement, Add	18.60	15.75	15.05	14.55	14.15	13.60	13.25	13.00	12.45	12.15	11.85
Unfinished Basement, Add	9.60	7.35	6.80	6.40	6.10	5.65	5.35	5.15	4.70	4.50	4.25

Modifications

Add to the total cost

Upgrade Kitchen Cabinets	$ + 714
Solid Surface Countertops	+ 696
Full Bath - including plumbing, wall and floor finishes	+ 3633
Half Bath - including plumbing, wall and floor finishes	+ 2208
One Car Attached Garage	+ 9356
One Car Detached Garage	+ 12,048
Fireplace & Chimney	+ 4435

Adjustments

For multi family - add to total cost

Additional Kitchen	$ + 2755
Additional Bath	+ 3633
Additional Entry & Exit	+ 1356
Separate Heating	+ 1302
Separate Electric	+ 908

For Townhouse/Rowhouse - Multiply cost per square foot by

Inner Unit	.95
End Unit	.97

Alternatives

Add to or deduct from the cost per square foot of living area

Composition Roll Roofing	– .50
Cedar Shake Roof	+ 2.10
Upgrade Walls and Ceilings to Skim Coat Plaster	+ .70
Upgrade Ceilings to Textured Finish	+ .42
Air Conditioning, in Heating Ductwork	+ 2.10
In Separate Ductwork	+ 4.70
Heating Systems, Hot Water	+ 1.40
Heat Pump	+ 1.95
Electric Heat	– 1.00
Not Heated	– 2.91

Additional upgrades or components

Kitchen Cabinets & Countertops	Page 77
Bathroom Vanities	78
Fireplaces & Chimneys	78
Windows, Skylights & Dormers	78
Appliances	79
Breezeways & Porches	79
Finished Attic	79
Garages	80
Site Improvements	80
Wings & Ells	24

Important: See the Reference Section for Location Factors (to adjust for your city) and Estimating Forms

Economy 1-1/2 Story

Living Area - 1600 S.F.
Perimeter - 135 L.F.

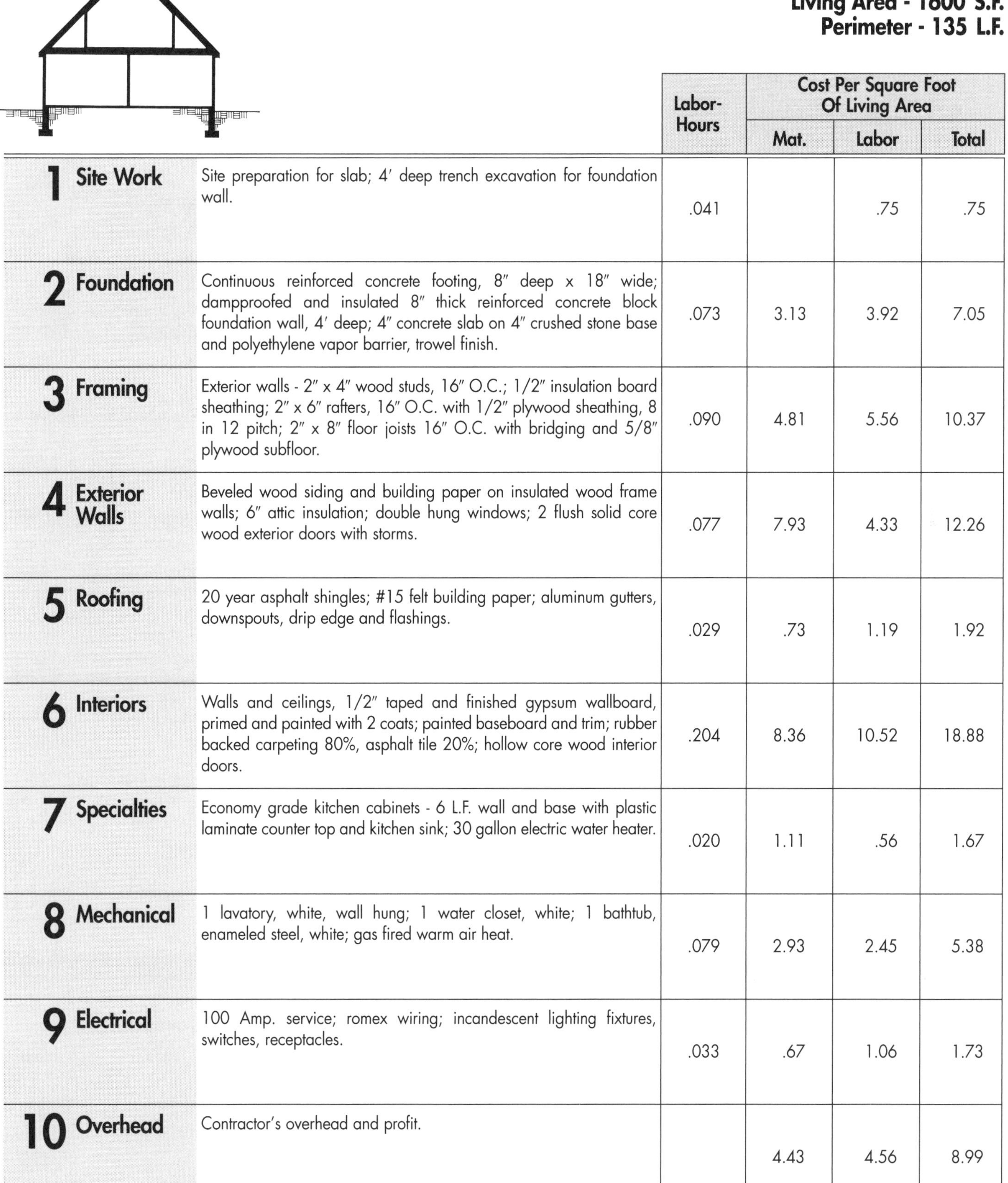

		Labor-Hours	Cost Per Square Foot Of Living Area		
			Mat.	Labor	Total
1 Site Work	Site preparation for slab; 4′ deep trench excavation for foundation wall.	.041		.75	.75
2 Foundation	Continuous reinforced concrete footing, 8″ deep x 18″ wide; dampproofed and insulated 8″ thick reinforced concrete block foundation wall, 4′ deep; 4″ concrete slab on 4″ crushed stone base and polyethylene vapor barrier, trowel finish.	.073	3.13	3.92	7.05
3 Framing	Exterior walls - 2″ x 4″ wood studs, 16″ O.C.; 1/2″ insulation board sheathing; 2″ x 6″ rafters, 16″ O.C. with 1/2″ plywood sheathing, 8 in 12 pitch; 2″ x 8″ floor joists 16″ O.C. with bridging and 5/8″ plywood subfloor.	.090	4.81	5.56	10.37
4 Exterior Walls	Beveled wood siding and building paper on insulated wood frame walls; 6″ attic insulation; double hung windows; 2 flush solid core wood exterior doors with storms.	.077	7.93	4.33	12.26
5 Roofing	20 year asphalt shingles; #15 felt building paper; aluminum gutters, downspouts, drip edge and flashings.	.029	.73	1.19	1.92
6 Interiors	Walls and ceilings, 1/2″ taped and finished gypsum wallboard, primed and painted with 2 coats; painted baseboard and trim; rubber backed carpeting 80%, asphalt tile 20%; hollow core wood interior doors.	.204	8.36	10.52	18.88
7 Specialties	Economy grade kitchen cabinets - 6 L.F. wall and base with plastic laminate counter top and kitchen sink; 30 gallon electric water heater.	.020	1.11	.56	1.67
8 Mechanical	1 lavatory, white, wall hung; 1 water closet, white; 1 bathtub, enameled steel, white; gas fired warm air heat.	.079	2.93	2.45	5.38
9 Electrical	100 Amp. service; romex wiring; incandescent lighting fixtures, switches, receptacles.	.033	.67	1.06	1.73
10 Overhead	Contractor's overhead and profit.		4.43	4.56	8.99
		Total	34.10	34.90	**69.00**

- **Mass produced from stock plans**
- **Single family — 1 full bath, 1 kitchen**
- **No basement**
- **Asphalt shingles on roof**
- **Hot air heat**
- **Gypsum wallboard interior finishes**
- **Materials and workmanship are sufficient to meet codes**

Note: The illustration shown may contain some optional components (for example: garages and/or fireplaces) whose costs are shown in the modifications, adjustments, & alternatives below or at the end of the square foot section.

Base cost per square foot of living area

	Living Area										
Exterior Wall	1000	1200	1400	1600	1800	2000	2200	2600	3000	3400	3800
Wood Siding - Wood Frame	86.50	78.15	74.45	71.95	69.35	66.30	64.50	60.80	57.00	55.45	54.05
Brick Veneer - Wood Frame	93.85	85.05	80.85	78.05	75.10	71.85	69.75	65.55	61.45	59.60	58.00
Stucco on Wood Frame	82.65	74.60	71.15	68.80	66.40	63.45	61.75	58.30	54.70	53.25	51.95
Painted Concrete Block	86.85	78.50	74.80	72.25	69.65	66.55	64.70	61.00	57.25	55.60	54.20
Finished Basement, Add	12.65	12.10	11.70	11.40	11.10	10.90	10.65	10.30	9.95	9.75	9.60
Unfinished Basement, Add	5.90	5.50	5.10	4.90	4.65	4.50	4.30	4.00	3.75	3.60	3.50

Modifications

Add to the total cost

Upgrade Kitchen Cabinets	$ + 714
Solid Surface Countertops	+ 696
Full Bath - including plumbing, wall and floor finishes	+ 3633
Half Bath - including plumbing, wall and floor finishes	+ 2208
One Car Attached Garage	+ 9356
One Car Detached Garage	+ 12,048
Fireplace & Chimney	+ 4900

Adjustments

For multi family - add to total cost

Additional Kitchen	$ + 2755
Additional Bath	+ 3633
Additional Entry & Exit	+ 1356
Separate Heating	+ 1302
Separate Electric	+ 908

For Townhouse/Rowhouse - Multiply cost per square foot by

Inner Unit	.93
End Unit	.96

Alternatives

Add to or deduct from the cost per square foot of living area

Composition Roll Roofing	− .35
Cedar Shake Roof	+ 1.45
Upgrade Walls and Ceilings to Skim Coat Plaster	+ .71
Upgrade Ceilings to Textured Finish	+ .42
Air Conditioning, in Heating Ductwork	+ 1.70
In Separate Ductwork	+ 4.30
Heating Systems, Hot Water	+ 1.36
Heat Pump	+ 2.07
Electric Heat	− .87
Not Heated	− 2.75

Additional upgrades or components

Kitchen Cabinets & Countertops	Page 77
Bathroom Vanities	78
Fireplaces & Chimneys	78
Windows, Skylights & Dormers	78
Appliances	79
Breezeways & Porches	79
Finished Attic	79
Garages	80
Site Improvements	80
Wings & Ells	24

Important: See the Reference Section for Location Factors (to adjust for your city) and Estimating Forms

Economy 2 Story

Living Area - 2000 S.F.
Perimeter - 135 L.F.

		Labor-Hours	Cost Per Square Foot Of Living Area		
			Mat.	Labor	Total
1 Site Work	Site preparation for slab; 4′ deep trench excavation for foundation wall.	.034		.60	.60
2 Foundation	Continuous reinforced concrete footing, 8″ deep x 18″ wide; dampproofed and insulated 8″ thick reinforced concrete block foundation wall, 4′ deep; 4″ concrete slab on 4″ crushed stone base and polyethylene vapor barrier, trowel finish.	.069	2.50	3.13	5.63
3 Framing	Exterior walls - 2″ x 4″ wood studs, 16″ O.C.; 1/2″ insulation board sheathing; wood truss roof frame, 24″ O.C. with 1/2″ plywood sheathing, 4 in 12 pitch; 2″ x 8″ floor joists 16″ O.C. with bridging and 5/8″ plywood subfloor.	.112	4.89	5.84	10.73
4 Exterior Walls	Beveled wood siding and building paper on insulated wood frame walls; 6″ attic insulation; double hung windows; 2 flush solid core wood exterior doors with storms.	.107	8.06	4.42	12.48
5 Roofing	20 year asphalt shingles; #15 felt building paper; aluminum gutters, downspouts, drip edge and flashings.	.024	.59	.95	1.54
6 Interiors	Walls and ceilings, 1/2″ taped and finished gypsum wallboard, primed and painted with 2 coats; painted baseboard and trim; rubber backed carpeting 80%, asphalt tile 20%; hollow core wood interior doors.	.219	8.34	10.52	18.86
7 Specialties	Economy grade kitchen cabinets - 6 L.F. wall and base with plastic laminate counter top and kitchen sink; 30 gallon electric water heater.	.017	.88	.44	1.32
8 Mechanical	1 lavatory, white, wall hung; 1 water closet, white; 1 bathtub, enameled steel, white; gas fired warm air heat.	.061	2.59	2.28	4.87
9 Electrical	100 Amp. service; romex wiring; incandescent lighting fixtures; switches, receptacles.	.030	.62	.99	1.61
10 Overhead	Contractor's overhead and profit		4.28	4.38	8.66
		Total	32.75	33.55	**66.30**

- **Mass produced from stock plans**
- **Single family — 1 full bath, 1 kitchen**
- **No basement**
- **Asphalt shingles on roof**
- **Hot air heat**
- **Gypsum wallboard interior finishes**
- **Materials and workmanship are sufficient to meet codes**

Note: The illustration shown may contain some optional components (for example: garages and/or fireplaces) whose costs are shown in the modifications, adjustments, & alternatives below or at the end of the square foot section.

Base cost per square foot of living area

Exterior Wall	Living Area 1000	1200	1400	1600	1800	2000	2200	2600	3000	3400	3800
Wood Siding - Wood Frame	80.60	72.65	69.30	67.05	64.75	61.85	60.25	56.95	53.50	52.05	50.85
Brick Veneer - Wood Frame	86.15	77.80	74.10	71.60	69.10	66.00	64.20	60.55	56.85	55.15	53.80
Stucco on Wood Frame	77.70	70.00	66.80	64.65	62.55	59.70	58.20	55.10	51.80	50.40	49.30
Painted Concrete Block	80.85	72.95	69.60	67.25	65.00	62.00	60.40	57.10	53.65	52.20	50.95
Finished Basement, Add	12.65	12.10	11.70	11.40	11.10	10.90	10.65	10.30	9.95	9.75	9.60
Unfinished Basement, Add	5.90	5.50	5.10	4.90	4.65	4.50	4.30	4.00	3.75	3.60	3.50

Modifications

Add to the total cost

Upgrade Kitchen Cabinets	$ + 714
Solid Surface Countertops	+ 696
Full Bath - including plumbing, wall and floor finishes	+ 3633
Half Bath - including plumbing, wall and floor finishes	+ 2208
One Car Attached Garage	+ 9356
One Car Detached Garage	+ 12,048
Fireplace & Chimney	+ 4435

Adjustments

For multi family - add to total cost

Additional Kitchen	$ + 2755
Additional Bath	+ 3633
Additional Entry & Exit	+ 1356
Separate Heating	+ 1302
Separate Electric	+ 908

For Townhouse/Rowhouse - Multiply cost per square foot by

Inner Unit	.94
End Unit	.97

Alternatives

Add to or deduct from the cost per square foot of living area

Composition Roll Roofing	– .35
Cedar Shake Roof	+ 1.45
Upgrade Walls and Ceilings to Skim Coat Plaster	+ .67
Upgrade Ceilings to Textured Finish	+ .42
Air Conditioning, in Heating Ductwork	+ 1.70
In Separate Ductwork	+ 4.30
Heating Systems, Hot Water	+ 1.36
Heat Pump	+ 2.07
Electric Heat	– .87
Not Heated	– 2.75

Additional upgrades or components

Kitchen Cabinets & Countertops	Page 77
Bathroom Vanities	78
Fireplaces & Chimneys	78
Windows, Skylights & Dormers	78
Appliances	79
Breezeways & Porches	79
Finished Attic	79
Garages	80
Site Improvements	80
Wings & Ells	24

Important: See the Reference Section for Location Factors (to adjust for your city) and Estimating Forms

Economy Bi-Level

Living Area - 2000 S.F.
Perimeter - 135 L.F.

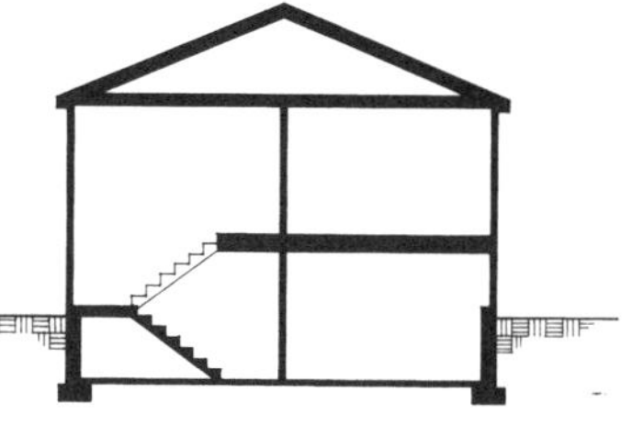

		Labor-Hours	Cost Per Square Foot Of Living Area		
			Mat.	Labor	Total
1 Site Work	Excavation for lower level, 4′ deep. Site preparation for slab.	.029		.60	.60
2 Foundation	Continuous reinforced concrete footing, 8″ deep x 18″ wide; dampproofed and insulated 8″ thick reinforced concrete block foundation wall, 4′ deep; 4″ concrete slab on 4″ crushed stone base and polyethylene vapor barrier, trowel finish.	.069	2.50	3.13	5.63
3 Framing	Exterior walls - 2″ x 4″ wood studs, 16″ O.C.; 1/2″ insulation board sheathing; wood truss roof frame, 24″ O.C. with 1/2″ plywood sheathing, 4 in 12 pitch; 2″ x 8″ floor joists 16″ O.C. with bridging and 5/8″ plywood subfloor.	.107	4.58	5.49	10.07
4 Exterior Walls	Beveled wood siding and building paper on insulated wood frame walls; 6″ attic insulation; double hung windows; 2 flush solid core wood exterior doors with storms.	.089	6.35	3.46	9.81
5 Roofing	20 year asphalt shingles; #15 felt building paper; aluminum gutters, downspouts, drip edge and flashings.	.024	.59	.95	1.54
6 Interiors	Walls and ceilings, 1/2″ taped and finished gypsum wallboard, primed and painted with 2 coats; painted baseboard and trim, rubber backed carpeting 80%, asphalt tile 20%; hollow core wood interior doors.	.213	8.14	10.19	18.33
7 Specialties	Economy grade kitchen cabinets - 6 L.F. wall and base with plastic laminate counter top and kitchen sink; 30 gallon electric water heater.	.018	.88	.44	1.32
8 Mechanical	1 lavatory, white, wall hung; 1 water closet, white; 1 bathtub, enameled steel, white; gas fired warm air heat.	.061	2.59	2.28	4.87
9 Electrical	100 Amp. service; romex wiring; incandescent lighting fixtures; switches, receptacles.	.030	.62	.99	1.61
10 Overhead	Contractor's overhead and profit.		3.95	4.12	8.07
		Total	30.20	31.65	**61.85**

- **Mass produced from stock plans**
- **Single family — 1 full bath, 1 kitchen**
- **No basement**
- **Asphalt shingles on roof**
- **Hot air heat**
- **Gypsum wallboard interior finishes**
- **Materials and workmanship are sufficient to meet codes**

Note: The illustration shown may contain some optional components (for example: garages and/or fireplaces) whose costs are shown in the modifications, adjustments, & alternatives below or at the end of the square foot section.

©Design Basics, Inc.

Base cost per square foot of living area

Exterior Wall	Living Area 1200	1500	1800	2000	2200	2400	2800	3200	3600	4000	4400
Wood Siding - Wood Frame	74.35	68.40	63.90	62.20	59.55	57.40	55.85	53.60	50.90	50.00	47.95
Brick Veneer - Wood Frame	79.45	73.00	68.05	66.15	63.35	60.95	59.30	56.75	53.90	52.90	50.65
Stucco on Wood Frame	71.70	66.00	61.75	60.15	57.60	55.55	54.10	51.95	49.40	48.55	46.60
Solid Masonry	74.60	68.60	64.15	62.35	59.75	57.60	56.00	53.80	51.10	50.15	48.05
Finished Basement, Add*	15.20	14.50	13.90	13.60	13.35	13.10	12.85	12.50	12.25	12.10	11.90
Unfinished Basement, Add*	6.55	6.00	5.50	5.35	5.15	4.90	4.70	4.45	4.25	4.15	4.00

*Basement under middle level only.

Modifications

Add to the total cost

Upgrade Kitchen Cabinets	$ + 714
Solid Surface Countertops	+ 696
Full Bath - including plumbing, wall and floor finishes	+ 3633
Half Bath - including plumbing, wall and floor finishes	+ 2208
One Car Attached Garage	+ 9356
One Car Detached Garage	+ 12,048
Fireplace & Chimney	+ 4435

Adjustments

For multi family - add to total cost

Additional Kitchen	$ + 2755
Additional Bath	+ 3633
Additional Entry & Exit	+ 1356
Separate Heating	+ 1302
Separate Electric	+ 908

For Townhouse/Rowhouse - Multiply cost per square foot by

Inner Unit	.93
End Unit	.96

Alternatives

Add to or deduct from the cost per square foot of living area

Composition Roll Roofing	– .50
Cedar Shake Roof	+ 2.10
Upgrade Walls and Ceilings to Skim Coat Plaster	+ .62
Upgrade Ceilings to Textured Finish	+ .42
Air Conditioning, in Heating Ductwork	+ 1.45
In Separate Ductwork	+ 4.00
Heating Systems, Hot Water	+ 1.30
Heat Pump	+ 2.15
Electric Heat	– .75
Not Heated	– 2.66

Additional upgrades or components

Kitchen Cabinets & Countertops	Page 77
Bathroom Vanities	78
Fireplaces & Chimneys	78
Windows, Skylights & Dormers	78
Appliances	79
Breezeways & Porches	79
Finished Attic	79
Garages	80
Site Improvements	80
Wings & Ells	24

Economy Tri-Level

Living Area - 2400 S.F.
Perimeter - 163 L.F.

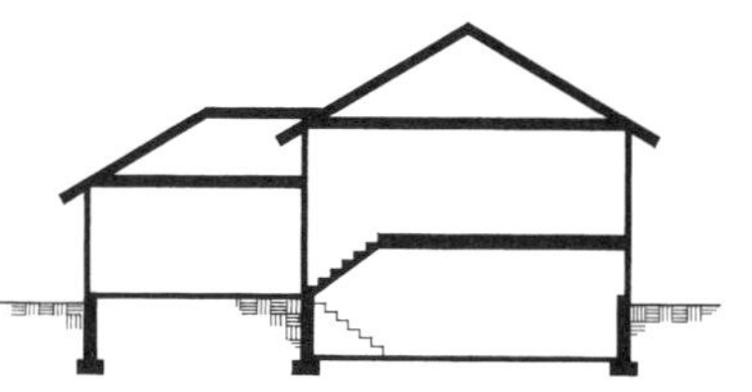

		Labor-Hours	Cost Per Square Foot Of Living Area		
			Mat.	Labor	Total
1 Site Work	Site preparation for slab; 4′ deep trench excavation for foundation wall, excavation for lower level, 4′ deep.	.027		.50	.50
2 Foundation	Continuous reinforced concrete footing, 8″ deep x 18″ wide; dampproofed and insulated 8″ thick reinforced concrete block foundation wall, 4′ deep; 4″ concrete slab on 4″ crushed stone base and polyethylene vapor barrier, trowel finish.	.071	2.82	3.40	6.22
3 Framing	Exterior walls - 2″ x 4″ wood studs, 16″ O.C.; 1/2″ insulation board sheathing; wood truss roof frame, 24″ O.C. with 1/2″ plywood sheathing, 4 in 12 pitch; 2″ x 8″ floor joists 16″ O.C. with bridging and 5/8″ plywood subfloor.	.094	4.36	4.93	9.29
4 Exterior Walls	Beveled wood siding and building paper on insulated wood frame walls; 6″ attic insulation; double hung windows; 2 flush solid core wood exterior doors with storms.	.081	5.51	3.01	8.52
5 Roofing	20 year asphalt shingles; #15 felt building paper; aluminum gutters, downspouts, drip edge and flashings.	.032	.78	1.27	2.05
6 Interiors	Walls and ceilings, 1/2″ taped and finished gypsum wallboard, primed and painted with 2 coats; painted baseboard and trim, rubber backed carpeting 80%, asphalt tile 20%; hollow core wood interior doors.	.177	7.16	9.03	16.19
7 Specialties	Economy grade kitchen cabinets - 6 L.F. wall and base with plastic laminate counter top and kitchen sink; 30 gallon electric water heater.	.014	.73	.37	1.10
8 Mechanical	1 lavatory, white, wall hung; 1 water closet, white; 1 bathtub, enameled steel, white; gas fired warm air heat.	.057	2.35	2.17	4.52
9 Electrical	100 Amp. service; romex wiring; incandescent lighting fixtures, switches, receptacles.	.029	.59	.95	1.54
10 Overhead	Contractor's overhead and profit		3.65	3.82	7.47
		Total	27.95	29.45	**57.40**

1 Story — Base cost per square foot of living area

Exterior Wall	Living Area 50	100	200	300	400	500	600	700
Wood Siding - Wood Frame	132.60	101.20	87.45	73.35	68.85	66.20	64.40	64.95
Brick Veneer - Wood Frame	149.85	113.55	97.80	80.20	75.05	71.95	69.85	70.25
Stucco on Wood Frame	123.65	94.80	82.20	69.80	65.70	63.20	61.55	62.25
Painted Concrete Block	133.50	101.80	88.05	73.65	69.15	66.50	64.65	65.25
Finished Basement, Add	36.95	30.15	27.30	22.60	21.60	21.10	20.70	20.45
Unfinished Basement, Add	20.65	15.40	13.20	9.55	8.80	8.40	8.10	7.90

1-1/2 Story — Base cost per square foot of living area

Exterior Wall	Living Area 100	200	300	400	500	600	700	800
Wood Siding - Wood Frame	104.15	83.60	71.05	63.90	60.10	58.40	56.05	55.40
Brick Veneer - Wood Frame	119.55	95.90	81.40	71.95	67.55	65.35	62.60	61.90
Stucco on Wood Frame	96.15	77.20	65.80	59.80	56.30	54.75	52.60	52.00
Painted Concrete Block	104.95	84.25	71.60	64.30	60.55	58.70	56.40	55.70
Finished Basement, Add	24.70	21.85	19.95	17.90	17.35	16.95	16.55	16.55
Unfinished Basement, Add	12.70	10.55	9.10	7.45	7.05	6.75	6.50	6.45

2 Story — Base cost per square foot of living area

Exterior Wall	Living Area 100	200	400	600	800	1000	1200	1400
Wood Siding - Wood Frame	105.95	78.80	66.85	55.60	51.70	49.40	47.85	48.60
Brick Veneer - Wood Frame	123.20	91.10	77.15	62.45	57.90	55.15	53.30	53.90
Stucco on Wood Frame	97.05	72.40	61.55	52.10	48.55	46.45	45.00	45.85
Painted Concrete Block	106.85	79.40	67.40	55.95	52.05	49.70	48.10	48.85
Finished Basement, Add	18.50	15.10	13.70	11.30	10.85	10.60	10.35	10.20
Unfinished Basement, Add	10.35	7.70	6.65	4.80	4.40	4.20	4.05	3.95

Base costs do not include bathroom or kitchen facilities. Use Modifications/Adjustments/Alternatives on pages 77-80 where appropriate.

1 Story

1-1/2 Story

2 Story

2-1/2 Story

Bi-Level

Tri-Level

- **Simple design from standard plans**
- **Single family — 1 full bath, 1 kitchen**
- **No basement**
- **Asphalt shingles on roof**
- **Hot air heat**
- **Gypsum wallboard interior finishes**
- **Materials and workmanship are average**

Note: The illustration shown may contain some optional components (for example: garages and/or fireplaces) whose costs are shown in the modifications, adjustments, & alternatives below or at the end of the square foot section.

Base cost per square foot of living area

Exterior Wall	Living Area 600	800	1000	1200	1400	1600	1800	2000	2400	2800	3200
Wood Siding - Wood Frame	117.95	107.30	99.20	92.70	87.15	83.60	81.70	79.30	74.40	70.90	68.55
Brick Veneer - Wood Frame	135.25	123.70	115.00	107.90	101.85	98.00	95.95	93.25	87.95	84.20	81.55
Stucco on Wood Frame	126.70	116.05	108.05	101.60	96.15	92.60	90.80	88.45	83.55	80.15	77.80
Solid Masonry	146.90	134.15	124.45	116.40	109.60	105.20	102.90	99.85	94.00	89.75	86.70
Finished Basement, Add	29.65	28.65	27.30	26.10	25.05	24.50	24.05	23.55	22.85	22.25	21.70
Unfinished Basement, Add	12.40	11.20	10.40	9.65	9.05	8.65	8.45	8.10	7.70	7.35	7.05

Modifications

Add to the total cost

Upgrade Kitchen Cabinets	$ + 3276
Solid Surface Countertops (Included)	
Full Bath - including plumbing, wall and floor finishes	+ 4559
Half Bath - including plumbing, wall and floor finishes	+ 2770
One Car Attached Garage	+ 10,056
One Car Detached Garage	+ 13,167
Fireplace & Chimney	+ 4420

Adjustments

For multi family - add to total cost

Additional Kitchen	$ + 5392
Additional Bath	+ 4559
Additional Entry & Exit	+ 1356
Separate Heating	+ 1302
Separate Electric	+ 1643

For Townhouse/Rowhouse - Multiply cost per square foot by

Inner Unit	.92
End Unit	.96

Alternatives

Add to or deduct from the cost per square foot of living area

Cedar Shake Roof	+ 2.65
Clay Tile Roof	+ 4.15
Slate Roof	+ 7.77
Upgrade Walls to Skim Coat Plaster	+ .38
Upgrade Ceilings to Textured Finish	+ .42
Air Conditioning, in Heating Ductwork	+ 2.90
In Separate Ductwork	+ 5.55
Heating Systems, Hot Water	+ 1.50
Heat Pump	+ 1.84
Electric Heat	– .70
Not Heated	– 2.63

Additional upgrades or components

Kitchen Cabinets & Countertops	Page 77
Bathroom Vanities	78
Fireplaces & Chimneys	78
Windows, Skylights & Dormers	78
Appliances	79
Breezeways & Porches	79
Finished Attic	79
Garages	80
Site Improvements	80
Wings & Ells	44

Important: See the Reference Section for Location Factors (to adjust for your city) and Estimating Forms

Average 1 Story

Living Area - 1600 S.F.
Perimeter - 163 L.F.

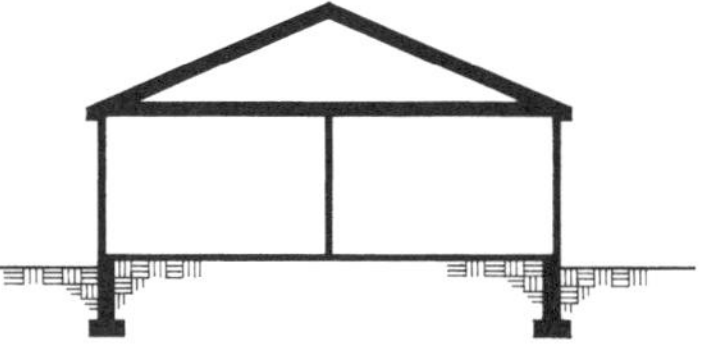

		Labor-Hours	Cost Per Square Foot Of Living Area		
			Mat.	Labor	Total
1 Site Work	Site preparation for slab; 4′ deep trench excavation for foundation wall.	.048		.77	.77
2 Foundation	Continuous reinforced concrete footing 8″ deep x 18″ wide; dampproofed and insulated reinforced concrete foundation wall, 8″ thick, 4′ deep; 4″ concrete slab on 4″ crushed stone base and polyethylene vapor barrier, trowel finish.	.113	4.37	5.27	9.64
3 Framing	Exterior walls - 2″ x 4″ wood studs, 16″ O.C.; 1/2″ plywood sheathing; 2″ x 6″ rafters 16″ O.C. with 1/2″ plywood sheathing, 4 in 12 pitch; 2″ x 6″ ceiling joists 16″ O.C.; 1/2″ plywood subfloor on 1″ x 2″ wood sleepers 16″ O.C.	.136	6.06	7.40	13.46
4 Exterior Walls	Beveled wood siding and building paper on insulated wood frame walls; 6″ attic insulation; double hung windows; 3 flush solid core wood exterior doors with storms.	.098	7.96	4.04	12.00
5 Roofing	25 year asphalt shingles; #15 felt building paper; aluminum gutters, downspouts, drip edge and flashings.	.047	1.20	1.95	3.15
6 Interiors	Walls and ceilings, 1/2″ taped and finished gypsum wallboard, primed and painted with 2 coats; painted baseboard and trim, finished hardwood floor 40%, carpet with 1/2″ underlayment 40%, vinyl tile with 1/2″ underlayment 15%, ceramic tile with 1/2″ underlayment 5%; hollow core and louvered doors.	.251	11.03	10.41	21.44
7 Specialties	Average grade kitchen cabinets - 14 L.F. wall and base with solid surface counter top and kitchen sink; 40 gallon electric water heater.	.009	1.96	.89	2.85
8 Mechanical	1 lavatory, white, wall hung; 1 water closet, white; 1 bathtub with shower, enameled steel, white; gas fired warm air heat.	.098	3.11	2.51	5.62
9 Electrical	200 Amp. service; romex wiring; incandescent lighting fixtures, switches, receptacles.	.041	1.05	1.45	2.50
10 Overhead	Contractor's overhead and profit and plans.		6.26	5.91	12.17
		Total	43.00	40.60	**83.60**

- **Simple design from standard plans**
- **Single family — 1 full bath, 1 kitchen**
- **No basement**
- **Asphalt shingles on roof**
- **Hot air heat**
- **Gypsum wallboard interior finishes**
- **Materials and workmanship are average**

Note: The illustration shown may contain some optional components (for example: garages and/or fireplaces) whose costs are shown in the modifications, adjustments, & alternatives below or at the end of the square foot section.

©By Designer

Base cost per square foot of living area

Exterior Wall	Living Area 600	800	1000	1200	1400	1600	1800	2000	2400	2800	3200
Wood Siding - Wood Frame	131.45	110.70	99.40	94.20	90.55	84.85	82.05	79.20	73.30	71.00	68.55
Brick Veneer - Wood Frame	142.00	118.40	106.50	100.90	96.90	90.65	87.55	84.40	77.85	75.40	72.60
Stucco on Wood Frame	130.05	109.70	98.45	93.40	89.75	84.15	81.40	78.50	72.70	70.50	68.05
Solid Masonry	158.30	130.20	117.45	111.10	106.60	99.50	95.95	92.35	84.95	82.05	78.75
Finished Basement, Add	24.25	21.15	20.25	19.55	19.05	18.25	17.85	17.50	16.70	16.40	15.95
Unfinished Basement, Add	10.65	8.40	7.80	7.35	7.05	6.60	6.30	6.10	5.65	5.40	5.15

Modifications

Add to the total cost

Upgrade Kitchen Cabinets	$ + 3276
Solid Surface Countertops (Included)	
Full Bath - including plumbing, wall and floor finishes	+ 4559
Half Bath - including plumbing, wall and floor finishes	+ 2770
One Car Attached Garage	+ 10,056
One Car Detached Garage	+ 13,167
Fireplace & Chimney	+ 4420

Adjustments

For multi family - add to total cost

Additional Kitchen	$ + 5392
Additional Bath	+ 4559
Additional Entry & Exit	+ 1356
Separate Heating	+ 1302
Separate Electric	+ 1643

For Townhouse/Rowhouse - Multiply cost per square foot by

Inner Unit	.92
End Unit	.96

Alternatives

Add to or deduct from the cost per square foot of living area

Cedar Shake Roof	+ 1.90
Clay Tile Roof	+ 3.00
Slate Roof	+ 5.61
Upgrade Walls to Skim Coat Plaster	+ .43
Upgrade Ceilings to Textured Finish	+ .42
Air Conditioning, in Heating Ductwork	+ 2.20
In Separate Ductwork	+ 4.85
Heating Systems, Hot Water	+ 1.42
Heat Pump	+ 2.04
Electric Heat	– .63
Not Heated	– 2.54

Additional upgrades or components

Important: See the Reference Section for Location Factors (to adjust for your city) and Estimating Forms

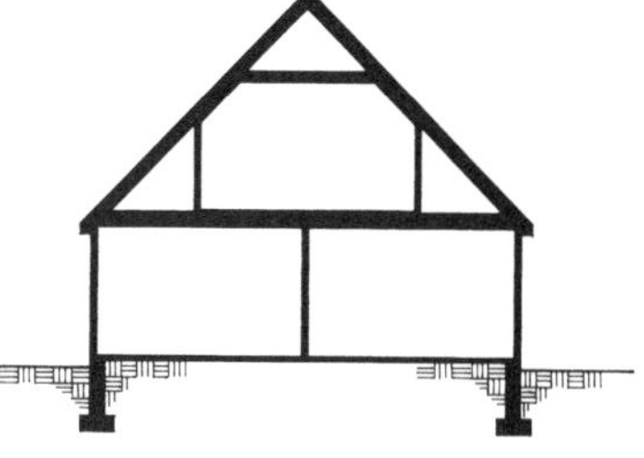

Average 1-1/2 Story

Living Area - 1800 S.F.
Perimeter - 144 L.F.

		Labor-Hours	Cost Per Square Foot Of Living Area		
			Mat.	Labor	Total
1 Site Work	Site preparation for slab; 4′ deep trench excavation for foundation wall.	.037		.69	.69
2 Foundation	Continuous reinforced concrete footing 8″ deep x 18″ wide; dampproofed and insulated reinforced concrete foundation wall, 8″ thick, 4′ deep; 4″ concrete slab on 4″ crushed stone base and polyethylene vapor barrier, trowel finish.	.073	3.11	3.85	6.96
3 Framing	Exterior walls - 2″ x 4″ wood studs, 16″ O.C.; 1/2″ plywood sheathing; 2″ x 6″ rafters 16″ O.C. with 1/2″ plywood sheathing, 8 in 12 pitch; 2″ x 8″ floor joists 16″ O.C. with 5/8″ plywood subfloor; 1/2″ plywood subfloor on 1″ x 2″ wood sleepers 16″ O.C.	.098	6.24	6.88	13.12
4 Exterior Walls	Beveled wood siding and building paper on insulated wood frame walls; 6″ attic insulation; double hung windows; 3 flush solid core wood exterior doors with storms.	.078	8.64	4.41	13.05
5 Roofing	25 year asphalt shingles; #15 felt building paper; aluminum gutters, downspouts, drip edge and flashings.	.029	.75	1.22	1.97
6 Interiors	Walls and ceilings, 1/2″ taped and finished gypsum wallboard, primed and painted with 2 coats; painted baseboard and trim, finished hardwood floor 40%, carpet with 1/2″ underlayment 40%, vinyl tile with 1/2″ underlayment 15%, ceramic tile with 1/2″ underlayment 5%; hollow core and louvered doors.	.225	12.36	11.81	24.17
7 Specialties	Average grade kitchen cabinets - 14 L.F. wall and base with solid surface counter top and kitchen sink; 40 gallon electric water heater.	.022	1.73	.78	2.51
8 Mechanical	1 lavatory, white, wall hung; 1 water closet, white; 1 bathtub with shower, enameled steel, white; gas fired warm air heat.	.049	2.90	2.41	5.31
9 Electrical	200 Amp. service; romex wiring; incandescent lighting fixtures, switches, receptacles.	.039	.99	1.38	2.37
10 Overhead	Contractor's overhead and profit and plans.		6.23	5.67	11.90
		Total	42.95	39.10	**82.05**

- **Simple design from standard plans**
- **Single family — 1 full bath, 1 kitchen**
- **No basement**
- **Asphalt shingles on roof**
- **Hot air heat**
- **Gypsum wallboard interior finishes**
- **Materials and workmanship are average**

Note: The illustration shown may contain some optional components (for example: garages and/or fireplaces) whose costs are shown in the modifications, adjustments, & alternatives below or at the end of the square foot section.

Base cost per square foot of living area

Exterior Wall	Living Area 1000	1200	1400	1600	1800	2000	2200	2600	3000	3400	3800
Wood Siding - Wood Frame	104.80	95.15	90.80	87.80	84.70	**81.25**	79.10	74.75	70.60	68.75	67.00
Brick Veneer - Wood Frame	113.05	102.75	97.90	94.60	91.10	87.40	85.00	80.05	75.45	73.35	71.45
Stucco on Wood Frame	103.80	94.20	89.90	86.95	83.90	80.50	78.35	74.10	69.90	68.15	66.45
Solid Masonry	125.65	114.40	108.80	105.00	100.90	96.85	93.90	88.20	82.95	80.50	78.25
Finished Basement, Add	17.10	16.95	16.35	15.95	15.55	15.30	15.05	14.50	14.05	13.80	13.55
Unfinished Basement, Add	6.80	6.30	5.95	5.75	5.45	5.30	5.15	4.80	4.55	4.45	4.25

Modifications

Add to the total cost

Upgrade Kitchen Cabinets	$ + 3276
Solid Surface Countertops (Included)	
Full Bath - including plumbing, wall and floor finishes	+ 4559
Half Bath - including plumbing, wall and floor finishes	+ 2770
One Car Attached Garage	+ 10,056
One Car Detached Garage	+ 13,167
Fireplace & Chimney	+ 4930

Adjustments

For multi family - add to total cost

Additional Kitchen	$ + 5392
Additional Bath	+ 4559
Additional Entry & Exit	+ 1356
Separate Heating	+ 1302
Separate Electric	+ 1643

For Townhouse/Rowhouse - Multiply cost per square foot by

Inner Unit	.90
End Unit	.95

Alternatives

Add to or deduct from the cost per square foot of living area

Cedar Shake Roof	+ 1.30
Clay Tile Roof	+ 2.10
Slate Roof	+ 3.89
Upgrade Walls to Skim Coat Plaster	+ .44
Upgrade Ceilings to Textured Finish	+ .42
Air Conditioning, in Heating Ductwork	+ 1.75
In Separate Ductwork	+ 4.40
Heating Systems, Hot Water	+ 1.39
Heat Pump	+ 2.15
Electric Heat	– .55
Not Heated	– 2.46

Additional upgrades or components

Kitchen Cabinets & Countertops	Page 77
Bathroom Vanities	78
Fireplaces & Chimneys	78
Windows, Skylights & Dormers	78
Appliances	79
Breezeways & Porches	79
Finished Attic	79
Garages	80
Site Improvements	80
Wings & Ells	44

Average 2 Story

Living Area - 2000 S.F.
Perimeter - 135 L.F.

		Labor-Hours	Cost Per Square Foot Of Living Area		
			Mat.	Labor	Total
1 Site Work	Site preparation for slab; 4′ deep trench excavation for foundation wall.	.034		.62	.62
2 Foundation	Continuous reinforced concrete footing 8″ deep x 18″ wide; dampproofed and insulated reinforced concrete foundation wall, 8″ thick, 4′ deep, 4″ concrete slab on 4″ crushed stone base and polyethylene vapor barrier, trowel finish.	.066	2.57	3.22	5.79
3 Framing	Exterior walls - 2″ x 4″ wood studs, 16″ O.C.; 1/2″ plywood sheathing; 2″ x 6″ rafters 16″ O.C. with 1/2″ plywood sheathing, 4 in 12 pitch; 2″ x 6″ ceiling joists 16″ O.C.; 2″ x 8″ floor joists 16″ O.C. with 5/8″ plywood subfloor; 1/2″ plywood subfloor on 1″ by 2″ wood sleepers 16″ O.C.	.131	6.46	7.13	13.59
4 Exterior Walls	Beveled wood siding and building paper on insulated wood frame walls; 6″ attic insulation; double hung windows; 3 flush solid core wood exterior doors with storms.	.111	9.40	4.78	14.18
5 Roofing	25 year asphalt shingles; #15 felt building paper; aluminum gutters, downspouts, drip edge and flashings.	.024	.61	.98	1.59
6 Interiors	Walls and ceilings, 1/2″ taped and finished gypsum wallboard, primed and painted with 2 coats; painted baseboard and trim, finished hardwood floor 40%, carpet with 1/2″ underlayment 40%, vinyl tile with 1/2″ underlayment 15%, ceramic tile with 1/2″ underlayment 5%; hollow core and louvered doors.	.232	12.26	11.81	24.07
7 Specialties	Average grade kitchen cabinets - 14 L.F. wall and base with solid surface counter top and kitchen sink; 40 gallon electric water heater.	.021	1.56	.71	2.27
8 Mechanical	1 lavatory, white, wall hung; 1 water closet, white; 1 bathtub with shower; enameled steel, white; gas fired warm air heat.	.060	2.74	2.33	5.07
9 Electrical	200 Amp. service; romex wiring; incandescent lighting fixtures, switches, receptacles.	.039	.95	1.33	2.28
10 Overhead	Contractor's overhead and profit and plans.		6.20	5.59	11.79
		Total	42.75	38.50	**81.25**

RESIDENTIAL | Average | 2-1/2 Story

- **Simple design from standard plans**
- **Single family — 1 full bath, 1 kitchen**
- **No basement**
- **Asphalt shingles on roof**
- **Hot air heat**
- **Gypsum wallboard interior finishes**
- **Materials and workmanship are average**

Note: The illustration shown may contain some optional components (for example: garages and/or fireplaces) whose costs are shown in the modifications, adjustments, & alternatives below or at the end of the square foot section.

Base cost per square foot of living area

Exterior Wall	Living Area 1200	1400	1600	1800	2000	2400	2800	3200	3600	4000	4400
Wood Siding - Wood Frame	104.05	98.05	89.60	88.05	85.05	80.35	76.50	**72.55**	70.65	67.10	65.95
Brick Veneer - Wood Frame	112.75	105.95	96.85	95.30	91.95	86.55	82.40	77.95	75.70	71.80	70.45
Stucco on Wood Frame	102.95	97.05	88.65	87.10	84.20	79.55	75.75	71.90	70.00	66.50	65.30
Solid Masonry	126.20	118.10	108.00	106.50	102.40	96.00	91.45	86.15	83.50	79.00	77.45
Finished Basement, Add	14.50	14.20	13.70	13.60	13.30	12.70	12.45	12.05	11.80	11.50	11.40
Unfinished Basement, Add	5.65	5.20	4.90	4.80	4.60	4.30	4.10	3.85	3.70	3.55	3.45

Modifications

Add to the total cost

Upgrade Kitchen Cabinets	$ + 3276
Solid Surface Countertops (Included)	
Full Bath - including plumbing, wall and floor finishes	+ 4559
Half Bath - including plumbing, wall and floor finishes	+ 2770
One Car Attached Garage	+ 10,056
One Car Detached Garage	+ 13,167
Fireplace & Chimney	+ 5600

Adjustments

For multi family - add to total cost

Additional Kitchen	$ + 5392
Additional Bath	+ 4559
Additional Entry & Exit	+ 1356
Separate Heating	+ 1302
Separate Electric	+ 1643

For Townhouse/Rowhouse - Multiply cost per square foot by

Inner Unit	.90
End Unit	.95

Alternatives

Add to or deduct from the cost per square foot of living area

Cedar Shake Roof	+ 1.15
Clay Tile Roof	+ 1.80
Slate Roof	+ 3.37
Upgrade Walls to Skim Coat Plaster	+ .41
Upgrade Ceilings to Textured Finish	+ .42
Air Conditioning, in Heating Ductwork	+ 1.60
In Separate Ductwork	+ 4.25
Heating Systems, Hot Water	+ 1.25
Heat Pump	+ 2.20
Electric Heat	– .97
Not Heated	– 2.87

Additional upgrades or components

Kitchen Cabinets & Countertops	Page 77
Bathroom Vanities	78
Fireplaces & Chimneys	78
Windows, Skylights & Dormers	78
Appliances	79
Breezeways & Porches	79
Finished Attic	79
Garages	80
Site Improvements	80
Wings & Ells	44

Average 2-1/2 Story

Living Area - 3200 S.F.
Perimeter - 150 L.F.

		Labor-Hours	Cost Per Square Foot Of Living Area		
			Mat.	Labor	Total
1 Site Work	Site preparation for slab; 4′ deep trench excavation for foundation wall.	.046		.39	.39
2 Foundation	Continuous reinforced concrete footing 8″ deep x 18″ wide, dampproofed and insulated reinforced concrete foundation wall, 8″ thick, 4′ deep; 4″ concrete slab on 4″ crushed stone base and polyethylene vapor barrier, trowel finish.	.061	1.86	2.27	4.13
3 Framing	Exterior walls - 2″ x 4″ wood studs, 16″ O.C.; 1/2″ plywood sheathing; 2″ x 6″ rafters 16″ O.C. with 1/2″ plywood sheathing, 4 in 12 pitch; 2″ x 6″ ceiling joists 16″ O.C.; 2″ x 8″ floor joists 16″ O.C. with 5/8″ plywood subfloor; 1/2″ plywood subfloor on 1″ by 2″ wood sleepers 16″ O.C.	.127	6.39	6.97	13.36
4 Exterior Walls	Beveled wood siding and building paper on insulated wood frame walls; 6″ attic insulation; double hung windows; 3 flush solid core wood exterior doors with storms.	.136	7.89	4.03	11.92
5 Roofing	25 year asphalt shingles; #15 felt building paper; aluminum gutters, downspouts, drip edge and flashings.	.018	.46	.75	1.21
6 Interiors	Walls and ceilings, 1/2″ taped and finished gypsum wallboard, primed and painted with 2 coats; painted baseboard and trim, finished hardwood floor 40%, carpet with 1/2″ underlayment 40%, vinyl tile with 1/2″ underlayment 15%, ceramic tile with 1/2″ underlayment 5%; hollow core and louvered doors.	.286	11.97	11.40	23.37
7 Specialties	Average grade kitchen cabinets - 14 L.F. wall and base with solid surface counter top and kitchen sink; 40 gallon electric water heater.	.030	.98	.44	1.42
8 Mechanical	1 lavatory, white, wall hung; 1 water closet, white; 1 bathtub with shower, enameled steel, white; gas fired warm air heat.	.072	2.18	2.07	4.25
9 Electrical	200 Amp. service; romex wiring; incandescent lighting fixtures, switches, receptacles.	.046	.81	1.17	1.98
10 Overhead	Contractor's overhead and profit and plans.		5.51	5.01	10.52
		Total	38.05	34.50	**72.55**

- **Simple design from standard plans**
- **Single family – 1 full bath, 1 kitchen**
- **No basement**
- **Asphalt shingles on roof**
- **Hot air heat**
- **Gypsum wallboard interior finishes**
- **Materials and workmanship are average**

Note: The illustration shown may contain some optional components (for example: garages and/or fireplaces) whose costs are shown in the modifications, adjustments, & alternatives below or at the end of the square foot section.

Base cost per square foot of living area

Exterior Wall	Living Area 1500	1800	2100	2500	3000	3500	4000	4500	5000	5500	6000
Wood Siding - Wood Frame	96.80	88.00	84.30	81.35	**75.80**	73.35	69.95	66.20	65.05	63.70	62.25
Brick Veneer - Wood Frame	105.00	95.60	91.40	88.15	81.95	79.15	75.20	71.05	69.80	68.25	66.55
Stucco on Wood Frame	95.75	87.05	83.40	80.50	75.00	72.65	69.25	65.50	64.40	63.10	61.75
Solid Masonry	117.60	107.30	102.30	98.55	91.35	88.10	83.35	78.55	77.00	75.25	73.10
Finished Basement, Add	12.45	12.35	11.95	11.75	11.30	11.05	10.70	10.40	10.35	10.20	10.00
Unfinished Basement, Add	4.55	4.25	4.05	3.90	3.65	3.45	3.25	3.10	3.05	2.95	2.85

Modifications

Add to the total cost

Upgrade Kitchen Cabinets	$ + 3276
Solid Surface Countertops (Included)	
Full Bath - including plumbing, wall and floor finishes	+ 4559
Half Bath - including plumbing, wall and floor finishes	+ 2770
One Car Attached Garage	+ 10,056
One Car Detached Garage	+ 13,167
Fireplace & Chimney	+ 5600

Adjustments

For multi family - add to total cost

Additional Kitchen	$ + 5392
Additional Bath	+ 4559
Additional Entry & Exit	+ 1356
Separate Heating	+ 1302
Separate Electric	+ 1643

For Townhouse/Rowhouse - Multiply cost per square foot by

Inner Unit	.88
End Unit	.94

Alternatives

Add to or deduct from the cost per square foot of living area

Cedar Shake Roof	+ .90
Clay Tile Roof	+ 1.40
Slate Roof	+ 2.59
Upgrade Walls to Skim Coat Plaster	+ .44
Upgrade Ceilings to Textured Finish	+ .42
Air Conditioning, in Heating Ductwork	+ 1.60
In Separate Ductwork	+ 4.25
Heating Systems, Hot Water	+ 1.25
Heat Pump	+ 2.20
Electric Heat	– .75
Not Heated	– 2.67

Additional upgrades or components

Important: See the Reference Section for Location Factors (to adjust for your city) and Estimating Forms

Average 3 Story

Living Area - 3000 S.F.
Perimeter - 135 L.F.

		Labor-Hours	Cost Per Square Foot Of Living Area		
			Mat.	Labor	Total
1 Site Work	Site preparation for slab; 4′ deep trench excavation for foundation wall.	.038		.41	.41
2 Foundation	Continuous reinforced concrete footing 8″ deep x 18″ wide, dampproofed and insulated reinforced concrete foundation wall, 8″ thick, 4′ deep; 4″ concrete slab on 4″ crushed stone base and polyethylene vapor barrier, trowel finish.	.053	1.73	2.15	3.88
3 Framing	Exterior walls - 2″ x 4″ wood studs, 16″ O.C.; 1/2″ plywood sheathing; 2″ x 6″ rafters 16″ O.C. with 1/2″ plywood sheathing, 4 in 12 pitch; 2″ x 6″ ceiling joists 16″ O.C.; 2″ x 8″ floor joists 16″ O.C. with 5/8″ plywood subfloor; 1/2″ plywood subfloor on 1″ by 2″ wood sleepers 16″ O.C.	.128	6.58	7.20	13.78
4 Exterior Walls	Horizontal beveled wood siding; building paper; 3-1/2″ batt insulation; wood double hung windows; 3 flush solid core wood exterior doors; storms and screens.	.139	8.99	4.59	13.58
5 Roofing	25 year asphalt shingles; #15 felt building paper; aluminum gutters, downspouts, drip edge and flashings.	.014	.40	.65	1.05
6 Interiors	Walls and ceilings, 1/2″ taped and finished gypsum wallboard, primed and painted with 2 coats; painted baseboard and trim, finished hardwood floor 40%, carpet with 1/2″ underlayment 40%, vinyl tile with 1/2″ underlayment 15%, ceramic tile with 1/2″ underlayment 5%; hollow core and louvered doors.	.280	12.37	11.85	24.22
7 Specialties	Average grade kitchen cabinets - 14 L.F. wall and base with solid surface counter top and kitchen sink; 40 gallon electric water heater.	.025	1.04	.46	1.50
8 Mechanical	1 lavatory, white, wall hung; 1 water closet, white; 1 bathtub with shower, enameled steel, white; gas fired warm air heat.	.065	2.25	2.10	4.35
9 Electrical	200 Amp. service; romex wiring; incandescent lighting fixtures, switches, receptacles.	.042	.82	1.18	2.00
10 Overhead	Contractor's overhead and profit and plans.		5.82	5.21	11.03
		Total	40.00	35.80	**75.80**

- **Simple design from standard plans**
- **Single family — 1 full bath, 1 kitchen**
- **No basement**
- **Asphalt shingles on roof**
- **Hot air heat**
- **Gypsum wallboard interior finishes**
- **Materials and workmanship are average**

Note: The illustration shown may contain some optional components (for example: garages and/or fireplaces) whose costs are shown in the modifications, adjustments, & alternatives below or at the end of the square foot section.

Base cost per square foot of living area

Exterior Wall	Living Area 1000	1200	1400	1600	1800	2000	2200	2600	3000	3400	3800
Wood Siding - Wood Frame	98.10	88.90	84.95	82.30	79.45	**76.25**	74.35	70.45	66.55	64.90	63.35
Brick Veneer - Wood Frame	104.25	94.65	90.25	87.40	84.25	80.85	78.70	74.40	70.20	68.40	66.70
Stucco on Wood Frame	97.35	88.20	84.25	81.65	78.85	75.60	73.75	69.95	66.05	64.45	62.95
Solid Masonry	113.75	103.40	98.50	95.20	91.65	87.90	85.45	80.45	75.85	73.70	71.75
Finished Basement, Add	17.10	16.95	16.35	15.95	15.55	15.30	15.05	14.50	14.05	13.80	13.55
Unfinished Basement, Add	6.80	6.30	5.95	5.75	5.45	5.30	5.15	4.80	4.55	4.45	4.25

Modifications

Add to the total cost

Upgrade Kitchen Cabinets	$ + 3276
Solid Surface Countertops (Included)	
Full Bath - including plumbing, wall and floor finishes	+ 4559
Half Bath - including plumbing, wall and floor finishes	+ 2770
One Car Attached Garage	+ 10,056
One Car Detached Garage	+ 13,167
Fireplace & Chimney	+ 4420

Adjustments

For multi family - add to total cost

Additional Kitchen	$ + 5392
Additional Bath	+ 4559
Additional Entry & Exit	+ 1356
Separate Heating	+ 1302
Separate Electric	+ 1643

For Townhouse/Rowhouse - Multiply cost per square foot by

Inner Unit	.91
End Unit	.96

Alternatives

Add to or deduct from the cost per square foot of living area

Cedar Shake Roof	+ 1.30
Clay Tile Roof	+ 2.10
Slate Roof	+ 3.89
Upgrade Walls to Skim Coat Plaster	+ .40
Upgrade Ceilings to Textured Finish	+ .42
Air Conditioning, in Heating Ductwork	+ 1.75
In Separate Ductwork	+ 4.40
Heating Systems, Hot Water	+ 1.39
Heat Pump	+ 2.15
Electric Heat	– .55
Not Heated	– 2.46

Additional upgrades or components

Kitchen Cabinets & Countertops	Page 77
Bathroom Vanities	78
Fireplaces & Chimneys	78
Windows, Skylights & Dormers	78
Appliances	79
Breezeways & Porches	79
Finished Attic	79
Garages	80
Site Improvements	80
Wings & Ells	44

Important: See the Reference Section for Location Factors (to adjust for your city) and Estimating Forms

Average Bi-Level

Living Area - 2000 S.F.
Perimeter - 135 L.F.

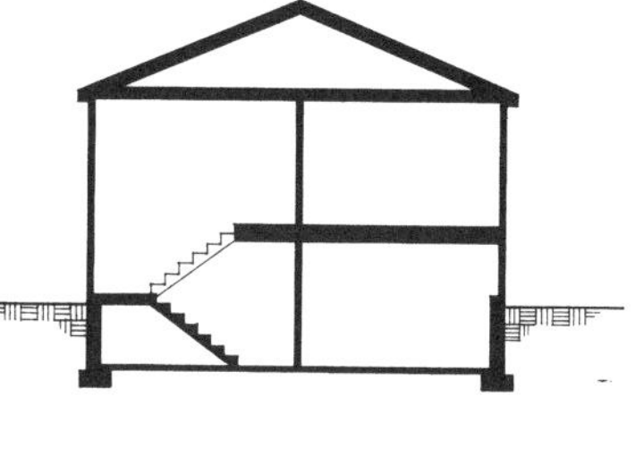

		Labor-Hours	Cost Per Square Foot Of Living Area		
			Mat.	Labor	Total
1 Site Work	Excavation for lower level, 4′ deep. Site preparation for slab.	.029		.62	.62
2 Foundation	Continuous reinforced concrete footing 8″ deep x 18″ wide, dampproofed and insulated reinforced concrete foundation wall, 8″ thick, 4′ deep; 4″ concrete slab on 4″ crushed stone base and polyethylene vapor barrier, trowel finish.	.066	2.57	3.22	5.79
3 Framing	Exterior walls - 2″ x 4″ wood studs, 16″ O.C.; 1/2″ plywood sheathing; 2″ x 6″ rafters 16″ O.C. with 1/2″ plywood sheathing, 4 in 12 pitch; 2″ x 6″ ceiling joists 16″ O.C.; 2″ x 8″ floor joists 16″ O.C. with 5/8″ plywood subfloor; 1/2″ plywood subfloor on 1″ by 2″ wood sleepers 16″ O.C.	.118	6.15	6.77	12.92
4 Exterior Walls	Horizontal beveled wood siding; building paper; 3-1/2″ batt insulation; wood double hung windows; 3 flush solid core wood exterior doors; storms and screens.	.091	7.36	3.72	11.08
5 Roofing	25 year asphalt shingles; #15 felt building paper; aluminum gutters, downspouts, drip edge and flashings.	.024	.61	.98	1.59
6 Interiors	Walls and ceilings, 1/2″ taped and finished gypsum wallboard, primed and painted with 2 coats; painted baseboard and trim, finished hardwood floor 40%, carpet with 1/2″ underlayment 40%, vinyl tile with 1/2″ underlayment 15%, ceramic tile with 1/2″ underlayment 5%; hollow core and louvered doors.	.217	12.06	11.47	23.53
7 Specialties	Average grade kitchen cabinets - 14 L.F. wall and base with solid surface counter top and kitchen sink; 40 gallon electric water heater.	.021	1.56	.71	2.27
8 Mechanical	1 lavatory, white, wall hung; 1 water closet, white; 1 bathtub with shower, enameled steel, white; gas fired warm air heat.	.061	2.74	2.33	5.07
9 Electrical	200 Amp. service; romex wiring; incandescent lighting fixtures, switches, receptacles.	.039	.95	1.33	2.28
10 Overhead	Contractor's overhead and profit and plans.		5.80	5.30	11.10
		Total	39.80	36.45	**76.25**

- **Simple design from standard plans**
- **Single family — 1 full bath, 1 kitchen**
- **No basement**
- **Asphalt shingles on roof**
- **Hot air heat**
- **Gypsum wallboard interior finishes**
- **Materials and workmanship are average**

Note: The illustration shown may contain some optional components (for example: garages and/or fireplaces) whose costs are shown in the modifications, adjustments, & alternatives below or at the end of the square foot section.

Base cost per square foot of living area

	Living Area										
Exterior Wall	1200	1500	1800	2100	2400	2700	3000	3400	3800	4200	4600
Wood Siding - Wood Frame	93.25	86.25	81.00	76.50	73.60	72.00	70.10	68.35	65.45	63.10	61.90
Brick Veneer - Wood Frame	98.90	91.35	85.65	80.70	77.55	75.80	73.65	71.80	68.65	66.15	64.80
Stucco on Wood Frame	92.50	85.60	80.45	75.95	73.10	71.50	69.65	67.90	65.00	62.70	61.50
Solid Masonry	107.55	99.25	92.75	87.15	83.60	81.65	79.15	77.10	73.60	70.75	69.30
Finished Basement, Add*	19.60	19.30	18.45	17.80	17.35	17.15	16.80	16.60	16.20	15.90	15.70
Unfinished Basement, Add*	7.55	7.00	6.50	6.10	5.85	5.70	5.50	5.35	5.10	4.95	4.90

*Basement under middle level only.

Modifications

Add to the total cost

Upgrade Kitchen Cabinets	$ + 3276
Solid Surface Countertops (Included)	
Full Bath - including plumbing, wall and floor finishes	+ 4559
Half Bath - including plumbing, wall and floor finishes	+ 2770
One Car Attached Garage	+ 10,056
One Car Detached Garage	+ 13,167
Fireplace & Chimney	+ 4420

Adjustments

For multi family - add to total cost

Additional Kitchen	$ + 5392
Additional Bath	+ 4559
Additional Entry & Exit	+ 1356
Separate Heating	+ 1302
Separate Electric	+ 1643

For Townhouse/Rowhouse - Multiply cost per square foot by

Inner Unit	.90
End Unit	.95

Alternatives

Add to or deduct from the cost per square foot of living area

Cedar Shake Roof	+ 1.90
Clay Tile Roof	+ 3.00
Slate Roof	+ 5.61
Upgrade Walls to Skim Coat Plaster	+ .35
Upgrade Ceilings to Textured Finish	+ .42
Air Conditioning, in Heating Ductwork	+ 1.48
In Separate Ductwork	+ 4.15
Heating Systems, Hot Water	+ 1.34
Heat Pump	+ 2.23
Electric Heat	− .45
Not Heated	− 2.37

Additional upgrades or components

Kitchen Cabinets & Countertops	Page 77
Bathroom Vanities	78
Fireplaces & Chimneys	78
Windows, Skylights & Dormers	78
Appliances	79
Breezeways & Porches	79
Finished Attic	79
Garages	80
Site Improvements	80
Wings & Ells	44

Average Tri-Level

Living Area - 2400 S.F.
Perimeter - 163 L.F.

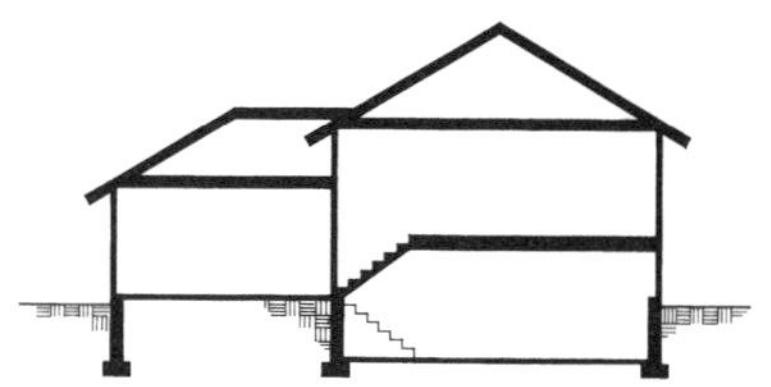

		Labor-Hours	Cost Per Square Foot Of Living Area		
			Mat.	Labor	Total
1 Site Work	Site preparation for slab; 4′ deep trench excavation for foundation wall, excavation for lower level, 4′ deep.	.029		.51	.51
2 Foundation	Continuous reinforced concrete footing 8″ deep x 18″ wide; dampproofed and insulated reinforced concrete foundation wall, 8″ thick, 4′ deep; 4″ concrete slab on 4″ crushed stone base and polyethylene vapor barrier, trowel finish.	.080	2.90	3.49	6.39
3 Framing	Exterior walls - 2″ x 4″ wood studs, 16″ O.C.; 1/2″ plywood sheathing; 2″ x 6″ rafters 16″ O.C. with 1/2″ plywood sheathing, 4 in 12 pitch; 2″ x 6″ ceiling joists 16″ O.C.; 2″ x 8″ floor joists 16″ O.C. with 5/8″ plywood subfloor; 1/2″ plywood subfloor on 1″ by 2″ wood sleepers 16″ O.C.	.124	5.81	6.37	12.18
4 Exterior Walls	Horizontal beveled wood siding: building paper; 3-1/2″ batt insulation; wood double hung windows; 3 flush solid core wood exterior doors; storms and screens.	.083	6.38	3.24	9.62
5 Roofing	25 year asphalt shingles; #15 felt building paper; aluminum gutters, downspouts, drip edge and flashings.	.032	.80	1.31	2.11
6 Interiors	Walls and ceilings, 1/2″ taped and finished gypsum wallboard, primed and painted with 2 coats; painted baseboard and trim, finished hardwood floor 40%, carpet with 1/2″ underlayment 40%, vinyl tile with 1/2″ underlayment 15%, ceramic tile with 1/2″ underlayment 5%; hollow core and louvered doors.	.186	12.23	11.18	23.41
7 Specialties	Average grade kitchen cabinets - 14 L.F. wall and base with solid surface counter top and kitchen sink; 40 gallon electric water heater.	.012	1.29	.60	1.89
8 Mechanical	1 lavatory, white, wall hung; 1 water closet, white; 1 bathtub with shower, enameled steel, white; gas fired warm air heat.	.059	2.42	2.22	4.64
9 Electrical	200 Amp. service; romex wiring; incandescent lighting fixtures, switches, receptacles.	.036	.89	1.26	2.15
10 Overhead	Contractor's overhead and profit and plans.		5.58	5.12	10.70
		Total	38.30	35.30	**73.60**

- **Post and beam frame**
- **Log exterior walls**
- **Simple design from standard plans**
- **Single family — 1 full bath, 1 kitchen**
- **No basement**
- **Asphalt shingles on roof**
- **Hot air heat**
- **Gypsum wallboard interior finishes**
- **Materials and workmanship are average**

Note: The illustration shown may contain some optional components (for example: garages and/or fireplaces) whose costs are shown in the modifications, adjustments, & alternatives below or at the end of the square foot section.

Base cost per square foot of living area

Exterior Wall	Living Area 600	800	1000	1200	1400	1600	1800	2000	2400	2800	3200
6" Log - Solid Wall	134.35	122.50	113.50	106.10	99.80	95.80	93.65	90.85	85.45	81.50	78.80
8" Log - Solid Wall	134.30	122.45	113.45	106.05	99.80	95.75	93.60	90.85	85.40	81.50	78.75
Finished Basement, Add	29.65	28.65	27.30	26.10	25.05	24.50	24.05	23.55	22.85	22.25	21.70
Unfinished Basement, Add	12.40	11.20	10.40	9.65	9.05	8.65	8.45	8.10	7.70	7.35	7.05

Modifications

Add to the total cost

Upgrade Kitchen Cabinets	$ + 3276
Solid Surface Countertops (Included)	
Full Bath - including plumbing, wall and floor finishes	+ 4559
Half Bath - including plumbing, wall and floor finishes	+ 2770
One Car Attached Garage	+ 10,056
One Car Detached Garage	+ 13,167
Fireplace & Chimney	+ 4420

Adjustments

For multi family - add to total cost

Additional Kitchen	$ + 5392
Additional Bath	+ 4559
Additional Entry & Exit	+ 1356
Separate Heating	+ 1302
Separate Electric	+ 1643

For Townhouse/Rowhouse - Multiply cost per square foot by

Inner Unit	.92
End Unit	.96

Alternatives

Add to or deduct from the cost per square foot of living area

Cedar Shake Roof	+ 1.30
Air Conditioning, in Heating Ductwork	+ 2.90
In Separate Ductwork	+ 5.54
Heating Systems, Hot Water	+ 1.50
Heat Pump	+ 1.83
Electric Heat	– .72
Not Heated	– 2.63

Additional upgrades or components

Important: See the Reference Section for Location Factors (to adjust for your city) and Estimating Forms

Solid Wall 1 Story

Living Area - 1600 S.F.
Perimeter - 163 L.F.

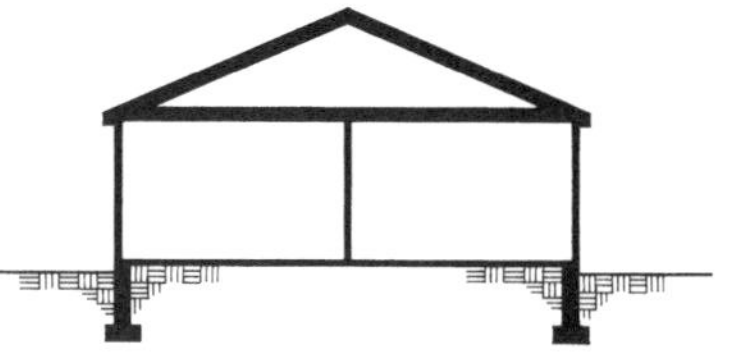

		Labor-Hours	Cost Per Square Foot Of Living Area		
			Mat.	Labor	Total
1 Site Work	Site preparation for slab; 4′ deep trench excavation for foundation wall.	.048		.77	.77
2 Foundation	Continuous reinforced concrete footing 8″ deep x 18″ wide; dampproofed and insulated reinforced concrete foundation wall, 8″ thick, 4′ deep; 4″ concrete slab on 4″ crushed stone base and polyethylene vapor barrier, trowel finish.	.113	4.37	5.27	9.64
3 Framing	Exterior walls - Precut traditional log home. Handcrafted white cedar or pine logs. Delivery included.	.136	19.22	11.38	30.60
4 Exterior Walls	Wood double-hung windows, solid wood exterior doors.	.098	4.72	2.27	6.99
5 Roofing	25 year asphalt shingles; #15 felt building paper; aluminum gutters, downspouts, drip edge and flashings.	.047	1.20	1.95	3.15
6 Interiors	Walls and ceilings, 1/2″ taped and finished gypsum wallboard, primed and painted with 2 coats; painted baseboard and trim, finished hardwood floor 40%, carpet with 1/2″ underlayment 40%, vinyl tile with 1/2″ underlayment 15%, ceramic tile with 1/2″ underlayment 5%; hollow core and louvered doors.	.251	10.39	9.36	19.75
7 Specialties	Average grade kitchen cabinets - 14 L.F. wall and base with solid surface counter top and kitchen sink; 40 gallon electric water heater.	.009	1.96	.89	2.85
8 Mechanical	1 lavatory, white, wall hung; 1 water closet, white; 1 bathtub with shower, enameled steel, white; gas fired warm air heat.	.098	3.11	2.51	5.62
9 Electrical	200 Amp. service; romex wiring; incandescent lighting fixtures, switches, receptacles.	.041	1.05	1.45	2.50
10 Overhead	Contractor's overhead and profit and plans.		7.83	6.10	13.93
		Total	53.85	41.95	**95.80**

- **Post and beam frame**
- **Log exterior walls**
- **Simple design from standard plans**
- **Single family — 1 full bath, 1 kitchen**
- **No basement**
- **Asphalt shingles on roof**
- **Hot air heat**
- **Gypsum wallboard interior finishes**
- **Materials and workmanship are average**

Note: The illustration shown may contain some optional components (for example: garages and/or fireplaces) whose costs are shown in the modifications, adjustments, & alternatives below or at the end of the square foot section.

Base cost per square foot of living area

Exterior Wall	Living Area 1000	1200	1400	1600	1800	2000	2200	2600	3000	3400	3800
6" Log-Solid	120.95	110.35	105.30	101.85	98.15	**94.35**	91.75	86.60	81.80	79.60	77.55
8" Log-Solid	104.80	95.15	90.80	87.80	84.70	81.25	79.10	74.75	70.60	68.75	67.00
Finished Basement, Add	20.45	18.65	17.60	16.95	16.35	15.95	15.55	15.30	14.70	14.25	13.90
Unfinished Basement, Add	8.90	7.40	6.80	6.30	5.95	5.75	5.45	5.30	5.00	4.65	4.45

Modifications

Add to the total cost

Upgrade Kitchen Cabinets	$ + 3276
Solid Surface Countertops (Included)	
Full Bath - including plumbing, wall and floor finishes	+ 4559
Half Bath - including plumbing, wall and floor finishes	+ 2770
One Car Attached Garage	+ 10,056
One Car Detached Garage	+ 13,167
Fireplace & Chimney	+ 4930

Adjustments

For multi family - add to total cost

Additional Kitchen	$ + 5392
Additional Bath	+ 4559
Additional Entry & Exit	+ 1356
Separate Heating	+ 1302
Separate Electric	+ 1643

For Townhouse/Rowhouse - Multiply cost per square foot by

Inner Unit	.92
End Unit	.96

Alternatives

Add to or deduct from the cost per square foot of living area

Cedar Shake Roof	+ 1.90
Air Conditioning, in Heating Ductwork	+ 1.75
In Separate Ductwork	+ 4.40
Heating Systems, Hot Water	+ 1.39
Heat Pump	+ 2.15
Electric Heat	− .55
Not Heated	− 2.46

Additional upgrades or components

Kitchen Cabinets & Countertops	Page 77
Bathroom Vanities	78
Fireplaces & Chimneys	78
Windows, Skylights & Dormers	78
Appliances	79
Breezeways & Porches	79
Finished Attic	79
Garages	80
Site Improvements	80
Wings & Ells	44

Solid Wall 2 Story

Living Area - 2000 S.F.
Perimeter - 135 L.F.

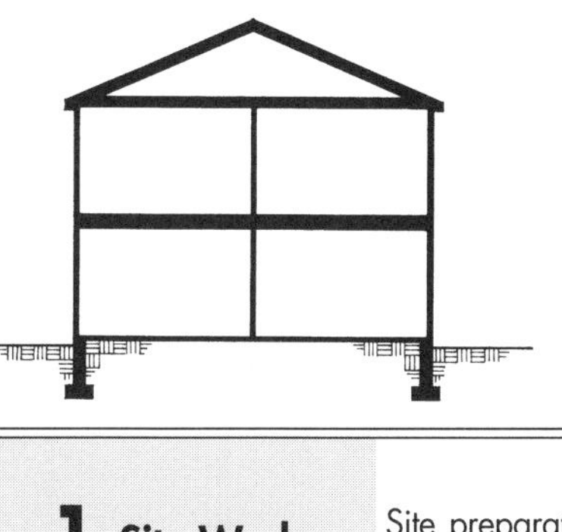

		Labor-Hours	Cost Per Square Foot Of Living Area		
			Mat.	Labor	Total
1 Site Work	Site preparation for slab; 4′ deep trench excavation for foundation wall.	.034		.62	.62
2 Foundation	Continuous reinforced concrete footing 8″ deep x 18″ wide; dampproofed and insulated reinforced concrete foundation wall, 8″ thick, 4′ deep, 4″ concrete slab on 4″ crushed stone base and polyethylene vapor barrier, trowel finish.	.066	2.57	3.22	5.79
3 Framing	Exterior walls - Precut traditional log home. Handcrafted white cedar or pine logs. Delivery included.	.131	21.70	11.80	33.50
4 Exterior Walls	Wood double-hung windows, solid wood exterior doors.	.111	5.19	2.45	7.64
5 Roofing	25 year asphalt shingles; #15 felt building paper; aluminum gutters, downspouts, drip edge and flashings.	.024	.61	.98	1.59
6 Interiors	Walls and ceilings, 1/2″ taped and finished gypsum wallboard, primed and painted with 2 coats; painted baseboard and trim, finished hardwood floor 40%, carpet with 1/2″ underlayment 40%, vinyl tile with 1/2″ underlayment 15%, ceramic tile with 1/2″ underlayment 5%; hollow core and louvered doors.	.232	11.43	10.45	21.88
7 Specialties	Average grade kitchen cabinets - 14 L.F. wall and base with solid surface counter top and kitchen sink; 40 gallon electric water heater.	.021	1.56	.71	2.27
8 Mechanical	1 lavatory, white, wall hung; 1 water closet, white; 1 bathtub with shower; enameled steel, white; gas fired warm air heat.	.060	2.74	2.33	5.07
9 Electrical	200 Amp. service; romex wiring; incandescent lighting fixtures, switches, receptacles.	.039	.95	1.33	2.28
10 Overhead	Contractor's overhead and profit and plans.		7.95	5.76	13.71
		Total	54.70	39.65	**94.35**

1 Story — Base cost per square foot of living area

Exterior Wall	Living Area 50	100	200	300	400	500	600	700
Wood Siding - Wood Frame	156.25	120.90	105.35	89.80	84.75	81.70	79.70	80.50
Brick Veneer - Wood Frame	165.75	124.90	107.10	87.75	81.95	78.40	76.10	76.70
Stucco on Wood Frame	153.60	118.90	103.75	88.65	83.70	80.70	78.70	79.55
Solid Masonry	204.75	155.55	134.25	109.05	102.10	97.85	96.40	96.50
Finished Basement, Add	47.60	39.45	35.45	28.85	27.50	26.75	26.20	25.80
Unfinished Basement, Add	22.35	17.00	14.75	10.95	10.25	9.75	9.50	9.25

1-1/2 Story — Base cost per square foot of living area

Exterior Wall	Living Area 100	200	300	400	500	600	700	800
Wood Siding - Wood Frame	125.80	101.45	86.85	78.80	74.35	72.35	69.65	68.90
Brick Veneer - Wood Frame	173.25	130.25	108.35	95.25	88.65	85.15	81.30	79.90
Stucco on Wood Frame	154.00	114.85	95.55	85.25	79.40	76.45	73.10	71.75
Solid Masonry	199.50	151.25	125.90	108.95	101.25	97.05	92.55	91.05
Finished Basement, Add	32.10	29.00	26.35	23.45	22.65	22.10	21.60	21.55
Unfinished Basement, Add	14.05	11.80	10.30	8.65	8.20	7.90	7.60	7.60

2 Story — Base cost per square foot of living area

Exterior Wall	Living Area 100	200	400	600	800	1000	1200	1400
Wood Siding - Wood Frame	125.20	94.25	80.60	68.05	63.60	61.00	59.25	60.20
Brick Veneer - Wood Frame	176.50	124.00	100.05	81.00	74.50	70.60	68.00	68.35
Stucco on Wood Frame	154.95	108.55	87.25	72.45	66.80	63.40	61.10	61.75
Solid Masonry	205.90	145.00	117.55	92.70	85.05	80.40	77.30	77.35
Finished Basement, Add	25.40	21.35	19.35	16.10	15.40	15.00	14.75	14.55
Unfinished Basement, Add	11.30	8.60	7.50	5.65	5.25	5.00	4.85	4.75

Base costs do not include bathroom or kitchen facilities. Use Modifications/Adjustments/Alternatives on pages 77-80 where appropriate.

1 Story

1-1/2 Story

2 Story

2-1/2 Story

Bi-Level

Tri-Level

- **A distinct residence from designer's plans**
- **Single family — 1 full bath, 1 half bath, 1 kitchen**
- **No basement**
- **Asphalt shingles on roof**
- **Forced hot air heat/air conditioning**
- **Gypsum wallboard interior finishes**
- **Materials and workmanship are above average**

Note: The illustration shown may contain some optional components (for example: garages and/or fireplaces) whose costs are shown in the modifications, adjustments, & alternatives below or at the end of the square foot section.

Base cost per square foot of living area

Exterior Wall	Living Area 800	1000	1200	1400	1600	1800	2000	2400	2800	3200	3600
Wood Siding - Wood Frame	141.80	128.75	118.50	110.05	104.40	101.40	97.70	90.60	85.60	81.95	78.30
Brick Veneer - Wood Frame	159.60	145.90	134.95	125.90	119.90	116.70	112.65	105.15	99.80	95.85	91.95
Stone Veneer - Wood Frame	164.65	150.50	139.10	129.70	123.40	120.10	115.90	108.05	102.50	98.35	94.25
Solid Masonry	166.65	152.25	140.70	131.20	124.80	121.45	117.10	109.20	103.55	99.35	95.20
Finished Basement, Add	45.05	44.85	42.80	41.25	40.15	39.55	38.75	37.55	36.55	35.85	35.10
Unfinished Basement, Add	19.10	18.05	17.05	16.25	15.80	15.45	15.10	14.50	14.05	13.60	13.35

Modifications

Add to the total cost

Upgrade Kitchen Cabinets	$ + 942
Solid Surface Countertops (Included)	
Full Bath - including plumbing, wall and floor finishes	+ 5402
Half Bath - including plumbing, wall and floor finishes	+ 3283
Two Car Attached Garage	+ 20,503
Two Car Detached Garage	+ 23,231
Fireplace & Chimney	+ 4600

Adjustments

For multi family - add to total cost

Additional Kitchen	$ + 11,941
Additional Full Bath & Half Bath	+ 8685
Additional Entry & Exit	+ 1356
Separate Heating & Air Conditioning	+ 5201
Separate Electric	+ 1643

For Townhouse/Rowhouse - Multiply cost per square foot by

Inner Unit	.90
End Unit	.95

Alternatives

Add to or deduct from the cost per square foot of living area

Cedar Shake Roof	+ 2.20
Clay Tile Roof	+ 3.70
Slate Roof	+ 7.43
Upgrade Ceilings to Textured Finish	+ .42
Air Conditioning, in Heating Ductwork	Base System
Heating Systems, Hot Water	+ 1.53
Heat Pump	+ 1.82
Electric Heat	– 2.03
Not Heated	– 3.30

Additional upgrades or components

Kitchen Cabinets & Countertops	Page 77
Bathroom Vanities	78
Fireplaces & Chimneys	78
Windows, Skylights & Dormers	78
Appliances	79
Breezeways & Porches	79
Finished Attic	79
Garages	80
Site Improvements	80
Wings & Ells	60

Living Area - 2400 S.F.
Perimeter - 207 L.F.

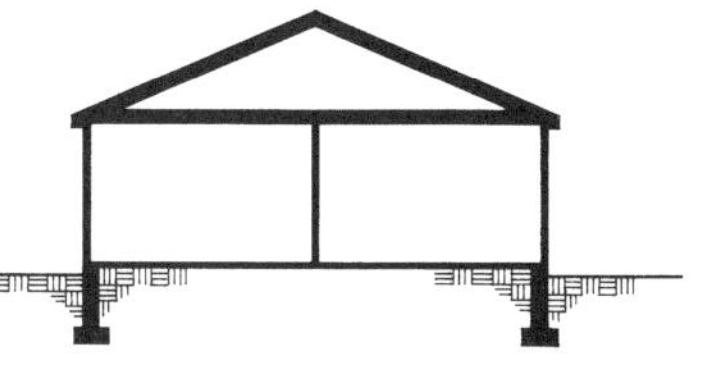

		Labor-Hours	Cost Per Square Foot Of Living Area Mat.	Labor	Total
1 Site Work	Site preparation for slab; 4′ deep trench excavation for foundation wall.	.028		.61	.61
2 Foundation	Continuous reinforced concrete footing 8″ deep x 18″ wide; dampproofed and insulated reinforced concrete foundation wall, 8″ thick, 4′ deep; 4″ concrete slab on 4″ crushed stone base and polyethylene vapor barrier, trowel finish.	.113	4.86	5.79	10.65
3 Framing	Exterior walls - 2″ x 6″ wood studs, 16″ O.C.; 1/2″ plywood sheathing; 2″ x 8″ rafters 16″ O.C. with 1/2″ plywood sheathing, 4 in 12 pitch; 2″ x 6″ ceiling joists 16″ O.C.; 5/8″ plywood subfloor on 1″ x 3″ wood sleepers 16″ O.C.	.190	4.07	5.37	9.44
4 Exterior Walls	Horizontal beveled wood siding; building paper; 6″ batt insulation; wood double hung windows; 3 solid core wood exterior doors; storms and screens.	.085	7.62	2.70	10.32
5 Roofing	30 year asphalt shingles; #15 felt building paper; aluminum gutters, downspouts and drip edge; copper flashings.	.082	3.08	2.83	5.91
6 Interiors	Walls and ceilings - 5/8″ gypsum wallboard, skim coat plaster, painted with primer and 2 coats; hardwood baseboard and trim, sanded and finished; hardwood floor 70%, ceramic tile with underlayment 20%, vinyl tile with underlayment 10%; wood panel interior doors, primed and painted with 2 coats.	.292	13.19	10.16	23.35
7 Specialties	Custom grade kitchen cabinets - 20 L.F. wall and base with solid surface counter top and kitchen sink; 4 L.F. bathroom vanity; 75 gallon electric water heater, medicine cabinet.	.019	4.12	1.00	5.12
8 Mechanical	Gas fired warm air heat/air conditioning; one full bath including: bathtub, corner shower, built in lavatory and water closet; one 1/2 bath including: built in lavatory and water closet.	.092	5.13	2.56	7.69
9 Electrical	200 Amp. service; romex wiring; fluorescent and incandescent lighting fixtures, switches, receptacles.	.039	.98	1.44	2.42
10 Overhead	Contractor's overhead and profit and design.		8.60	6.49	15.09
		Total	51.65	38.95	**90.60**

- **A distinct residence from designer's plans**
- **Single family — 1 full bath, 1 half bath, 1 kitchen**
- **No basement**
- **Asphalt shingles on roof**
- **Forced hot air heat/air conditioning**
- **Gypsum wallboard interior finishes**
- **Materials and workmanship are above average**

Note: The illustration shown may contain some optional components (for example: garages and/or fireplaces) whose costs are shown in the modifications, adjustments, & alternatives below or at the end of the square foot section.

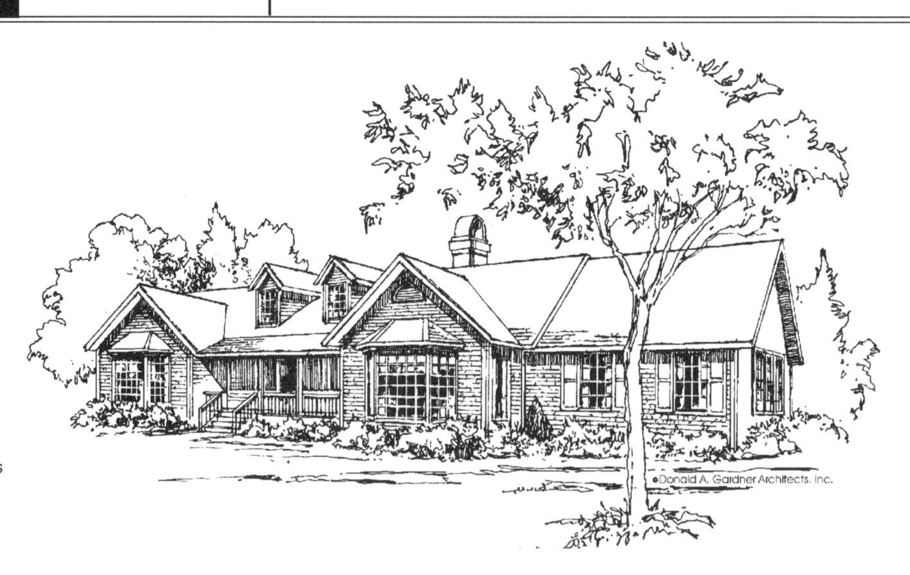

Base cost per square foot of living area

	Living Area										
Exterior Wall	1000	1200	1400	1600	1800	2000	2400	2800	3200	3600	4000
Wood Siding - Wood Frame	128.30	119.80	113.65	106.10	101.90	97.75	89.70	**86.05**	82.75	80.15	76.45
Brick Veneer - Wood Frame	136.50	127.40	120.95	112.70	108.20	103.65	94.90	91.00	87.40	84.60	80.55
Stone Veneer - Wood Frame	141.75	132.45	125.70	117.05	112.30	107.55	98.40	94.25	90.35	87.50	83.25
Solid Masonry	143.90	134.35	127.55	118.75	113.85	109.10	99.70	95.55	91.50	88.60	84.30
Finished Basement, Add	30.00	30.10	29.30	28.05	27.30	26.70	25.45	24.85	24.25	23.85	23.30
Unfinished Basement, Add	12.80	12.25	11.85	11.25	10.95	10.60	10.00	9.70	9.40	9.25	9.00

Modifications

Add to the total cost

Upgrade Kitchen Cabinets	$ + 942
Solid Surface Countertops (Included)	
Full Bath - including plumbing, wall and floor finishes	+ 5402
Half Bath - including plumbing, wall and floor finishes	+ 3283
Two Car Attached Garage	+ 20,503
Two Car Detached Garage	+ 23,231
Fireplace & Chimney	+ 4600

Adjustments

For multi family - add to total cost

Additional Kitchen	$ + 11,941
Additional Full Bath & Half Bath	+ 8685
Additional Entry & Exit	+ 1356
Separate Heating & Air Conditioning	+ 5201
Separate Electric	+ 1643

For Townhouse/Rowhouse - Multiply cost per square foot by

Inner Unit	.90
End Unit	.95

Alternatives

Add to or deduct from the cost per square foot of living area

Cedar Shake Roof	+ 1.62
Clay Tile Roof	+ 2.70
Slate Roof	+ 5.37
Upgrade Ceilings to Textured Finish	+ .42
Air Conditioning, in Heating Ductwork	Base System
Heating Systems, Hot Water	+ 1.45
Heat Pump	+ 1.92
Electric Heat	− 1.80
Not Heated	− 3.04

Additional upgrades or components

Kitchen Cabinets & Countertops	Page 77
Bathroom Vanities	78
Fireplaces & Chimneys	78
Windows, Skylights & Dormers	78
Appliances	79
Breezeways & Porches	79
Finished Attic	79
Garages	80
Site Improvements	80
Wings & Ells	60

Important: See the Reference Section for Location Factors (to adjust for your city) and Estimating Forms

Custom 1-1/2 Story

Living Area - 2800 S.F.
Perimeter - 175 L.F.

SQUARE FOOT COSTS

		Labor-Hours	Cost Per Square Foot Of Living Area		
			Mat.	Labor	Total
1 Site Work	Site preparation for slab; 4′ deep trench excavation for foundation wall.	.028		.52	.52
2 Foundation	Continuous reinforced concrete footing 8″ deep x 18″ wide; dampproofed and insulated reinforced concrete foundation wall, 8″ thick, 4′ deep; 4″ concrete slab on 4″ crushed stone base and polyethylene vapor barrier, trowel finish.	.065	3.40	4.14	7.54
3 Framing	Exterior walls - 2″ x 6″ wood studs, 16″ O.C.; 1/2″ plywood sheathing; 2″ x 8″ rafters 16″ O.C. with 1/2″ plywood sheathing, 8 in 12 pitch; 2″ x 10″ floor joists 16″ O.C. with 5/8″ plywood subfloor; 5/8″ plywood subfloor on 1″ x 3″ wood sleepers 16″ O.C.	.192	4.98	5.66	10.64
4 Exterior Walls	Horizontal beveled wood siding; building paper; 6″ batt insulation; wood double hung windows; 3 solid core wood exterior doors; storms and screens.	.064	7.60	2.73	10.33
5 Roofing	30 year asphalt shingles; #15 felt building paper; aluminum gutters, downspouts and drip edge; copper flashings.	.048	1.92	1.77	3.69
6 Interiors	Walls and ceilings - 5/8″ gypsum wallboard, skim coat plaster, painted with primer and 2 coats; hardwood baseboard and trim, sanded and finished; hardwood floor 70%, ceramic tile with underlayment 20%, vinyl tile with underlayment 10%; wood panel interior doors, primed and painted with 2 coats.	.259	14.27	11.15	25.42
7 Specialties	Custom grade kitchen cabinets - 20 L.F. wall and base with solid surface counter top and kitchen sink; 4 L.F. bathroom vanity; 75 gallon electric water heater, medicine cabinet.	.030	3.53	.86	4.39
8 Mechanical	Gas fired warm air heat/air conditioning; one full bath including: bathtub, corner shower, built in lavatory and water closet; one 1/2 bath including: built in lavatory and water closet.	.084	4.48	2.39	6.87
9 Electrical	200 Amp. service; romex wiring; fluorescent and incandescent lighting fixtures, switches, receptacles.	.038	.93	1.37	2.30
10 Overhead	Contractor's overhead and profit and design.		8.24	6.11	14.35
		Total	49.35	36.70	**86.05**

RESIDENTIAL | Custom | 2 Story

- **A distinct residence from designer's plans**
- **Single family — 1 full bath, 1 half bath, 1 kitchen**
- **No basement**
- **Asphalt shingles on roof**
- **Forced hot air heat/air conditioning**
- **Gypsum wallboard interior finishes**
- **Materials and workmanship are above average**

Note: The illustration shown may contain some optional components (for example: garages and/or fireplaces) whose costs are shown in the modifications, adjustments, & alternatives below or at the end of the square foot section.

Base cost per square foot of living area

Exterior Wall	Living Area										
	1200	1400	1600	1800	2000	2400	2800	3200	3600	4000	4400
Wood Siding - Wood Frame	120.60	113.75	108.80	104.55	99.80	92.95	**87.10**	83.35	81.15	78.75	76.65
Brick Veneer - Wood Frame	129.35	121.90	116.65	111.90	106.85	99.30	92.90	88.80	86.40	83.65	81.45
Stone Veneer - Wood Frame	135.05	127.20	121.70	116.65	111.45	103.50	96.65	92.30	89.80	86.85	84.55
Solid Masonry	137.30	129.30	123.70	118.55	113.25	105.10	98.20	93.70	91.15	88.10	85.75
Finished Basement, Add	24.15	24.20	23.60	22.90	22.45	21.40	20.65	20.15	19.80	19.35	19.10
Unfinished Basement, Add	10.25	9.85	9.50	9.20	9.00	8.50	8.15	7.85	7.75	7.50	7.40

Modifications

Add to the total cost

Upgrade Kitchen Cabinets	$ + 942
Solid Surface Countertops (Included)	
Full Bath - including plumbing, wall and floor finishes	+ 5402
Half Bath - including plumbing, wall and floor finishes	+ 3283
Two Car Attached Garage	+ 20,503
Two Car Detached Garage	+ 23,231
Fireplace & Chimney	+ 5190

Adjustments

For multi family - add to total cost

Additional Kitchen	$ + 11,941
Additional Full Bath & Half Bath	+ 8685
Additional Entry & Exit	+ 1356
Separate Heating & Air Conditioning	+ 5201
Separate Electric	+ 1643

For Townhouse/Rowhouse - Multiply cost per square foot by

Inner Unit	.87
End Unit	.93

Alternatives

Add to or deduct from the cost per square foot of living area

Cedar Shake Roof	+ 1.10
Clay Tile Roof	+ 1.85
Slate Roof	+ 3.72
Upgrade Ceilings to Textured Finish	+ .42
Air Conditioning, in Heating Ductwork	Base System
Heating Systems, Hot Water	+ 1.42
Heat Pump	+ 2.14
Electric Heat	– 1.80
Not Heated	– 2.88

Additional upgrades or components

Kitchen Cabinets & Countertops	Page 77
Bathroom Vanities	78
Fireplaces & Chimneys	78
Windows, Skylights & Dormers	78
Appliances	79
Breezeways & Porches	79
Finished Attic	79
Garages	80
Site Improvements	80
Wings & Ells	60

Important: See the Reference Section for Location Factors (to adjust for your city) and Estimating Forms

Custom 2 Story

Living Area - 2800 S.F.
Perimeter - 156 L.F.

		Labor-Hours	Cost Per Square Foot Of Living Area		
			Mat.	Labor	Total
1 Site Work	Site preparation for slab; 4′ deep trench excavation for foundation wall.	.024		.52	.52
2 Foundation	Continuous reinforced concrete footing 8″ deep x 18″ wide; dampproofed and insulated reinforced concrete foundation wall, 8″ thick, 4′ deep; 4″ concrete slab on 4″ crushed stone base and polyethylene vapor barrier, trowel finish.	.058	2.88	3.54	6.42
3 Framing	Exterior walls - 2″ x 6″ wood studs, 16″ O.C.; 1/2″ plywood sheathing; 2″ x 8″ rafters 16″ O.C. with 1/2″ plywood sheathing, 6 in 12 pitch; 2″ x 8″ ceiling joists 16″ O.C.; 2″ x 10″ floor joists 16″ O.C. with 5/8″ plywood subfloor; 5/8″ plywood subfloor on 1″ x 3″ wood sleepers 16″ O.C.	.159	5.42	5.85	11.27
4 Exterior Walls	Horizontal beveled wood siding; building paper; 6″ batt insulation; wood double hung windows; 3 solid core wood exterior doors; storms and screens.	.091	8.65	3.12	11.77
5 Roofing	30 year asphalt shingles; #15 felt building paper; aluminum gutters, downspouts and drip edge; copper flashings.	.042	1.55	1.42	2.97
6 Interiors	Walls and ceilings - 5/8″ gypsum wallboard, skim coat plaster, painted with primer and 2 coats; hardwood baseboard and trim, sanded and finished; hardwood floor 70%, ceramic tile with underlayment 20%, vinyl tile with underlayment 10%; wood panel interior doors, primed and painted with 2 coats.	.271	14.50	11.42	25.92
7 Specialties	Custom grade kitchen cabinets - 20 L.F. wall and base with solid surface counter top and kitchen sink; 4 L.F. bathroom vanity; 75 gallon electric water heater, medicine cabinet.	.028	3.53	.86	4.39
8 Mechanical	Gas fired warm air heat/air conditioning; one full bath including: bathtub, corner shower; built in lavatory and water closet; one 1/2 bath including: built in lavatory and water closet.	.078	4.60	2.43	7.03
9 Electrical	200 Amp. service; romex wiring; fluorescent and incandescent lighting fixtures, switches, receptacles.	.038	.93	1.37	2.30
10 Overhead	Contractor's overhead and profit and design.		8.39	6.12	14.51
		Total	50.45	36.65	**87.10**

- **A distinct residence from designer's plans**
- **Single family — 1 full bath, 1 half bath, 1 kitchen**
- **No basement**
- **Asphalt shingles on roof**
- **Forced hot air heat/air conditioning**
- **Gypsum wallboard interior finishes**
- **Materials and workmanship are above average**

Note: The illustration shown may contain some optional components (for example: garages and/or fireplaces) whose costs are shown in the modifications, adjustments, & alternatives below or at the end of the square foot section.

Base cost per square foot of living area

Exterior Wall	Living Area 1500	1800	2100	2400	2800	3200	3600	4000	4500	5000	5500
Wood Siding - Wood Frame	119.05	107.40	100.70	96.70	91.40	86.30	83.65	79.15	76.90	74.85	72.70
Brick Veneer - Wood Frame	128.05	115.80	108.15	103.80	98.15	92.50	89.45	84.60	82.05	79.65	77.35
Stone Veneer - Wood Frame	134.00	121.20	113.00	108.40	102.60	96.45	93.30	88.10	85.35	82.85	80.40
Solid Masonry	136.35	123.35	114.90	110.20	104.35	98.00	94.70	89.55	86.65	84.10	81.55
Finished Basement, Add	19.15	19.05	18.05	17.60	17.15	16.45	16.10	15.60	15.30	15.05	14.80
Unfinished Basement, Add	8.20	7.75	7.30	7.10	6.90	6.55	6.40	6.10	6.00	5.85	5.75

Modifications

Add to the total cost

Upgrade Kitchen Cabinets	$ + 942
Solid Surface Countertops (Included)	
Full Bath - including plumbing, wall and floor finishes	+ 5402
Half Bath - including plumbing, wall and floor finishes	+ 3283
Two Car Attached Garage	+ 20,503
Two Car Detached Garage	+ 23,231
Fireplace & Chimney	+ 5865

Adjustments

For multi family - add to total cost

Additional Kitchen	$ + 11,941
Additional Full Bath & Half Bath	+ 8685
Additional Entry & Exit	+ 1356
Separate Heating & Air Conditioning	+ 5201
Separate Electric	+ 1643

For Townhouse/Rowhouse - Multiply cost per square foot by

Inner Unit	.87
End Unit	.94

Alternatives

Add to or deduct from the cost per square foot of living area

Cedar Shake Roof	+ .95
Clay Tile Roof	+ 1.60
Slate Roof	+ 3.22
Upgrade Ceilings to Textured Finish	+ .42
Air Conditioning, in Heating Ductwork	Base System
Heating Systems, Hot Water	+ 1.28
Heat Pump	+ 2.20
Electric Heat	– 3.15
Not Heated	– 2.88

Additional upgrades or components

Kitchen Cabinets & Countertops	Page 77
Bathroom Vanities	78
Fireplaces & Chimneys	78
Windows, Skylights & Dormers	78
Appliances	79
Breezeways & Porches	79
Finished Attic	79
Garages	80
Site Improvements	80
Wings & Ells	60

Important: See the Reference Section for Location Factors (to adjust for your city) and Estimating Forms

Custom 2-1/2 Story

Living Area - 3200 S.F.
Perimeter - 150 L.F.

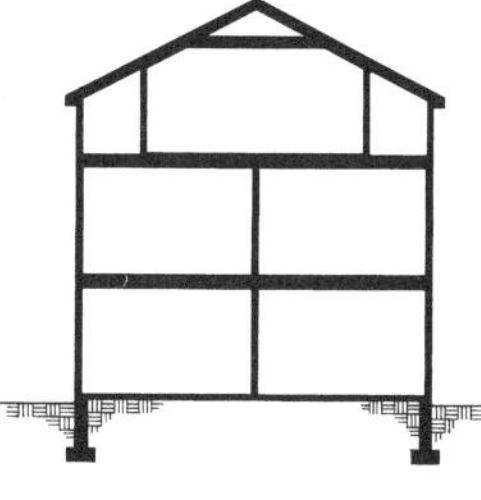

		Labor-Hours	Cost Per Square Foot Of Living Area		
			Mat.	Labor	Total
1 Site Work	Site preparation for slab; 4′ deep trench excavation for foundation wall.	.048		.46	.46
2 Foundation	Continuous reinforced concrete footing 8″ deep x 18″ wide; dampproofed and insulated reinforced concrete foundation wall, 8″ thick, 4′ deep; 4″ concrete slab on 4″ crushed stone base and polyethylene vapor barrier, trowel finish.	.063	2.37	2.94	5.31
3 Framing	Exterior walls - 2″ x 6″ wood studs, 16″ O.C.; 1/2″ plywood sheathing; 2″ x 8″ rafters 16″ O.C. with 1/2″ plywood sheathing, 6 in 12 pitch; 2″ x 8″ ceiling joists 16″ O.C.; 2″ x 10″ floor joists 16″ O.C. with 5/8″ plywood subfloor; 5/8″ plywood subfloor on 1″ x 3″ wood sleepers 16″ O.C.	.177	5.80	6.08	11.88
4 Exterior Walls	Horizontal beveled wood siding; building paper; 6″ batt insulation; wood double hung windows; 3 solid core wood exterior doors; storms and screens.	.134	8.93	3.25	12.18
5 Roofing	30 year asphalt shingles; #15 felt building paper; aluminum gutters, downspouts and drip edge; copper flashings.	.032	1.18	1.09	2.27
6 Interiors	Walls and ceilings - 5/8″ gypsum wallboard, skim coat plaster, painted with primer and 2 coats; hardwood baseboard and trim, sanded and finished; hardwood floor 70%, ceramic tile with underlayment 20%, vinyl tile with underlayment 10%; wood panel interior doors, primed and painted with 2 coats.	.354	15.09	12.14	27.23
7 Specialties	Custom grade kitchen cabinets - 20 L.F. wall and base with solid surface counter top and kitchen sink; 4 L.F. bathroom vanity; 75 gallon electric water heater, medicine cabinet.	.053	3.08	.75	3.83
8 Mechanical	Gas fired warm air heat/air conditioning; one full bath including: bathtub, corner shower; built in lavatory and water closet; one 1/2 bath including: built in lavatory and water closet.	.104	4.20	2.34	6.54
9 Electrical	200 Amp. service; romex wiring; fluorescent and incandescent lighting fixtures, switches, receptacles.	.048	.90	1.34	2.24
10 Overhead	Contractor's overhead and profit and design.		8.30	6.06	14.36
		Total	49.85	36.45	**86.30**

- **A distinct residence from designer's plans**
- **Single family — 1 full bath, 1 half bath, 1 kitchen**
- **No basement**
- **Asphalt shingles on roof**
- **Forced hot air heat/air conditioning**
- **Gypsum wallboard interior finishes**
- **Materials and workmanship are above average**

Note: The illustration shown may contain some optional components (for example: garages and/or fireplaces) whose costs are shown in the modifications, adjustments, & alternatives below or at the end of the square foot section.

Base cost per square foot of living area

Exterior Wall	Living Area 1500	1800	2100	2500	3000	3500	4000	4500	5000	5500	6000
Wood Siding - Wood Frame	118.55	107.05	101.60	97.10	89.95	86.45	82.15	77.45	75.85	74.10	72.30
Brick Veneer - Wood Frame	127.95	115.75	109.75	104.90	97.00	93.15	88.20	83.10	81.30	79.35	77.20
Stone Veneer - Wood Frame	134.05	121.45	115.05	110.00	101.60	97.50	92.15	86.80	84.80	82.70	80.45
Solid Masonry	136.50	123.70	117.15	111.95	103.45	99.20	93.75	88.20	86.20	84.05	81.65
Finished Basement, Add	16.85	16.80	16.15	15.65	14.95	14.50	13.95	13.60	13.40	13.15	12.90
Unfinished Basement, Add	7.20	6.85	6.55	6.35	6.00	5.80	5.55	5.35	5.25	5.15	5.05

Modifications

Add to the total cost

Upgrade Kitchen Cabinets	$ + 942
Solid Surface Countertops (Included)	
Full Bath - including plumbing, wall and floor finishes	+ 5402
Half Bath - including plumbing, wall and floor finishes	+ 3283
Two Car Attached Garage	+ 20,503
Two Car Detached Garage	+ 23,231
Fireplace & Chimney	+ 5865

Adjustments

For multi family - add to total cost

Additional Kitchen	$ + 11,941
Additional Full Bath & Half Bath	+ 8685
Additional Entry & Exit	+ 1356
Separate Heating & Air Conditioning	+ 5201
Separate Electric	+ 1643

For Townhouse/Rowhouse - Multiply cost per square foot by

Inner Unit	.85
End Unit	.93

Alternatives

Add to or deduct from the cost per square foot of living area

Cedar Shake Roof	+ .75
Clay Tile Roof	+ 1.25
Slate Roof	+ 2.48
Upgrade Ceilings to Textured Finish	+ .42
Air Conditioning, in Heating Ductwork	Base System
Heating Systems, Hot Water	+ 1.28
Heat Pump	+ 2.20
Electric Heat	– 3.15
Not Heated	– 2.78

Additional upgrades or components

Kitchen Cabinets & Countertops	Page 77
Bathroom Vanities	78
Fireplaces & Chimneys	78
Windows, Skylights & Dormers	78
Appliances	79
Breezeways & Porches	79
Finished Attic	79
Garages	80
Site Improvements	80
Wings & Ells	60

Custom 3 Story

Living Area - 3000 S.F.
Perimeter - 135 L.F.

		Labor-Hours	Cost Per Square Foot Of Living Area		
			Mat.	Labor	Total
1 Site Work	Site preparation for slab; 4′ deep trench excavation for foundation wall.	.048		.49	.49
2 Foundation	Continuous reinforced concrete footing 8″ deep x 18″ wide; dampproofed and insulated reinforced concrete foundation wall, 8″ thick, 4′ deep; 4″ concrete slab on 4″ crushed stone base and polyethylene vapor barrier, trowel finish.	.060	2.22	2.78	5.00
3 Framing	Exterior walls - 2″ x 6″ wood studs, 16″ O.C.; 1/2″ plywood sheathing; 2″ x 8″ rafters 16″ O.C. with 1/2″ plywood sheathing, 6 in 12 pitch; 2″ x 8″ ceiling joists 16″ O.C.; 2″ x 10″ floor joists 16″ O.C. with 5/8″ plywood subfloor; 5/8″ plywood subfloor on 1″ x 3″ wood sleepers 16″ O.C.	.191	6.09	6.27	12.36
4 Exterior Walls	Horizontal beveled wood siding; building paper; 6″ batt insulation; wood double hung windows; 3 solid core wood exterior doors; storms and screens.	.150	10.18	3.73	13.91
5 Roofing	30 year asphalt shingles; #15 felt building paper; aluminum gutters, downspouts and drip edge; copper flashings.	.028	1.02	.94	1.96
6 Interiors	Walls and ceilings - 5/8″ gypsum wallboard, skim coat plaster, painted with primer and 2 coats; hardwood baseboard and trim, sanded and finished; hardwood floor 70%, ceramic tile with underlayment 20%, vinyl tile with underlayment 10%; wood panel interior doors, primed and painted with 2 coats.	.409	15.55	12.58	28.13
7 Specialties	Custon grade kitchen cabinets - 20 L.F. wall and base with solid surface counter top and kitchen sink; 4 L.F. bathroom vanity; 75 gallon electric water heater, medicine cabinet.	.053	3.29	.80	4.09
8 Mechanical	Gas fired warm air heat/air conditioning; one full bath including: bathtub, corner shower; built in lavatory and water closet; one 1/2 bath including: built in lavatory and water closet.	.105	4.37	2.40	6.77
9 Electrical	200 Amp. service; romex wiring; fluorescent and incandescent lighting fixtures, switches, receptacles.	.048	.91	1.35	2.26
10 Overhead	Contractor's overhead and profit and design.		8.72	6.26	14.98
		Total	52.35	37.60	**89.95**

RESIDENTIAL | Custom | Bi-Level

- **A distinct residence from designer's plans**
- **Single family — 1 full bath, 1 half bath, 1 kitchen**
- **No basement**
- **Asphalt shingles on roof**
- **Forced hot air heat/air conditioning**
- **Gypsum wallboard interior finishes**
- **Materials and workmanship are above average**

Note: The illustration shown may contain some optional components (for example: garages and/or fireplaces) whose costs are shown in the modifications, adjustments, & alternatives below or at the end of the square foot section.

Base cost per square foot of living area

Exterior Wall	Living Area 1200	1400	1600	1800	2000	2400	2800	3200	3600	4000	4400
Wood Siding - Wood Frame	114.15	107.70	103.05	99.15	94.50	88.15	**82.75**	79.25	77.25	75.05	73.15
Brick Veneer - Wood Frame	120.70	113.80	108.85	104.60	99.80	92.95	87.05	83.30	81.15	78.75	76.70
Stone Veneer - Wood Frame	124.95	117.80	112.70	108.20	103.25	96.05	89.90	86.00	83.75	81.15	79.05
Solid Masonry	126.65	119.30	114.10	109.60	104.60	97.25	91.00	87.00	84.70	82.05	79.95
Finished Basement, Add	24.15	24.20	23.60	22.90	22.45	21.40	20.65	20.15	19.80	19.35	19.10
Unfinished Basement, Add	10.25	9.85	9.50	9.20	9.00	8.50	8.15	7.85	7.75	7.50	7.40

Modifications

Add to the total cost

Upgrade Kitchen Cabinets	$ + 942
Solid Surface Countertops (Included)	
Full Bath - including plumbing, wall and floor finishes	+ 5402
Half Bath - including plumbing, wall and floor finishes	+ 3283
Two Car Attached Garage	+ 20,503
Two Car Detached Garage	+ 23,231
Fireplace & Chimney	+ 4600

Adjustments

For multi family - add to total cost

Additional Kitchen	$ + 11,951
Additional Full Bath & Half Bath	+ 8685
Additional Entry & Exit	+ 1356
Separate Heating & Air Conditioning	+ 5201
Separate Electric	+ 1643

For Townhouse/Rowhouse - Multiply cost per square foot by

Inner Unit	.89
End Unit	.95

Alternatives

Add to or deduct from the cost per square foot of living area

Cedar Shake Roof	+ 1.10
Clay Tile Roof	+ 1.85
Slate Roof	+ 3.72
Upgrade Ceilings to Textured Finish	+ .42
Air Conditioning, in Heating Ductwork	Base System
Heating Systems, Hot Water	+ 1.42
Heat Pump	+ 2.14
Electric Heat	– 1.80
Not Heated	– 2.78

Additional upgrades or components

Kitchen Cabinets & Countertops	Page 77
Bathroom Vanities	78
Fireplaces & Chimneys	78
Windows, Skylights & Dormers	78
Appliances	79
Breezeways & Porches	79
Finished Attic	79
Garages	80
Site Improvements	80
Wings & Ells	60

Important: See the Reference Section for Location Factors (to adjust for your city) and Estimating Forms

Custom Bi-Level

Living Area - 2800 S.F.
Perimeter - 156 L.F.

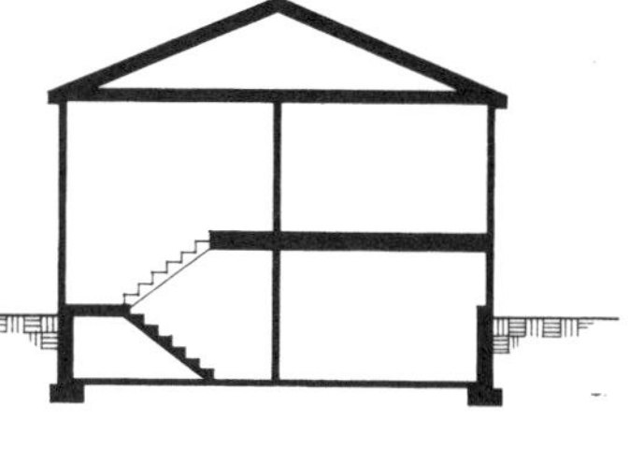

		Labor-Hours	Cost Per Square Foot Of Living Area		
			Mat.	Labor	Total
1 Site Work	Excavation for lower level, 4′ deep. Site preparation for slab.	.024		.52	.52
2 Foundation	Continuous reinforced concrete footing 8″ deep x 18″ wide; dampproofed and insulated reinforced concrete foundation wall, 8″ thick, 4′ deep; 4″ concrete slab on 4″ crushed stone base and polyethylene vapor barrier, trowel finish.	.058	2.88	3.54	6.42
3 Framing	Exterior walls - 2″ x 6″ wood studs, 16″ O.C.; 1/2″ plywood sheathing; 2″ x 8″ rafters 16″ O.C. with 1/2″ plywood sheathing, 6 in 12 pitch; 2″ x 8″ ceiling joists 16″ O.C.; 2″ x 10″ floor joists 16″ O.C. with 5/8″ plywood subfloor; 5/8″ plywood subfloor on 1″ x 3″ wood sleepers 16″ O.C.	.147	5.13	5.58	10.71
4 Exterior Walls	Horizontal beveled wood siding; building paper; 6″ batt insulation; wood double hung windows; 3 solid core wood exterior doors; storms and screens.	.079	6.75	2.42	9.17
5 Roofing	30 year asphalt shingles; #15 felt building paper; aluminum gutters, downspouts and drip edge; copper flashings.	.033	1.55	1.42	2.97
6 Interiors	Walls and ceilings - 5/8″ gypsum wallboard, skim coat plaster, painted with primer and 2 coats; hardwood baseboard and trim, sanded and finished; hardwood floor 70%, ceramic tile with underlayment 20%, vinyl tile with underlayment 10%; wood panel interior doors, primed and painted with 2 coats.	.257	14.32	11.12	25.44
7 Specialties	Custom grade kitchen cabinets - 20 L.F. wall and base with solid surface counter top and kitchen sink; 4 L.F. bathroom vanity; 75 gallon electric water heater, medicine cabinet.	.028	3.53	.86	4.39
8 Mechanical	Gas fired warm air heat/air conditioning; one full bath including: bathtub, corner shower, built in lavatory and water closet; one 1/2 bath including: built in lavatory and water closet.	.078	4.60	2.43	7.03
9 Electrical	200 Amp. service; romex wiring; fluorescent and incandescent lighting fixtures, switches, receptacles.	.038	.93	1.37	2.30
10 Overhead	Contractor's overhead and profit and design.		7.96	5.84	13.80
		Total	47.65	35.10	**82.75**

- **A distinct residence from designer's plans**
- **Single family — 1 full bath, 1 half bath, 1 kitchen**
- **No basement**
- **Asphalt shingles on roof**
- **Forced hot air heat/air conditioning**
- **Gypsum wallboard interior finishes**
- **Materials and workmanship are above average**

Note: The illustration shown may contain some optional components (for example: garages and/or fireplaces) whose costs are shown in the modifications, adjustments, & alternatives below or at the end of the square foot section.

Base cost per square foot of living area

Exterior Wall	Living Area 1200	1500	1800	2100	2400	2800	3200	3600	4000	4500	5000
Wood Siding - Wood Frame	117.40	107.00	99.20	92.60	88.30	85.20	81.45	77.45	75.70	72.00	69.90
Brick Veneer - Wood Frame	123.85	112.85	104.50	97.40	92.85	89.55	85.50	81.15	79.35	75.30	73.05
Stone Veneer - Wood Frame	128.10	116.70	107.95	100.60	95.80	92.40	88.10	83.65	81.70	77.55	75.15
Solid Masonry	129.75	118.25	109.35	101.80	96.95	93.50	89.15	84.55	82.65	78.35	75.95
Finished Basement, Add*	30.00	29.85	28.55	27.45	26.75	26.25	25.60	24.95	24.70	24.10	23.70
Unfinished Basement, Add*	12.75	12.00	11.35	10.85	10.50	10.30	9.95	9.65	9.50	9.25	9.05

*Basement under middle level only.

Modifications

Add to the total cost

Upgrade Kitchen Cabinets	$ + 942
Solid Surface Countertops (Included)	
Full Bath - including plumbing, wall and floor finishes	+ 5402
Half Bath - including plumbing, wall and floor finishes	+ 3283
Two Car Attached Garage	+ 20,503
Two Car Detached Garage	+ 23,231
Fireplace & Chimney	+ 4600

Adjustments

For multi family - add to total cost

Additional Kitchen	$ + 11,494
Additional Full Bath & Half Bath	+ 8685
Additional Entry & Exit	+ 1356
Separate Heating & Air Conditioning	+ 5201
Separate Electric	+ 1643

For Townhouse/Rowhouse - Multiply cost per square foot by

Inner Unit	.87
End Unit	.94

Alternatives

Add to or deduct from the cost per square foot of living area

Cedar Shake Roof	+ 1.60
Clay Tile Roof	+ 2.70
Slate Roof	+ 5.37
Upgrade Ceilings to Textured Finish	+ .42
Air Conditioning, in Heating Ductwork	Base System
Heating Systems, Hot Water	+ 1.37
Heat Pump	+ 2.22
Electric Heat	– 1.60
Not Heated	– 2.78

Additional upgrades or components

Important: See the Reference Section for Location Factors (to adjust for your city) and Estimating Forms

Custom Tri-Level

Living Area - 3200 S.F.
Perimeter - 198 L.F.

		Labor-Hours	Cost Per Square Foot Of Living Area		
			Mat.	Labor	Total
1 Site Work	Site preparation for slab; 4′ deep trench excavation for foundation wall, excavation for lower level, 4′ deep.	.023		.46	.46
2 Foundation	Continuous reinforced concrete footing 8″ deep x 18″ wide; dampproofed and insulated reinforced concrete foundation wall, 8″ thick, 4′ deep; 4″ concrete slab on 4″ crushed stone base and polyethylene vapor barrier, trowel finish.	.073	3.39	4.09	7.48
3 Framing	Exterior walls - 2″ x 6″ wood studs, 16″ O.C.; 1/2″ plywood sheathing; 2″ x 8″ rafters 16″ O.C. with 1/2″ plywood sheathing, 6 in 12 pitch; 2″ x 8″ ceiling joists 16″ O.C.; 2″ x 10″ floor joists 16″ O.C. with 5/8″ plywood subfloor; 5/8″ plywood subfloor on 1″ x 3″ wood sleepers 16″ O.C.	.162	4.73	5.46	10.19
4 Exterior Walls	Horizontal beveled wood siding; building paper; 6″ batt insulation; wood double hung windows; 3 solid core wood exterior doors; storms and screens.	.076	6.55	2.33	8.88
5 Roofing	30 year asphalt shingles; #15 felt building paper; aluminum gutters, downspouts and drip edge; copper flashings.	.045	2.06	1.89	3.95
6 Interiors	Walls and ceilings - 5/8″ gypsum wallboard, skim coat plaster, painted with primer and 2 coats; hardwood baseboard and trim, sanded and finished; hardwood floor 70%, ceramic tile with underlayment 20%, vinyl tile with underlayment 10%; wood panel interior doors, primed and painted with 2 coats.	.242	13.68	10.64	24.32
7 Specialties	Custom grade kitchen cabinets - 20 L.F. wall and base with solid surface counter top and kitchen sink; 4 L.F. bathroom vanity; 75 gallon electric water heater, medicine cabinet.	.026	3.08	.75	3.83
8 Mechanical	Gas fired warm air heat/air conditioning; one full bath including: bathtub, corner shower, built in lavatory and water closet; one 1/2 bath including: built in lavatory and water closet.	.073	4.20	2.34	6.54
9 Electrical	200 Amp. service; romex wiring; fluorescent and incandescent lighting fixtures, switches, receptacles.	.036	.90	1.34	2.24
10 Overhead	Contractor's overhead and profit and design.		7.71	5.85	13.56
		Total	46.30	35.15	**81.45**

1 Story — Base cost per square foot of living area

Exterior Wall	Living Area 50	100	200	300	400	500	600	700
Wood Siding - Wood Frame	183.25	142.60	124.90	106.30	100.55	97.00	94.70	95.55
Brick Veneer - Wood Frame	205.25	158.30	138.00	115.00	108.40	104.35	101.70	102.25
Stone Veneer - Wood Frame	219.55	168.55	146.55	120.70	113.50	109.15	106.20	106.70
Solid Masonry	225.20	172.60	149.85	122.95	115.50	111.00	108.00	108.35
Finished Basement, Add	71.15	60.70	55.05	45.50	43.65	42.50	41.75	41.15
Unfinished Basement, Add	51.80	35.85	28.50	21.95	20.20	19.20	18.50	18.00

1-1/2 Story — Base cost per square foot of living area

Exterior Wall	Living Area 100	200	300	400	500	600	700	800
Wood Siding - Wood Frame	146.25	119.95	104.05	94.70	89.90	87.65	84.65	83.80
Brick Veneer - Wood Frame	165.85	135.65	117.10	104.95	99.35	96.55	93.05	92.15
Stone Veneer - Wood Frame	178.65	145.90	125.60	111.55	105.45	102.35	98.55	97.60
Solid Masonry	183.65	149.90	129.00	114.20	107.90	104.60	100.75	99.75
Finished Basement, Add	47.55	44.00	40.20	36.00	34.85	34.10	33.45	33.30
Unfinished Basement, Add	31.05	23.70	20.15	17.25	16.20	15.50	14.90	14.70

2 Story — Base cost per square foot of living area

Exterior Wall	Living Area 100	200	400	600	800	1000	1200	1400
Wood Siding - Wood Frame	146.00	111.60	96.60	82.35	77.50	74.45	72.50	73.60
Brick Veneer - Wood Frame	168.00	127.30	109.65	91.10	85.30	81.80	79.50	80.30
Stone Veneer - Wood Frame	182.25	137.50	118.15	96.75	90.45	86.60	84.05	84.70
Solid Masonry	187.90	141.55	121.55	99.00	92.45	88.45	85.80	86.40
Finished Basement, Add	35.60	30.40	27.55	22.80	21.85	21.25	20.90	20.65
Unfinished Basement, Add	25.95	17.95	14.30	11.00	10.10	9.60	9.25	9.00

Base costs do not include bathroom or kitchen facilities. Use Modifications/Adjustments/Alternatives on pages 77-80 where appropriate.

1 Story

1-1/2 Story

2 Story

2-1/2 Story

Bi-Level

Tri-Level

- **Unique residence built from an architect's plan**
- **Single family — 1 full bath, 1 half bath, 1 kitchen**
- **No basement**
- **Cedar shakes on roof**
- **Forced hot air heat/air conditioning**
- **Gypsum wallboard interior finishes**
- **Many special features**
- **Extraordinary materials and workmanship**

Note: The illustration shown may contain some optional components (for example: garages and/or fireplaces) whose costs are shown in the modifications, adjustments, & alternatives below or at the end of the square foot section.

Base cost per square foot of living area

	Living Area										
Exterior Wall	1000	1200	1400	1600	1800	2000	2400	2800	3200	3600	4000
Wood Siding - Wood Frame	163.65	151.45	141.35	134.65	130.90	126.55	118.30	112.40	108.25	103.90	100.40
Brick Veneer - Wood Frame	171.75	158.70	147.95	140.80	136.90	132.15	123.50	117.10	112.65	108.05	104.25
Solid Brick	183.50	169.35	157.65	149.85	145.65	140.35	130.95	124.00	119.05	113.95	109.75
Solid Stone	183.40	169.20	157.55	149.80	145.60	140.30	130.90	124.00	118.95	113.85	109.75
Finished Basement, Add	44.50	47.80	45.85	44.60	43.75	42.75	41.25	40.05	39.15	38.25	37.55
Unfinished Basement, Add	19.85	18.70	17.70	17.10	16.75	16.25	15.55	15.00	14.50	14.10	13.70

Modifications

Add to the total cost

Upgrade Kitchen Cabinets	$ + 1243
Solid Surface Countertops (Included)	
Full Bath - including plumbing, wall and floor finishes	+ 6267
Half Bath - including plumbing, wall and floor finishes	+ 3808
Two Car Attached Garage	+ 23,502
Two Car Detached Garage	+ 26,461
Fireplace & Chimney	+ 6445

Adjustments

For multi family - add to total cost

Additional Kitchen	$ + 15,425
Additional Full Bath & Half Bath	+ 10,075
Additional Entry & Exit	+ 2275
Separate Heating & Air Conditioning	+ 5201
Separate Electric	+ 1643

For Townhouse/Rowhouse - Multiply cost per square foot by

Inner Unit	.90
End Unit	.95

Alternatives

Add to or deduct from the cost per square foot of living area

Heavyweight Asphalt Shingles	– 2.20
Clay Tile Roof	+ 1.50
Slate Roof	+ 5.65
Upgrade Ceilings to Textured Finish	+ .42
Air Conditioning, in Heating Ductwork	Base System
Heating Systems, Hot Water	+ 1.63
Heat Pump	+ 1.96
Electric Heat	– 1.78
Not Heated	– 3.57

Additional upgrades or components

Kitchen Cabinets & Countertops	Page 77
Bathroom Vanities	78
Fireplaces & Chimneys	78
Windows, Skylights & Dormers	78
Appliances	79
Breezeways & Porches	79
Finished Attic	79
Garages	80
Site Improvements	80
Wings & Ells	76

Important: See the Reference Section for Location Factors (to adjust for your city) and Estimating Forms

Luxury 1 Story

Living Area - 2800 S.F.
Perimeter - 219 L.F.

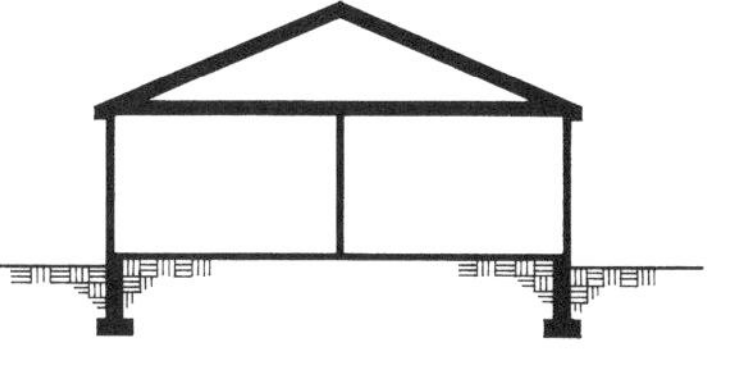

		Labor-Hours	Cost Per Square Foot Of Living Area		
			Mat.	Labor	Total
1 Site Work	Site preparation for slab; 4′ deep trench excavation for foundation wall.	.028		.57	.57
2 Foundation	Continuous reinforced concrete footing 8″ deep x 18″ wide; dampproofed and insulated reinforced concrete foundation wall, 12″ thick, 4′ deep; 4″ concrete slab on 4″ crushed stone base and polyethylene vapor barrier, trowel finish.	.098	5.74	6.01	11.75
3 Framing	Exterior walls - 2″ x 6″ wood studs, 16″ O.C.; 5/8″ plywood sheathing; 2″ x 10″ rafters 16″ O.C. with 5/8″ plywood sheathing, 6 in 12 pitch; 2″ x 8″ ceiling joists 16″ O.C.; 5/8″ plywood subfloor on 1″ x 3″ wood sleepers 16″ O.C.	.260	11.17	11.03	22.20
4 Exterior Walls	Horizontal beveled wood siding; building paper; 6″ batt insulation; wood double hung windows; 3 solid core wood exterior doors; storms and screens.	.204	7.40	2.67	10.07
5 Roofing	Red cedar shingles; #15 felt building paper; aluminum gutters, downspouts and drip edge; copper flashings.	.082	3.61	3.31	6.92
6 Interiors	Walls and ceilings - 5/8″ gypsum wallboard, skim coat plaster, painted with primer and 2 coats; hardwood baseboard and trim, sanded and finished; hardwood floor 70%, ceramic tile with underlayment 20%, vinyl tile with underlayment 10%; wood panel interior doors, primed and painted with 2 coats.	.287	11.60	11.19	22.79
7 Specialties	Luxury grade kitchen cabinets - 25 L.F. wall and base with solid surface counter top and kitchen sink; 6 L.F. bathroom vanity; 75 gallon electric water heater; medicine cabinet.	.052	4.59	1.09	5.68
8 Mechanical	Gas fired warm air heat/air conditioning; one full bath including: bathtub, corner shower; built in lavatory and water closet; one 1/2 bath including: built in lavatory and water closet.	.078	5.22	2.68	7.90
9 Electrical	200 Amp. service; romex wiring; fluorescent and incandescent lighting fixtures; intercom, switches, receptacles.	.044	1.11	1.64	2.75
10 Overhead	Contractor's overhead and profit and architect's fees.		12.11	9.66	21.77
		Total	62.55	49.85	**112.40**

- **Unique residence built from an architect's plan**
- **Single family — 1 full bath, 1 half bath, 1 kitchen**
- **No basement**
- **Cedar shakes on roof**
- **Forced hot air heat/air conditioning**
- **Gypsum wallboard interior finishes**
- **Many special features**
- **Extraordinary materials and workmanship**

Note: The illustration shown may contain some optional components (for example: garages and/or fireplaces) whose costs are shown in the modifications, adjustments, & alternatives below or at the end of the square foot section.

Base cost per square foot of living area

	Living Area										
Exterior Wall	**1000**	**1200**	**1400**	**1600**	**1800**	**2000**	**2400**	**2800**	**3200**	**3600**	**4000**
Wood Siding - Wood Frame	150.85	140.45	132.95	123.95	119.00	114.00	104.55	100.20	96.35	93.15	88.95
Brick Veneer - Wood Frame	160.20	149.20	141.25	131.55	126.20	120.85	110.60	105.90	101.60	98.25	93.70
Solid Brick	173.75	162.00	153.35	142.60	136.65	130.75	119.30	114.25	109.25	105.65	100.60
Solid Stone	173.65	161.80	153.25	142.50	136.50	130.65	119.25	114.15	109.15	105.55	100.50
Finished Basement, Add	31.40	34.10	33.05	31.45	30.60	29.90	28.25	27.55	26.75	26.35	25.65
Unfinished Basement, Add	14.25	13.55	13.10	12.35	12.00	11.65	10.85	10.55	10.15	9.95	9.65

Modifications

Add to the total cost

Upgrade Kitchen Cabinets	$ + 1243
Solid Surface Countertops (Included)	
Full Bath - including plumbing, wall and floor finishes	+ 6267
Half Bath - including plumbing, wall and floor finishes	+ 3808
Two Car Attached Garage	+ 23,502
Two Car Detached Garage	+ 26,461
Fireplace & Chimney	+ 6445

Adjustments

For multi family - add to total cost

Additional Kitchen	$ + 15,425
Additional Full Bath & Half Bath	+ 10,075
Additional Entry & Exit	+ 2275
Separate Heating & Air Conditioning	+ 5201
Separate Electric	+ 1643

For Townhouse/Rowhouse - Multiply cost per square foot by

Inner Unit	.90
End Unit	.95

Alternatives

Add to or deduct from the cost per square foot of living area

Heavyweight Asphalt Shingles	– 1.60
Clay Tile Roof	+ 1.10
Slate Roof	+ 4.14
Upgrade Ceilings to Textured Finish	+ .42
Air Conditioning, in Heating Ductwork	Base System
Heating Systems, Hot Water	+ 1.56
Heat Pump	+ 2.17
Electric Heat	– 1.78
Not Heated	– 3.29

Additional upgrades or components

Kitchen Cabinets & Countertops	Page 77
Bathroom Vanities	78
Fireplaces & Chimneys	78
Windows, Skylights & Dormers	78
Appliances	79
Breezeways & Porches	79
Finished Attic	79
Garages	80
Site Improvements	80
Wings & Ells	76

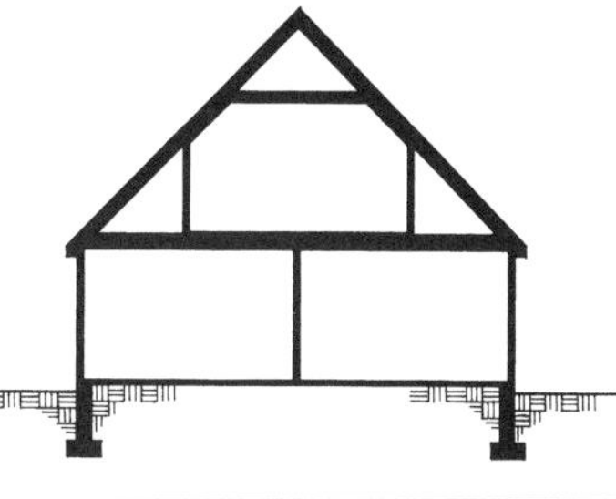

Luxury 1-1/2 Story

Living Area - 2800 S.F.
Perimeter - 175 L.F.

		Labor-Hours	Cost Per Square Foot Of Living Area		
			Mat.	Labor	Total
1 Site Work	Site preparation for slab; 4′ deep trench excavation for foundation wall.	.025		.57	.57
2 Foundation	Continuous reinforced concrete footing 8″ deep x 18″ wide; dampproofed and insulated reinforced concrete foundation wall, 12″ thick, 4′ deep; 4″ concrete slab on 4″ crushed stone base and polyethylene vapor barrier, trowel finish.	.066	4.13	4.54	8.67
3 Framing	Exterior walls - 2″ x 6″ wood studs, 16″ O.C.; 5/8″ plywood sheathing; 2″ x 10″ rafters 16″ O.C. with 5/8″ plywood sheathing, 8 in 12 pitch; 2″ x 8″ ceiling joists 16″ O.C.; 2″ x 12″ floor joists 16″ O.C. with 5/8″ plywood subfloor; 5/8″ plywood subfloor on 1″ x 3″ wood sleepers 16″ O.C.	.189	6.81	7.15	13.96
4 Exterior Walls	Horizontal beveled wood siding; building paper; 6″ batt insulation; wood double hung windows; 3 solid core wood exterior doors; storms and screens.	.174	8.14	2.93	11.07
5 Roofing	Red cedar shingles; #15 felt building paper; aluminum gutters, downspouts and drip edge; copper flashings.	.065	2.25	2.07	4.32
6 Interiors	Walls and ceilings - 5/8″ gypsum wallboard, skim coat plaster, painted with primer and 2 coats; hardwood baseboard and trim, sanded and finished; hardwood floor 70%, ceramic tile with underlayment 20%, vinyl tile with underlayment 10%; wood panel interior doors, primed and painted with 2 coats.	.260	13.14	12.75	25.89
7 Specialties	Luxury grade kitchen cabinets - 25 L.F. wall and base with solid surface counter top and kitchen sink; 6 L.F. bathroom vanity; 75 gallon electric water heater; medicine cabinet.	.062	4.59	1.09	5.68
8 Mechanical	Gas fired warm air heat/air conditioning; one full bath including: bathtub, corner shower; built in lavatory and water closet; one 1/2 bath including: built in lavatory and water closet.	.080	5.22	2.68	7.90
9 Electrical	200 Amp. service; romex wiring; fluorescent and incandescent lighting fixtures; intercom, switches, receptacles.	.044	1.11	1.64	2.75
10 Overhead	Contractor's overhead and profit and architect's fees.		10.91	8.48	19.39
	Total		56.30	43.90	**100.20**

- **Unique residence built from an architect's plan**
- **Single family — 1 full bath, 1 half bath, 1 kitchen**
- **No basement**
- **Cedar shakes on roof**
- **Forced hot air heat/air conditioning**
- **Gypsum wallboard interior finishes**
- **Many special features**
- **Extraordinary materials and workmanship**

Note: The illustration shown may contain some optional components (for example: garages and/or fireplaces) whose costs are shown in the modifications, adjustments, & alternatives below or at the end of the square foot section.

Base cost per square foot of living area

	Living Area										
Exterior Wall	**1200**	**1400**	**1600**	**1800**	**2000**	**2400**	**2800**	**3200**	**3600**	**4000**	**4400**
Wood Siding - Wood Frame	140.10	131.70	125.80	120.75	115.15	107.10	100.25	**95.90**	93.20	90.45	87.95
Brick Veneer - Wood Frame	150.05	141.05	134.70	129.10	123.20	114.40	106.90	102.05	99.20	96.00	93.40
Solid Brick	164.60	154.65	147.70	141.30	135.00	125.05	116.60	111.10	107.95	104.25	101.35
Solid Stone	164.40	154.45	147.55	141.20	134.85	124.90	116.50	111.05	107.90	104.15	101.30
Finished Basement, Add	25.25	27.45	26.60	25.75	25.20	24.05	22.95	22.35	22.00	21.40	21.10
Unfinished Basement, Add	11.40	10.85	10.55	10.10	9.90	9.30	8.85	8.55	8.40	8.15	7.95

Modifications

Add to the total cost

Upgrade Kitchen Cabinets	$ + 1243
Solid Surface Countertops (Included)	
Full Bath - including plumbing, wall and floor finishes	+ 6267
Half Bath - including plumbing, wall and floor finishes	+ 3808
Two Car Attached Garage	+ 23,502
Two Car Detached Garage	+ 26,461
Fireplace & Chimney	+ 7070

Adjustments

For multi family - add to total cost

Additional Kitchen	$ + 15,425
Additional Full Bath & Half Bath	+ 10,075
Additional Entry & Exit	+ 2275
Separate Heating & Air Conditioning	+ 5201
Separate Electric	+ 1643

For Townhouse/Rowhouse - Multiply cost per square foot by

Inner Unit	.86
End Unit	.93

Alternatives

Add to or deduct from the cost per square foot of living area

Heavyweight Asphalt Shingles	− 1.10
Clay Tile Roof	+ .75
Slate Roof	+ 2.83
Upgrade Ceilings to Textured Finish	+ .42
Air Conditioning, in Heating Ductwork	Base System
Heating Systems, Hot Water	+ 1.52
Heat Pump	+ 2.29
Electric Heat	− 1.60
Not Heated	− 3.11

Additional upgrades or components

Kitchen Cabinets & Countertops	Page 77
Bathroom Vanities	78
Fireplaces & Chimneys	78
Windows, Skylights & Dormers	78
Appliances	79
Breezeways & Porches	79
Finished Attic	79
Garages	80
Site Improvements	80
Wings & Ells	76

Luxury 2 Story

Living Area - 3200 S.F.
Perimeter - 163 L.F.

		Labor-Hours	Cost Per Square Foot Of Living Area		
			Mat.	Labor	Total
1 Site Work	Site preparation for slab; 4′ deep trench excavation for foundation wall.	.024		.50	.50
2 Foundation	Continuous reinforced concrete footing 8″ deep x 18″ wide; dampproofed and insulated reinforced concrete foundation wall, 12″ thick, 4′ deep; 4″ concrete slab on 4″ crushed stone base and polyethylene vapor barrier, trowel finish.	.058	3.35	3.69	7.04
3 Framing	Exterior walls - 2″ x 6″ wood studs, 16″ O.C.; 5/8″ plywood sheathing; 2″ x 10″ rafters 16″ O.C. with 5/8″ plywood sheathing, 6 in 12 pitch; 2″ x 8″ ceiling joists 16″ O.C.; 2″ x 12″ floor joists 16″ O.C. with 5/8″ plywood subfloor; 5/8″ plywood subfloor on 1″ x 3″ wood sleepers 16″ O.C.	.193	6.83	7.11	13.94
4 Exterior Walls	Horizontal beveled wood siding, building paper; 6″ batt insulation; wood double hung windows; 3 solid core wood exterior doors; storms and screens.	.247	8.58	3.13	11.71
5 Roofing	Red cedar shingles; #15 felt building paper; aluminum gutters, downspouts and drip edge; copper flashings.	.049	1.80	1.65	3.45
6 Interiors	Walls and ceilings - 5/8″ gypsum wallboard, skim coat plaster, painted with primer and 2 coats; hardwood baseboard and trim, sanded and finished; hardwood floor 70%, ceramic tile with underlayment 20%, vinyl tile with underlayment 10%; wood panel interior doors, primed and painted with 2 coats.	.252	12.98	12.67	25.65
7 Specialties	Luxury grade kitchen cabinets - 25 L.F. wall and base with solid surface counter top and kitchen sink; 6 L.F. bathroom vanity; 75 gallon electric water heater; medicine cabinet.	.057	4.03	.96	4.99
8 Mechanical	Gas fired warm air heat/air conditioning; one full bath including: bathtub, corner shower; built in lavatory and water closet; one 1/2 bath including: built in lavatory and water closet.	.071	4.78	2.59	7.37
9 Electrical	200 Amp. service; romex wiring; fluorescent and incandescent lighting fixtures; intercom, switches, receptacles.	.042	1.07	1.61	2.68
10 Overhead	Contractor's overhead and profit and architect's fee.		10.43	8.14	18.57
		Total	53.85	42.05	**95.90**

- **Unique residence built from an architect's plan**
- **Single family — 1 full bath, 1 half bath, 1 kitchen**
- **No basement**
- **Cedar shakes on roof**
- **Forced hot air heat/air conditioning**
- **Gypsum wallboard interior finishes**
- **Many special features**
- **Extraordinary materials and workmanship**

Note: The illustration shown may contain some optional components (for example: garages and/or fireplaces) whose costs are shown in the modifications, adjustments, & alternatives below or at the end of the square foot section.

Base cost per square foot of living area

Exterior Wall	Living Area 1500	1800	2100	2500	3000	3500	4000	4500	5000	5500	6000
Wood Siding - Wood Frame	136.60	123.20	115.20	109.50	101.45	95.70	90.15	87.40	84.90	82.55	79.80
Brick Veneer - Wood Frame	146.95	132.75	123.75	117.55	108.80	102.40	96.25	93.25	90.50	87.80	84.95
Solid Brick	162.05	146.70	136.20	129.35	119.50	112.10	105.30	101.80	98.65	95.55	92.35
Solid Stone	161.90	146.50	136.10	129.20	119.35	111.95	105.20	101.65	98.55	95.40	92.20
Finished Basement, Add	20.15	21.60	20.40	19.70	18.75	17.95	17.35	16.90	16.55	16.30	16.05
Unfinished Basement, Add	9.15	8.65	8.10	7.75	7.30	6.95	6.70	6.50	6.35	6.20	6.05

Modifications

Add to the total cost

Upgrade Kitchen Cabinets	$ + 1243
Solid Surface Countertops (Included)	
Full Bath - including plumbing, wall and floor finishes	+ 6267
Half Bath - including plumbing, wall and floor finishes	+ 3808
Two Car Attached Garage	+ 23,502
Two Car Detached Garage	+ 26,461
Fireplace & Chimney	+ 7730

Adjustments

For multi family - add to total cost

Additional Kitchen	$ + 15,425
Additional Full Bath & Half Bath	+ 10,075
Additional Entry & Exit	+ 2275
Separate Heating & Air Conditioning	+ 5201
Separate Electric	+ 1643

For Townhouse/Rowhouse - Multiply cost per square foot by

Inner Unit	.86
End Unit	.93

Alternatives

Add to or deduct from the cost per square foot of living area

Heavyweight Asphalt Shingles	– .95
Clay Tile Roof	+ .65
Slate Roof	+ 2.47
Upgrade Ceilings to Textured Finish	+ .42
Air Conditioning, in Heating Ductwork	Base System
Heating Systems, Hot Water	+ 1.37
Heat Pump	+ 2.36
Electric Heat	– 3.15
Not Heated	– 3.11

Additional upgrades or components

Kitchen Cabinets & Countertops	Page 77
Bathroom Vanities	78
Fireplaces & Chimneys	78
Windows, Skylights & Dormers	78
Appliances	79
Breezeways & Porches	79
Finished Attic	79
Garages	80
Site Improvements	80
Wings & Ells	76

Important: See the Reference Section for Location Factors (to adjust for your city) and Estimating Forms

Luxury 2-1/2 Story

Living Area - 3000 S.F.
Perimeter - 148 L.F.

		Labor-Hours	Cost Per Square Foot Of Living Area		
			Mat.	Labor	Total
1 Site Work	Site preparation for slab; 4′ deep trench excavation for foundation wall.	.055		.53	.53
2 Foundation	Continuous reinforced concrete footing 8″ deep x 18″ wide; dampproofed and insulated reinforced concrete foundation wall, 12″ thick, 4′ deep; 4″ concrete slab on 4″ crushed stone base and polyethylene vapor barrier, trowel finish.	.067	2.93	3.34	6.27
3 Framing	Exterior walls - 2″ x 6″ wood studs, 16″ O.C.; 5/8″ plywood sheathing; 2″ x 10″ rafters 16″ O.C. with 5/8″ plywood sheathing, 6 in 12 pitch; 2″ x 8″ ceiling joists 16″ O.C.; 2″ x 12″ floor joists 16″ O.C. with 5/8″ plywood subfloor; 5/8″ plywood subfloor on 1″ x 3″ wood sleepers 16″ O.C.	.209	7.08	7.29	14.37
4 Exterior Walls	Horizontal beveled wood siding; building paper; 6″ batt insulation; wood double hung windows; 3 solid core wood exterior doors; storms and screens.	.405	10.01	3.67	13.68
5 Roofing	Red cedar shingles; #15 felt building paper; aluminum gutters, downspouts and drip edge; copper flashings.	.039	1.38	1.27	2.65
6 Interiors	Walls and ceilings - 5/8″ gypsum wallboard, skim coat plaster, painted with primer and 2 coats; hardwood baseboard and trim, sanded and finished; hardwood floor 70%, ceramic tile with underlayment 20%, vinyl tile with underlayment 10%; wood panel interior doors, primed and painted with 2 coats.	.341	14.49	14.19	28.68
7 Specialties	Luxury grade kitchen cabinets - 25 L.F. wall and base with solid surface counter top and kitchen sink; 6 L.F. bathroom vanity; 75 gallon electric water heater; medicine cabinet.	.119	4.29	1.02	5.31
8 Mechanical	Gas fired warm air heat/air conditioning; one full bath including: bathtub, corner shower; built in lavatory and water closet; one 1/2 bath including: built in lavatory and water closet.	.103	4.99	2.65	7.64
9 Electrical	200 Amp. service; romex wiring; fluorescent and incandescent lighting fixtures; intercom, switches, receptacles.	.054	1.08	1.62	2.70
10 Overhead	Contractor's overhead and profit and architect's fee.		11.10	8.52	19.62
		Total	57.35	44.10	**101.45**

- **Unique residence built from an architect's plan**
- **Single family — 1 full bath, 1 half bath, 1 kitchen**
- **No basement**
- **Cedar shakes on roof**
- **Forced hot air heat/air conditioning**
- **Gypsum wallboard interior finishes**
- **Many special features**
- **Extraordinary materials and workmanship**

Note: The illustration shown may contain some optional components (for example: garages and/or fireplaces) whose costs are shown in the modifications, adjustments, & alternatives below or at the end of the square foot section.

Base cost per square foot of living area

Exterior Wall	Living Area 1500	1800	2100	2500	3000	3500	4000	4500	5000	5500	6000
Wood Siding - Wood Frame	135.70	122.35	115.85	110.45	102.20	98.00	93.00	87.70	85.70	83.60	81.55
Brick Veneer - Wood Frame	146.45	132.35	125.15	119.35	110.20	105.65	99.95	94.05	91.90	89.60	87.15
Solid Brick	162.15	146.90	138.75	132.35	122.00	116.85	110.10	103.45	101.00	98.30	95.35
Solid Stone	161.95	146.70	138.65	132.20	121.85	116.65	110.00	103.30	100.85	98.20	95.25
Finished Basement, Add	17.75	19.00	18.20	17.65	16.80	16.30	15.60	15.05	14.85	14.60	14.25
Unfinished Basement, Add	8.00	7.60	7.25	7.00	6.60	6.35	6.05	5.80	5.70	5.60	5.40

Modifications

Add to the total cost

Upgrade Kitchen Cabinets	$ + 1243
Solid Surface Countertops (Included)	
Full Bath - including plumbing, wall and floor finishes	+ 6267
Half Bath - including plumbing, wall and floor finishes	+ 3808
Two Car Attached Garage	+ 23,502
Two Car Detached Garage	+ 26,461
Fireplace & Chimney	+ 7730

Adjustments

For multi family - add to total cost

Additional Kitchen	$ + 15,425
Additional Full Bath & Half Bath	+ 10,075
Additional Entry & Exit	+ 2275
Separate Heating & Air Conditioning	+ 5201
Separate Electric	+ 1643

For Townhouse/Rowhouse - Multiply cost per square foot by

Inner Unit	.84
End Unit	.92

Alternatives

Add to or deduct from the cost per square foot of living area

Heavyweight Asphalt Shingles	– .75
Clay Tile Roof	+ .50
Slate Roof	+ 1.88
Upgrade Ceilings to Textured Finish	+ .42
Air Conditioning, in Heating Ductwork	Base System
Heating Systems, Hot Water	+ 1.37
Heat Pump	+ 2.36
Electric Heat	– 3.15
Not Heated	– 3.01

Additional upgrades or components

Kitchen Cabinets & Countertops	Page 77
Bathroom Vanities	78
Fireplaces & Chimneys	78
Windows, Skylights & Dormers	78
Appliances	79
Breezeways & Porches	79
Finished Attic	79
Garages	80
Site Improvements	80
Wings & Ells	76

Important: See the Reference Section for Location Factors (to adjust for your city) and Estimating Forms

Luxury 3 Story

Living Area - 3000 S.F.
Perimeter - 135 L.F.

		Labor-Hours	Cost Per Square Foot Of Living Area		
			Mat.	Labor	Total
1 Site Work	Site preparation for slab; 4′ deep trench excavation for foundation wall.	.055		.53	.53
2 Foundation	Continuous reinforced concrete footing 8″ deep x 18″ wide; dampproofed and insulated reinforced concrete foundation wall, 12″ thick, 4′ deep; 4″ concrete slab on 4″ crushed stone base and polyethylene vapor barrier, trowel finish.	.063	2.64	3.04	5.68
3 Framing	Exterior walls - 2″ x 6″ wood studs, 16″ O.C.; 5/8″ plywood sheathing; 2″ x 10″ rafters 16″ O.C. with 5/8″ plywood sheathing, 6 in 12 pitch; 2″ x 8″ ceiling joists 16″ O.C.; 2″ x 12″ floor joists 16″ O.C. with 5/8″ plywood subfloor; 5/8″ plywood subfloor on 1″ x 3″ wood sleepers 16″ O.C.	.225	7.14	7.39	14.53
4 Exterior Walls	Horizontal beveled wood siding; building paper; 6″ batt insulation; wood double hung windows; 3 solid core wood exterior doors; storms and screens.	.454	10.91	3.99	14.90
5 Roofing	Red cedar shingles; #15 felt building paper; aluminum gutters, downspouts and drip edge; copper flashings.	.034	1.20	1.11	2.31
6 Interiors	Walls and ceilings - 5/8″ gypsum wallboard, skim coat plaster, painted with primer and 2 coats; hardwood baseboard and trim, sanded and finished; hardwood floor 70%, ceramic tile with underlayment 20%, vinyl tile with underlayment 10%; wood panel interior doors, primed and painted with 2 coats.	.390	14.52	14.28	28.80
7 Specialties	Luxury grade kitchen cabinets - 25 L.F. wall and base with solid surface counter top and kitchen sink; 6 L.F. bathroom vanity; 75 gallon electric water heater; medicine cabinet.	.119	4.29	1.02	5.31
8 Mechanical	Gas fired warm air heat/air conditioning; one full bath including: bathtub, corner shower; built in lavatory and water closet; one 1/2 bath including: built in lavatory and water closet.	.103	4.99	2.65	7.64
9 Electrical	200 Amp. service; romex wiring; fluorescent and incandescent lighting fixtures; intercom, switches, receptacles.	.053	1.08	1.62	2.70
10 Overhead	Contractor's overhead and profit and architect's fees.		11.23	8.57	19.80
		Total	58.00	44.20	**102.20**

- **Unique residence built from an architect's plan**
- **Single family — 1 full bath, 1 half bath, 1 kitchen**
- **No basement**
- **Cedar shakes on roof**
- **Forced hot air heat/air conditioning**
- **Gypsum wallboard interior finishes**
- **Many special features**
- **Extraordinary materials and workmanship**

Note: The illustration shown may contain some optional components (for example: garages and/or fireplaces) whose costs are shown in the modifications, adjustments, & alternatives below or at the end of the square foot section.

Base cost per square foot of living area

Exterior Wall	Living Area 1200	1400	1600	1800	2000	2400	2800	3200	3600	4000	4400
Wood Siding - Wood Frame	132.65	124.80	119.20	114.55	109.10	101.60	95.35	**91.25**	88.80	86.25	83.95
Brick Veneer - Wood Frame	140.10	131.80	125.85	120.80	115.20	107.05	100.25	95.90	93.30	90.45	88.05
Solid Brick	151.00	142.00	135.65	130.00	124.00	115.10	107.55	102.75	99.85	96.60	93.95
Solid Stone	150.85	141.85	135.50	129.90	123.90	114.95	107.45	102.65	99.75	96.50	93.90
Finished Basement, Add	25.25	27.45	26.60	25.75	25.20	24.05	22.95	22.35	22.00	21.40	21.10
Unfinished Basement, Add	11.40	10.85	10.55	10.10	9.90	9.30	8.85	8.55	8.40	8.15	7.95

Modifications

Add to the total cost

Upgrade Kitchen Cabinets	$ + 1243
Solid Surface Countertops (Included)	
Full Bath - including plumbing, wall and floor finishes	+ 6267
Half Bath - including plumbing, wall and floor finishes	+ 3808
Two Car Attached Garage	+ 23,502
Two Car Detached Garage	+ 26,461
Fireplace & Chimney	+ 6445

Adjustments

For multi family - add to total cost

Additional Kitchen	$ + 15,425
Additional Full Bath & Half Bath	+ 10,075
Additional Entry & Exit	+ 2275
Separate Heating & Air Conditioning	+ 5201
Separate Electric	+ 1643

For Townhouse/Rowhouse - Multiply cost per square foot by

Inner Unit	.89
End Unit	.94

Alternatives

Add to or deduct from the cost per square foot of living area

Heavyweight Asphalt Shingles	– 1.10
Clay Tile Roof	+ .75
Slate Roof	+ 5.65
Upgrade Ceilings to Textured Finish	+ .42
Air Conditioning, in Heating Ductwork	Base System
Heating Systems, Hot Water	+ 1.52
Heat Pump	+ 2.29
Electric Heat	– 1.60
Not Heated	– 3.11

Additional upgrades or components

Kitchen Cabinets & Countertops	Page 77
Bathroom Vanities	78
Fireplaces & Chimneys	78
Windows, Skylights & Dormers	78
Appliances	79
Breezeways & Porches	79
Finished Attic	79
Garages	80
Site Improvements	80
Wings & Ells	76

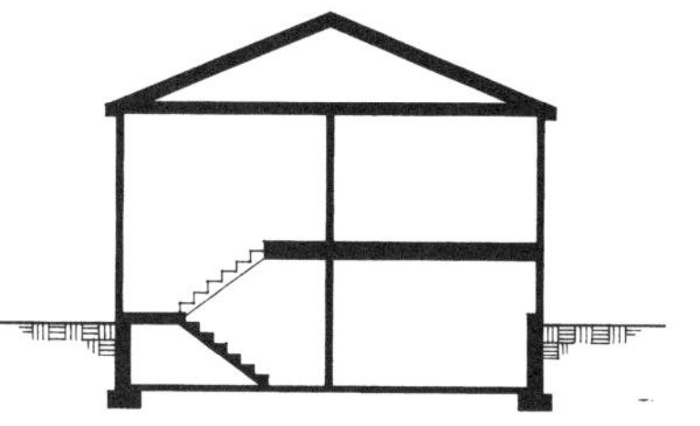

Luxury Bi-Level

Living Area - 3200 S.F.
Perimeter - 163 L.F.

		Labor-Hours	Cost Per Square Foot Of Living Area		
			Mat.	Labor	Total
1 Site Work	Excavation for lower level, 4′ deep. Site preparation for slab.	.024		.50	.50
2 Foundation	Continuous reinforced concrete footing 8″ deep x 18″ wide; dampproofed and insulated reinforced concrete foundation wall, 12″ thick, 4′ deep; 4″ concrete slab on 4″ crushed stone base and polyethylene vapor barrier, trowel finish.	.058	3.35	3.69	7.04
3 Framing	Exterior walls - 2″ x 6″ wood studs, 16″ O.C.; 5/8″ plywood sheathing; 2″ x 10″ rafters 16″ O.C. with 5/8″ plywood sheathing, 6 in 12 pitch; 2″ x 8″ ceiling joists 16″ O.C.; 2″ x 12″ floor joists 16″ O.C. with 5/8″ plywood subfloor; 5/8″ plywood subfloor on 1″ x 3″ wood sleepers 16″ O.C.	.232	6.49	6.76	13.25
4 Exterior Walls	Horizontal beveled wood siding; building paper; 6″ batt insulation; wood double hung windows; 3 solid core wood exterior doors; storms and screens.	.185	6.69	2.44	9.13
5 Roofing	Red cedar shingles: #15 felt building paper; aluminum gutters, downspouts and drip edge; copper flashings.	.042	1.80	1.65	3.45
6 Interiors	Walls and ceilings - 5/8″ gypsum wallboard, skim coat plaster, painted with primer and 2 coats; hardwood baseboard and trim, sanded and finished; hardwood floor 70%, ceramic tile with underlayment 20%; vinyl tile with underlayment 10%; wood panel interior doors, primed and painted with 2 coats.	.238	12.80	12.38	25.18
7 Specialties	Luxury grade kitchen cabinets - 25 L.F. wall and base with solid surface counter top and kitchen sink; 6 L.F. bathroom vanity; 75 gallon electric water heater; medicine cabinet.	.056	4.03	.96	4.99
8 Mechanical	Gas fired warm air heat/air conditioning; one full bath including: bathtub, corner shower; built in lavatory and water closet; one 1/2 bath including: built in lavatory and water closet.	.071	4.78	2.59	7.37
9 Electrical	200 Amp. service; romex wiring; fluorescent and incandescent lighting fixtures; intercom, switches, receptacles.	.042	1.07	1.61	2.68
10 Overhead	Contractor's overhead and profit and architect's fees.		9.84	7.82	17.66
		Total	50.85	40.40	**91.25**

- **Unique residence built from an architect's plan**
- **Single family — 1 full bath, 1 half bath, 1 kitchen**
- **No basement**
- **Cedar shakes on roof**
- **Forced hot air heat/air conditioning**
- **Gypsum wallboard interior finishes**
- **Many special features**
- **Extraordinary materials and workmanship**

Note: The illustration shown may contain some optional components (for example: garages and/or fireplaces) whose costs are shown in the modifications, adjustments, & alternatives below or at the end of the square foot section.

©Home Planners, Inc.

Base cost per square foot of living area

Exterior Wall	Living Area 1500	1800	2100	2400	2800	3200	3600	4000	4500	5000	5500
Wood Siding - Wood Frame	125.15	115.85	108.15	103.05	99.30	94.95	90.25	88.25	83.95	81.50	78.75
Brick Veneer - Wood Frame	131.90	121.95	113.65	108.20	104.25	99.55	94.55	92.45	87.75	85.05	82.15
Solid Brick	141.70	130.75	121.70	115.75	111.50	106.20	100.75	98.45	93.35	90.30	87.20
Solid Stone	141.60	130.70	121.60	115.65	111.40	106.20	100.70	98.40	93.30	90.25	87.10
Finished Basement, Add*	29.65	31.90	30.60	29.70	29.05	28.30	27.50	27.20	26.45	26.00	25.55
Unfinished Basement, Add*	13.20	12.45	11.80	11.40	11.15	10.75	10.35	10.20	9.85	9.60	9.45

*Basement under middle level only.

Modifications

Add to the total cost

Upgrade Kitchen Cabinets	$ + 1243
Solid Surface Countertops (Included)	
Full Bath - including plumbing, wall and floor finishes	+ 6267
Half Bath - including plumbing, wall and floor finishes	+ 3808
Two Car Attached Garage	+ 23,502
Two Car Detached Garage	+ 26,461
Fireplace & Chimney	+ 6445

Adjustments

For multi family - add to total cost

Additional Kitchen	$ + 15,425
Additional Full Bath & Half Bath	+ 10,075
Additional Entry & Exit	+ 2275
Separate Heating & Air Conditioning	+ 5201
Separate Electric	+ 1643

For Townhouse/Rowhouse - Multiply cost per square foot by

Inner Unit	.86
End Unit	.93

Alternatives

Add to or deduct from the cost per square foot of living area

Heavyweight Asphalt Shingles	– 1.60
Clay Tile Roof	+ 1.10
Slate Roof	+ 4.14
Upgrade Ceilings to Textured Finish	+ .42
Air Conditioning, in Heating Ductwork	Base System
Heating Systems, Hot Water	+ 1.47
Heat Pump	+ 2.38
Electric Heat	– 1.42
Not Heated	– 3.01

Additional upgrades or components

Kitchen Cabinets & Countertops	Page 77
Bathroom Vanities	78
Fireplaces & Chimneys	78
Windows, Skylights & Dormers	78
Appliances	79
Breezeways & Porches	79
Finished Attic	79
Garages	80
Site Improvements	80
Wings & Ells	76

Luxury Tri-Level

Living Area - 3600 S.F.
Perimeter - 207 L.F.

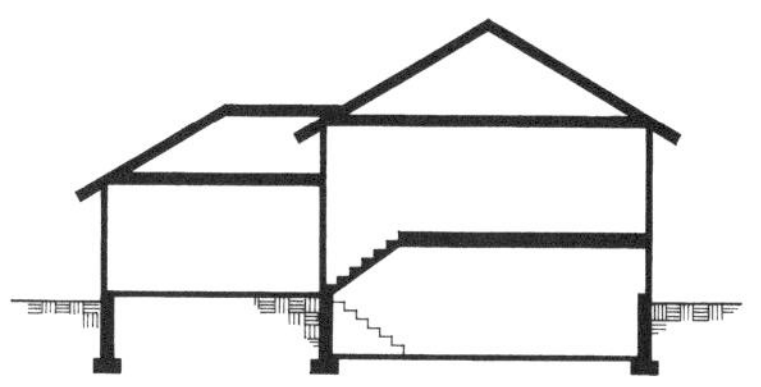

		Labor-Hours	Cost Per Square Foot Of Living Area		
			Mat.	Labor	Total
1 Site Work	Site preparation for slab; 4′ deep trench excavation for foundation wall, excavation for lower level, 4′ deep.	.021		.44	.44
2 Foundation	Continuous reinforced concrete footing 8″ deep x 18″ wide; dampproofed and insulated reinforced concrete foundation wall, 12″ thick, 4′ deep; 4″ concrete slab on 4″ crushed stone base and polyethylene vapor barrier, trowel finish.	.109	4.00	4.25	8.25
3 Framing	Exterior walls - 2″ x 6″ wood studs, 16″ O.C.; 5/8″ plywood sheathing; 2″ x 10″ rafters 16″ O.C. with 5/8″ plywood sheathing, 6 in 12 pitch; 2″ x 8″ ceiling joists 16″ O.C.; 2″ x 12″ floor joists 16″ O.C. with 5/8″ plywood subfloor; 5/8″ plywood subfloor on 1″ x 3″ wood sleepers 16″ O.C.	.204	6.47	6.75	13.22
4 Exterior Walls	Horizontal beveled wood siding; building paper; 6″ batt insulation; wood double hung windows; 3 solid core wood exterior doors; storms and screens.	.181	6.46	2.33	8.79
5 Roofing	Red cedar shingles; #15 felt building paper; aluminum gutters, downspouts and drip edge; copper flashings.	.056	2.40	2.20	4.60
6 Interiors	Walls and ceilings - 5/8″ gypsum wallboard, skim coat plaster, painted with primer and 2 coats; hardwood baseboard and trim, sanded and finished; hardwood floor 70%, ceramic tile with underlayment 20%, vinyl tile with underlayment 10%; wood panel interior doors, primed and painted with 2 coats.	.217	12.00	11.55	23.55
7 Specialties	Luxury grade kitchen cabinets - 25 L.F. wall and base with solid surface counter top and kitchen sink; 6 L.F. bathroom vanity; 75 gallon electric water heater; medicine cabinet.	.048	3.58	.86	4.44
8 Mechanical	Gas fired warm air heat/air conditioning; one full bath including: bathtub, corner shower; built in lavatory and water closet; one 1/2 bath including: built in lavatory and water closet.	.057	4.42	2.49	6.91
9 Electrical	200 Amp. service; romex wiring; fluorescent and incandescent lighting fixtures; intercom, switches, receptacles.	.039	1.04	1.56	2.60
10 Overhead	Contractor's overhead and profit and architect's fees.		9.68	7.77	17.45
		Total	50.05	40.20	**90.25**

1 Story — Base cost per square foot of living area

Exterior Wall	Living Area 50	100	200	300	400	500	600	700
Wood Siding - Wood Frame	203.70	157.80	137.80	116.80	110.20	106.25	103.65	104.55
Brick Veneer - Wood Frame	228.85	175.75	152.80	126.75	119.20	114.65	111.65	112.25
Solid Brick	265.50	201.90	174.60	141.30	132.30	126.80	123.30	123.50
Solid Stone	265.00	201.60	174.35	141.10	132.10	126.75	123.15	123.35
Finished Basement, Add	78.20	70.95	63.70	51.70	49.25	47.80	46.75	46.10
Unfinished Basement, Add	40.50	31.35	27.55	21.25	20.05	19.30	18.75	18.40

1-1/2 Story — Base cost per square foot of living area

Exterior Wall	Living Area 100	200	300	400	500	600	700	800
Wood Siding - Wood Frame	161.70	132.05	114.10	103.50	98.05	95.50	92.15	91.20
Brick Veneer - Wood Frame	184.15	150.00	129.05	115.15	108.80	105.70	101.70	100.70
Solid Brick	216.90	176.15	150.85	132.15	124.55	120.55	115.75	114.60
Solid Stone	216.45	175.90	150.60	131.95	124.35	120.35	115.65	114.40
Finished Basement, Add	51.50	50.95	46.15	40.80	39.35	38.35	37.45	37.35
Unfinished Basement, Add	25.85	22.05	19.55	16.75	16.05	15.50	15.05	15.00

2 Story — Base cost per square foot of living area

Exterior Wall	Living Area 100	200	400	600	800	1000	1200	1400
Wood Siding - Wood Frame	159.30	120.50	103.50	87.30	81.75	78.35	76.15	77.40
Brick Veneer - Wood Frame	184.45	138.45	118.40	97.25	90.70	86.75	84.15	85.05
Solid Brick	221.10	164.60	140.25	111.80	103.85	98.95	95.80	96.30
Solid Stone	220.65	164.25	140.00	111.65	103.65	98.85	95.60	96.15
Finished Basement, Add	39.15	35.55	31.95	25.90	24.70	23.95	23.45	23.15
Unfinished Basement, Add	20.20	15.70	13.80	10.65	10.00	9.60	9.35	9.20

Base costs do not include bathroom or kitchen facilities. Use Modifications/Adjustments/Alternatives on pages 77-80 where appropriate.

Important: See the Reference Section for Location Factors (to adjust for your city) and Estimating Forms

Kitchen cabinets - Base units, hardwood *(Cost per Unit)*

	Economy	Average	Custom	Luxury
24" deep, 35" high,				
One top drawer,				
One door below				
12" wide	$135	$180	$240	$315
15" wide	176	235	315	410
18" wide	191	254	340	445
21" wide	198	264	350	460
24" wide	225	300	400	525
Four drawers				
12" wide	293	390	520	685
15" wide	229	305	405	535
18" wide	255	340	450	595
24" wide	274	365	485	640
Two top drawers,				
Two doors below				
27" wide	248	330	440	580
30" wide	263	350	465	615
33" wide	274	365	485	640
36" wide	285	380	505	665
42" wide	304	405	540	710
48" wide	323	430	570	755
Range or sink base				
(Cost per unit)				
Two doors below				
30" wide	222	296	395	520
33" wide	236	315	420	550
36" wide	248	330	440	580
42" wide	263	350	465	615
48" wide	274	365	485	640
Corner Base Cabinet				
(Cost per unit)				
36" wide	356	475	630	830
Lazy Susan *(Cost per unit)*				
With revolving door	345	460	610	805

Kitchen cabinets - Wall cabinets, hardwood *(Cost per Unit)*

	Economy	Average	Custom	Luxury
12" deep, 2 doors				
12" high				
30" wide	$146	$194	$260	$ 340
36" wide	166	221	295	385
15" high				
30" wide	152	202	270	355
33" wide	168	224	300	390
36" wide	170	227	300	395
24" high				
30" wide	184	245	325	430
36" wide	202	269	360	470
42" wide	221	295	390	515
30" high, 1 door				
12" wide	133	177	235	310
15" wide	148	197	260	345
18" wide	161	214	285	375
24" wide	179	238	315	415
30" high, 2 doors				
27" wide	215	286	380	500
30" wide	209	279	370	490
36" wide	236	315	420	550
42" wide	255	340	450	595
48" wide	281	375	500	655
Corner wall, 30" high				
24" wide	154	205	275	360
30" wide	179	238	315	415
36" wide	192	256	340	450
Broom closet				
84" high, 24" deep				
18" wide	386	515	685	900
Oven Cabinet				
84" high, 24" deep				
27" wide	551	735	980	1285

Kitchen countertops *(Cost per L.F.)*

	Economy	Average	Custom	Luxury
Solid Surface				
24" wide, no backsplash	88	117	155	205
with backsplash	95	127	170	220
Stock plastic laminate, 24" wide				
with backsplash	15	21	25	35
Custom plastic laminate, no splash				
7/8" thick, alum. molding	23	31	40	55
1-1/4" thick, no splash	25	34	45	60
Marble				
1/2" - 3/4" thick w/splash	42	56	75	100
Maple, laminated				
1-1/2" thick w/splash	61	81	110	140
Stainless steel				
(per S.F.)	110	147	195	255
Cutting blocks, recessed				
16" x 20" x 1" (each)	83	110	145	195

Vanity bases *(Cost per Unit)*

	Economy	Average	Custom	Luxury
2 door, 30" high, 21" deep				
24" wide	177	236	315	415
30" wide	206	275	365	480
36" wide	270	360	480	630
48" wide	323	430	570	755

Solid surface vanity tops *(Cost Each)*

	Economy	Average	Custom	Luxury
Center bowl				
22" x 25"	$247	$267	$288	$311
22" x 31"	283	306	330	356
22" x 37"	325	351	379	409
22" x 49"	400	432	467	504

Fireplaces & Chimneys *(Cost per Unit)*

	1-1/2 Story	2 Story	3 Story
Economy (prefab metal)			
Exterior chimney & 1 fireplace	$4435	$4900	$5370
Interior chimney & 1 fireplace	4245	4725	4945
Average (masonry)			
Exterior chimney & 1 fireplace	4420	4930	5600
Interior chimney & 1 fireplace	4140	4645	5060
For more than 1 flue, add	320	545	905
For more than 1 fireplace, add	3130	3130	3130
Custom (masonry)			
Exterior chimney & 1 fireplace	4600	5190	5865
Interior chimney & 1 fireplace	4315	4880	5260
For more than 1 flue, add	360	625	850
For more than 1 fireplace, add	3305	3305	3305
Luxury (masonry)			
Exterior chimney & 1 fireplace	6645	7070	7730
Interior chimney & 1 fireplace	6150	6725	7115
For more than 1 flue, add	535	885	1240
For more than 1 fireplace, add	5080	5080	5080

Windows and Skylights *(Cost Each)*

	Economy	Average	Custom	Luxury
Fixed Picture Windows				
3'-6" x 4'-0"	$ 476	$ 514	$ 555	$ 599
4'-0" x 6'-0"	686	741	800	864
5'-0" x 6'-0"	960	1037	1120	1210
6'-0" x 6'-0"	986	1065	1150	1242
Bay/Bow Windows				
8'-0" x 5'-0"	1265	1366	1475	1593
10'-0" x 5'-0"	1779	1921	2075	2241
10'-0" x 6'-0"	1865	2014	2175	2349
12'-0" x 6'-0"	2229	2407	2600	2808
Palladian Windows				
3'-2" x 6'-4"		1736	1875	2025
4'-0" x 6'-0"		2106	2275	2457
5'-5" x 6'-10"		2523	2725	2943
8'-0" x 6'-0"		2917	3150	3402
Skylights				
46" x 21-1/2"	385	416	531	573
46" x 28"	423	457	517	558
57" x 44"	522	564	699	755

Dormers *(Cost/S.F. of plan area)*

	Economy	Average	Custom	Luxury
Framing and Roofing Only				
Gable dormer, 2" x 6" roof frame	$22	$25	$28	$45
2" x 8" roof frame	24	27	30	48
Shed dormer, 2" x 6" roof frame	15	16	18	29
2" x 8" roof frame	16	17	19	31
2" x 10" roof frame	17	19	21	32

Appliances *(Cost per Unit)*

	Economy	Average	Custom	Luxury
Range				
30" free standing, 1 oven	$ 325	$1075	$1450	$1825
2 oven	1725	1775	1800	1825
30" built-in, 1 oven	565	1208	1529	1850
2 oven	1225	1625	1825	2025
21" free standing				
1 oven	330	380	405	430
Counter Top Ranges				
4 burner standard	272	491	601	710
As above with griddle	590	718	781	845
Microwave Oven	186	408	519	630
Combination Range, Refrigerator, Sink				
30" wide	1050	1588	1856	2125
60" wide	3075	3537	3767	3998
72" wide	3475	3997	4257	4518
Comb. Range, Refrig., Sink, Microwave Oven & Ice Maker	5169	5945	6332	6720
Compactor				
4 to 1 compaction	540	593	619	645
Deep Freeze				
15 to 23 C.F.	495	588	634	680
30 C.F.	915	1020	1073	1125
Dehumidifier, portable, auto.				
15 pint	170	196	208	221
30 pint	189	218	232	246
Washing Machine, automatic	440	938	1254	1525
Water Heater				
Electric, glass lined				
30 gal.	380	468	511	555
80 gal.	780	1003	1114	1225
Water Heater, Gas, glass lined				
30 gal.	770	935	1018	1100
50 gal.	950	1175	1288	1400
Water Softener, automatic				
30 grains/gal.	640	736	784	832
100 grains/gal.	835	961	1023	1086
Dishwasher, built-in				
2 cycles	400	480	520	560
4 or more cycles	415	463	486	510
Dryer, automatic	510	843	1009	1175
Garage Door Opener	360	433	469	505
Garbage Disposal	95	154	183	212
Heater, Electric, built-in				
1250 watt ceiling type	172	214	235	256
1250 watt wall type	208	242	259	276
Wall type w/blower				
1500 watt	236	326	370	415
3000 watt	415	457	477	498
Hood For Range, 2 speed				
30" wide	129	477	651	825
42" wide	340	568	681	795
Humidifier, portable				
7 gal. per day	170	196	208	221
15 gal. per day	204	235	250	265
Ice Maker, automatic				
13 lb. per day	465	535	570	605
51 lb. per day	1325	1524	1624	1723
Refrigerator, no frost				
10-12 C.F.	530	693	774	855
14-16 C.F.	560	603	624	645
18-20 C.F.	640	858	966	1075
21-29 C.F.	805	1653	2076	2500
Sump Pump, 1/3 H.P.	222	306	348	390

Breezeway *(Cost per S.F.)*

Class	Type	Area (S.F.)			
		50	100	150	200
Economy	Open	$ 20.15	$ 17.15	$14.40	$14.15
	Enclosed	96.75	74.75	62.05	54.35
Average	Open	24.95	21.95	19.20	17.50
	Enclosed	106.00	79.00	64.65	56.85
Custom	Open	35.60	31.30	27.25	24.95
	Enclosed	146.20	108.75	88.90	78.15
Luxury	Open	36.75	32.20	29.05	28.15
	Enclosed	147.95	109.70	88.75	79.45

Porches *(Cost per S.F.)*

Class	Type	Area (S.F.)				
		25	50	100	200	300
Economy	Open	$ 57.80	$ 38.70	$30.20	$25.60	$21.85
	Enclosed	115.55	80.50	60.85	47.45	40.65
Average	Open	70.15	44.65	34.25	28.50	28.50
	Enclosed	138.65	93.95	71.05	55.00	46.60
Custom	Open	90.90	60.30	45.40	39.65	35.55
	Enclosed	180.70	123.45	93.95	73.15	63.00
Luxury	Open	97.30	63.60	46.95	42.20	37.60
	Enclosed	190.40	133.75	99.25	77.05	66.35

Finished attic *(Cost per S.F.)*

Class	Area (S.F.)				
	400	500	600	800	1000
Economy	$15.45	$14.90	$14.30	$14.05	$13.55
Average	23.75	23.20	22.65	22.35	21.75
Custom	29.00	28.30	27.70	27.25	26.70
Luxury	36.50	35.65	34.80	34.00	33.40

Alarm system *(Cost per System)*

	Burglar Alarm	Smoke Detector
Economy	$ 370	$ 58
Average	425	91
Custom	722	142
Luxury	1075	175

Sauna, prefabricated
(Cost per unit, including heater and controls—7' high)

Size	Cost
6' x 4'	$ 4475
6' x 5'	4975
6' x 6'	5325
6' x 9'	6625
8' x 10'	8700
8' x 12'	10,200
10' x 12'	10,800

Garages *

(Costs include exterior wall systems comparable with the quality of the residence. Included in the cost is an allowance for one personnel door, manual overhead door(s) and electrical fixture.)

Class	Type									
	Detached			Attached			Built-in		Basement	
	One Car	Two Car	Three Car	One Car	Two Car	Three Car	One Car	Two Car	One Car	Two Car
Economy										
Wood	$12,048	$18,428	$24,808	$ 9356	$16,154	$22,534	$-1548	$-3097	$1293	$1702
Masonry	16,722	24,277	31,831	12,280	20,254	27,809	-2122	-4245		
Average										
Wood	13,167	19,827	26,488	10,056	17,135	23,796	-1686	-3371	1472	2060
Masonry	16,876	24,469	32,063	12,377	20,389	27,983	-2141	-3663		
Custom										
Wood	15,170	23,231	31,293	11,858	20,503	28,564	-3207	-3371	2189	3494
Masonry	18,889	27,886	36,883	14,185	23,765	32,762	-3664	-4284		
Luxury										
Wood	17,015	26,461	35,907	13,473	23,502	32,949	-3283	-3522	2924	4629
Masonry	22,114	32,843	43,571	16,664	27,975	38,074	-3909	-4774		

*See the Introduction to this section for definitions of garage types.

Swimming pools *(Cost per S.F.)*

Residential	(includes equipment)
In-ground	$ 21.00 - 51.00
Deck equipment	1.30
Paint pool, preparation & 3 coats (epoxy)	3.16
Rubber base paint	2.93
Pool Cover	0.73
Swimming Pool Heaters	(Cost per unit)
(not including wiring, external piping, base or pad)	
Gas	
155 MBH	$ 2775
190 MBH	3250
500 MBH	11,600
Electric	
15 KW 7200 gallon pool	2525
24 KW 9600 gallon pool	3325
54 KW 24,000 gallon pool	4975

Wood and coal stoves

Wood Only	
Free Standing (minimum)	$ 1525
Fireplace Insert (minimum)	1537
Coal Only	
Free Standing	$ 1737
Fireplace Insert	1902
Wood and Coal	
Free Standing	$ 3573
Fireplace Insert	3660

Sidewalks *(Cost per S.F.)*

Concrete, 3000 psi with wire mesh	4" thick	$ 3.09
	5" thick	3.77
	6" thick	4.23
Precast concrete patio blocks (natural)	2" thick	8.90
Precast concrete patio blocks (colors)	2" thick	9.40
Flagstone, bluestone	1" thick	12.60
Flagstone, bluestone	1-1/2" thick	16.50
Slate (natural, irregular)	3/4" thick	12.30
Slate (random rectangular)	1/2" thick	18.25
Seeding		
Fine grading & seeding includes lime, fertilizer & seed	per S.Y.	1.99
Lawn Sprinkler System	per S.F.	.78

Fencing *(Cost per L.F.)*

Chain Link, 4' high, galvanized	$ 14.60
Gate, 4' high (each)	151.00
Cedar Picket, 3' high, 2 rail	10.65
Gate (each)	150.00
3 Rail, 4' high	12.80
Gate (each)	160.00
Cedar Stockade, 3 Rail, 6' high	13.00
Gate (each)	160.00
Board & Battens, 2 sides 6' high, pine	18.60
6' high, cedar	26.50
No. 1 Cedar, basketweave, 6' high	14.80
Gate, 6' high (each)	184.00

Carport *(Cost per S.F.)*

Economy	$ 7.28
Average	11.02
Custom	16.44
Luxury	18.67

Assemblies Section

Table of Contents

How to Use the Assemblies Section

Illustration
Each building assembly system is accompanied by a detailed, illustrated description. Each individual component is labeled. Every element involved in the total system function is shown.

Quantities
Each material in a system is shown with the quantity required for the system unit. For example, the rafters in this system have 1.170 L.F. per S.F. of ceiling area.

Unit of Measure
In the three right-hand columns, each cost figure is adjusted to agree with the unit of measure for the entire system. In this case, COST PER SQUARE FOOT (S.F.) is the common unit of measure. NOTE: In addition, under the UNIT heading, all the elements of each system are defined in relation to the product as a selling commodity. For example, "fascia board" is defined in linear feet, instead of in board feet.

Description
Each page includes a brief outline of any special conditions to be used when pricing a system. All units of measure are defined here.

System Definition
Not only are all components broken down for each system, but alternative components can be found on the opposite page. Simply insert any chosen new element into the chart to develop a custom system.

Labor-hours
Total labor-hours for a system can be found by simply multiplying the quantity of the system required times LABOR-HOURS. The resulting figure is the total labor-hours needed to complete the system.
(QUANTITY OF SYSTEM x LABOR-HOURS = TOTAL SYSTEM LABOR-HOURS)

Materials
This column contains the MATERIAL COST of each element. These cost figures include 10% for profit.

Installation
Labor rates include both the INSTALLATION COST of the contractor and the standard contractor's O&P. On the average, the LABOR COST will be 70.6% over the above BARE LABOR COST.

Totals
This row provides the necessary system cost totals. TOTAL SYSTEM COST can be derived by multiplying the TOTAL times each system's SQUARE FOOT ESTIMATE (TOTAL x SQUARE FEET = TOTAL SYSTEM COST).

Work Sheet
Using the SELECTIVE PRICE SHEET on the page opposite each system, it is possible to create estimates with alternative items for any number of systems.

Note:
Throughout this section, the words assembly and system are used interchangeably.

Total
MATERIAL COST + INSTALLATION COST = TOTAL. Work on the table from left to right across cost columns to derive totals.

3 | FRAMING — 12 | Gable End Roof Framing Systems

Ridge Board
Sheathing
Rafters
Rafter Tie
Fascia Board
Soffit Nailer
Ceiling Joists
Furring Strips

System Description	QUAN.	UNIT	LABOR HOURS	COST PER S.F. MAT.	INST.	TOTAL
2" X 6" RAFTERS, 16" O.C., 4/12 PITCH						
Rafters, 2" x 6", 16" O.C., 4/12 pitch	1.170	L.F.	.019	76	.80	1.56
Ceiling joists, 2" x 4", 16" O.C.	1.000	L.F.	.013	41	.54	95
Ridge board, 2" x 6"	.050	L.F.	.002	03	.07	10
Fascia board, 2" x 6"	.100	L.F.	.005	07	.23	30
Rafter tie, 1" x 4", 4' O.C.	.060	L.F.	.001	02	.05	07
Soffit nailer (outrigger), 2" x 4", 24" O.C.	.170	L.F.	.004	07	.19	26
Sheathing, exterior, plywood, CDX, 1/2" thick	1.170	S.F.	.013	80	.56	1.36
Furring strips, 1" x 3", 16" O.C.	1.000	L.F.	.023	31	.97	1.28
TOTAL		S.F.	.080	2.47	3.41	5.88
2" X 8" RAFTERS, 16" O.C., 4/12 PITCH						
Rafters, 2" x 8", 16" O.C., 4/12 pitch	1.170	L.F.	.020	1.10	.83	1.93
Ceiling joists, 2" x 6", 16" O.C.	1.000	L.F.	.013	65	.54	1.19
Ridge board, 2" x 8"	.050	L.F.	.002	.05	.08	.13
Fascia board, 2" x 8"	.100	L.F.	.007	.09	.30	.39
Rafter tie, 1" x 4", 4' O.C.	.060	L.F.	.001	.02	.05	.07
Soffit nailer (outrigger), 2" x 4", 24" O.C.	.170	L.F.	.004	07	.19	26
Sheathing, exterior, plywood, CDX, 1/2" thick	1.170	S.F.	.013	80	.56	1.36
Furring strips, 1" x 3", 16" O.C.	1.000	L.F.	.023	31	.97	1.28
TOTAL		S.F.	.083	3.09	3.52	6.61

The cost of this system is based on the square foot of plan area.
All quantities have been adjusted accordingly.

Description	QUAN.	UNIT	LABOR HOURS	COST PER S.F. MAT.	INST.	TOTAL

120 **Important: See the Reference Section for critical supporting data - Reference Nos., Crews & Location Factors**

Division 1
Site Work

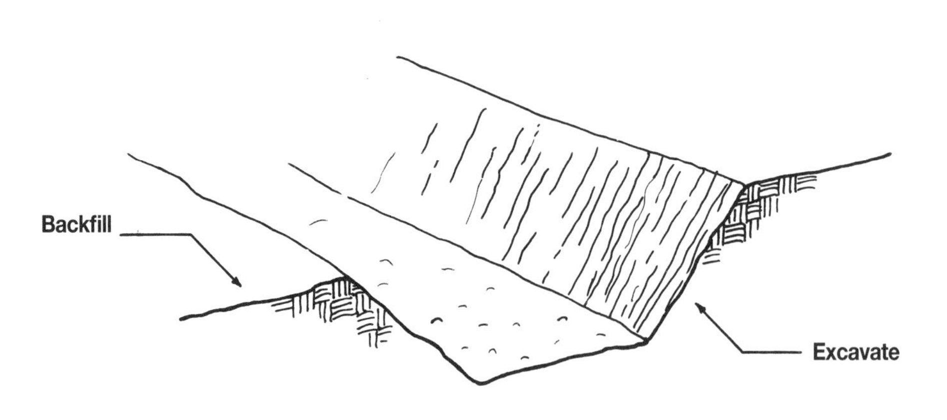

System Description	QUAN.	UNIT	LABOR HOURS	COST EACH		
				MAT.	INST.	TOTAL
BUILDING, 24′ X 38′, 4′ DEEP						
Cut & chip light trees to 6″ diam.	.190	Acre	9.120		513	513
Excavate, backhoe	174.000	C.Y.	4.641		349.74	349.74
Backfill, dozer, 4″ lifts, no compaction	87.000	C.Y.	.580		97.44	97.44
Rough grade, dozer, 30′ from building	87.000	C.Y.	.580		97.44	97.44
TOTAL		Ea.	14.921		1,057.62	1,057.62
BUILDING, 26′ X 46′, 4′ DEEP						
Cut & chip light trees to 6″ diam.	.210	Acre	10.080		567	567
Excavate, backhoe	201.000	C.Y.	5.361		404.01	404.01
Backfill, dozer, 4″ lifts, no compaction	100.000	C.Y.	.667		112	112
Rough grade, dozer, 30′ from building	100.000	C.Y.	.667		112	112
TOTAL		Ea.	16.775		1,195.01	1,195.01
BUILDING, 26′ X 60′, 4′ DEEP						
Cut & chip light trees to 6″ diam.	.240	Acre	11.520		648	648
Excavate, backhoe	240.000	C.Y.	6.401		482.40	482.40
Backfill, dozer, 4″ lifts, no compaction	120.000	C.Y.	.800		134.40	134.40
Rough grade, dozer, 30′ from building	120.000	C.Y.	.800		134.40	134.40
TOTAL		Ea.	19.521		1,399.20	1,399.20
BUILDING, 30′ X 66′, 4′ DEEP						
Cut & chip light trees to 6″ diam.	.260	Acre	12.480		702	702
Excavate, backhoe	268.000	C.Y.	7.148		538.68	538.68
Backfill, dozer, 4″ lifts, no compaction	134.000	C.Y.	.894		150.08	150.08
Rough grade, dozer, 30′ from building	134.000	C.Y.	.894		150.08	150.08
TOTAL		Ea.	21.416		1,540.84	1,540.84

The costs in this system are on a cost each basis.
Quantities are based on 1′-0″ clearance on each side of footing.

Description	QUAN.	UNIT	LABOR HOURS	COST EACH		
				MAT.	INST.	TOTAL

Important: See the Reference Section for critical supporting data - Reference Nos., Crews & Location Factors

Footing Excavation Price Sheet	QUAN.	UNIT	LABOR HOURS	COST EACH		
				MAT.	INST.	TOTAL
Clear and grub, medium brush, 30′ from building, 24′ x 38′	.190	Acre	9.120		515	515
26′ x 46′	.210	Acre	10.080		565	565
26′ x 60′	.240	Acre	11.520		650	650
30′ x 66′	.260	Acre	12.480		705	705
Light trees, to 6″ dia. cut & chip, 24′ x 38′	.190	Acre	9.120		515	515
26′ x 46′	.210	Acre	10.080		565	565
26′ x 60′	.240	Acre	11.520		650	650
30′ x 66′	.260	Acre	12.480		705	705
Medium trees, to 10″ dia. cut & chip, 24′ x 38′	.190	Acre	13.029		740	740
26′ x 46′	.210	Acre	14.400		815	815
26′ x 60′	.240	Acre	16.457		930	930
30′ x 66′	.260	Acre	17.829		1,000	1,000
Excavation, footing, 24′ x 38′, 2′ deep	68.000	C.Y.	.906		137	137
4′ deep	174.000	C.Y.	2.319		350	350
8′ deep	384.000	C.Y.	5.119		770	770
26′ x 46′, 2′ deep	79.000	C.Y.	1.053		159	159
4′ deep	201.000	C.Y.	2.679		405	405
8′ deep	404.000	C.Y.	5.385		810	810
26′ x 60′, 2′ deep	94.000	C.Y.	1.253		189	189
4′ deep	240.000	C.Y.	3.199		480	480
8′ deep	483.000	C.Y.	6.438		970	970
30′ x 66′, 2′ deep	105.000	C.Y.	1.400		211	211
4′ deep	268.000	C.Y.	3.572		540	540
8′ deep	539.000	C.Y.	7.185		1,075	1,075
Backfill, 24′ x 38′, 2″ lifts, no compaction	34.000	C.Y.	.227		38	38
Compaction, air tamped, add	34.000	C.Y.	2.267		330	330
4″ lifts, no compaction	87.000	C.Y.	.580		97.50	97.50
Compaction, air tamped, add	87.000	C.Y.	5.800		850	850
8″ lifts, no compaction	192.000	C.Y.	1.281		215	215
Compaction, air tamped, add	192.000	C.Y.	12.801		1,875	1,875
26′ x 46′, 2″ lifts, no compaction	40.000	C.Y.	.267		44.50	44.50
Compaction, air tamped, add	40.000	C.Y.	2.667		390	390
4″ lifts, no compaction	100.000	C.Y.	.667		112	112
Compaction, air tamped, add	100.000	C.Y.	6.667		975	975
8″ lifts, no compaction	202.000	C.Y.	1.347		227	227
Compaction, air tamped, add	202.000	C.Y.	13.467		1,975	1,975
26′ x 60′, 2″ lifts, no compaction	47.000	C.Y.	.313		52.50	52.50
Compaction, air tamped, add	47.000	C.Y.	3.133		455	455
4″ lifts, no compaction	120.000	C.Y.	.800		135	135
Compaction, air tamped, add	120.000	C.Y.	8.000		1,175	1,175
8″ lifts, no compaction	242.000	C.Y.	1.614		271	271
Compaction, air tamped, add	242.000	C.Y.	16.134		2,350	2,350
30′ x 66′, 2″ lifts, no compaction	53.000	C.Y.	.354		59.50	59.50
Compaction, air tamped, add	53.000	C.Y.	3.534		520	520
4″ lifts, no compaction	134.000	C.Y.	.894		151	151
Compaction, air tamped, add	134.000	C.Y.	8.934		1,300	1,300
8″ lifts, no compaction	269.000	C.Y.	1.794		300	300
Compaction, air tamped, add	269.000	C.Y.	17.934		2,625	2,625
Rough grade, 30′ from building, 24′ x 38′	87.000	C.Y.	.580		97.50	97.50
26′ x 46′	100.000	C.Y.	.667		112	112
26′ x 60′	120.000	C.Y.	.800		135	135
30′ x 66′	134.000	C.Y.	.894		151	151

Backfill

Excavate

System Description	QUAN.	UNIT	LABOR HOURS	COST EACH MAT.	COST EACH INST.	COST EACH TOTAL
BUILDING, 24′ X 38′, 8′ DEEP						
Clear & grub, dozer, medium brush, 30′ from building	.190	Acre	2.027		201.40	201.40
Excavate, track loader, 1-1/2 C.Y. bucket	550.000	C.Y.	7.860		671	671
Backfill, dozer, 8″ lifts, no compaction	180.000	C.Y.	1.201		201.60	201.60
Rough grade, dozer, 30′ from building	280.000	C.Y.	1.868		313.60	313.60
TOTAL		Ea.	12.956		1,387.60	1,387.60
BUILDING, 26′ X 46′, 8′ DEEP						
Clear & grub, dozer, medium brush, 30′ from building	.210	Acre	2.240		222.60	222.60
Excavate, track loader, 1-1/2 C.Y. bucket	672.000	C.Y.	9.603		819.84	819.84
Backfill, dozer, 8″ lifts, no compaction	220.000	C.Y.	1.467		246.40	246.40
Rough grade, dozer, 30′ from building	340.000	C.Y.	2.268		380.80	380.80
TOTAL		Ea.	15.578		1,669.64	1,669.64
BUILDING, 26′ X 60′, 8′ DEEP						
Clear & grub, dozer, medium brush, 30′ from building	.240	Acre	2.560		254.40	254.40
Excavate, track loader, 1-1/2 C.Y. bucket	829.000	C.Y.	11.846		1,011.38	1,011.38
Backfill, dozer, 8″ lifts, no compaction	270.000	C.Y.	1.801		302.40	302.40
Rough grade, dozer, 30′ from building	420.000	C.Y.	2.801		470.40	470.40
TOTAL		Ea.	19.008		2,038.58	2,038.58
BUILDING, 30′ X 66′, 8′ DEEP						
Clear & grub, dozer, medium brush, 30′ from building	.260	Acre	2.773		275.60	275.60
Excavate, track loader, 1-1/2 C.Y. bucket	990.000	C.Y.	14.147		1,207.80	1,207.80
Backfill dozer, 8″ lifts, no compaction	320.000	C.Y.	2.134		358.40	358.40
Rough grade, dozer, 30′ from building	500.000	C.Y.	3.335		560	560
TOTAL		Ea.	22.389		2,401.80	2,401.80

The costs in this system are on a cost each basis.
Quantities are based on 1′-0″ clearance beyond footing projection.

Description	QUAN.	UNIT	LABOR HOURS	COST EACH MAT.	COST EACH INST.	COST EACH TOTAL

 Important: See the Reference Section for critical supporting data - Reference Nos., Crews & Location Factors

Foundation Excavation Price Sheet	QUAN.	UNIT	LABOR HOURS	COST EACH		
				MAT.	INST.	TOTAL
Clear & grub, medium brush, 30′ from building, 24′ x 38′	.190	Acre	2.027		201	201
26′ x 46′	.210	Acre	2.240		223	223
26′ x 60′	.240	Acre	2.560		255	255
30′ x 66′	.260	Acre	2.773		276	276
Light trees, to 6″ dia. cut & chip, 24′ x 38′	.190	Acre	9.120		515	515
26′ x 46′	.210	Acre	10.080		565	565
26′ x 60′	.240	Acre	11.520		650	650
30′ x 66′	.260	Acre	12.480		705	705
Medium trees, to 10″ dia. cut & chip, 24′ x 38′	.190	Acre	13.029		740	740
26′ x 46′	.210	Acre	14.400		815	815
26′ x 60′	.240	Acre	16.457		930	930
30′ x 66′	.260	Acre	17.829		1,000	1,000
Excavation, basement, 24′ x 38′, 2′ deep	98.000	C.Y.	1.400		120	120
4′ deep	220.000	C.Y.	3.144		268	268
8′ deep	550.000	C.Y.	7.860		670	670
26′ x 46′, 2′ deep	123.000	C.Y.	1.758		151	151
4′ deep	274.000	C.Y.	3.915		335	335
8′ deep	672.000	C.Y.	9.603		820	820
26′ x 60′, 2′ deep	157.000	C.Y.	2.244		192	192
4′ deep	345.000	C.Y.	4.930		420	420
8′ deep	829.000	C.Y.	11.846		1,000	1,000
30′ x 66′, 2′ deep	192.000	C.Y.	2.744		234	234
4′ deep	419.000	C.Y.	5.988		510	510
8′ deep	990.000	C.Y.	14.147		1,200	1,200
Backfill, 24′ x 38′, 2″ lifts, no compaction	32.000	C.Y.	.213		36	36
Compaction, air tamped, add	32.000	C.Y.	2.133		315	315
4″ lifts, no compaction	72.000	C.Y.	.480		80.50	80.50
Compaction, air tamped, add	72.000	C.Y.	4.800		705	705
8″ lifts, no compaction	180.000	C.Y.	1.201		202	202
Compaction, air tamped, add	180.000	C.Y.	12.001		1,750	1,750
26′ x 46′, 2″ lifts, no compaction	40.000	C.Y.	.267		44.50	44.50
Compaction, air tamped, add	40.000	C.Y.	2.667		390	390
4″ lifts, no compaction	90.000	C.Y.	.600		101	101
Compaction, air tamped, add	90.000	C.Y.	6.000		880	880
8″ lifts, no compaction	220.000	C.Y.	1.467		247	247
Compacton, air tamped, add	220.000	C.Y.	14.667		2,150	2,150
26′ x 60′, 2″ lifts, no compaction	50.000	C.Y.	.334		56	56
Compaction, air tamped, add	50.000	C.Y.	3.334		490	490
4″ lifts, no compaction	110.000	C.Y.	.734		124	124
Compaction, air tamped, add	110.000	C.Y.	7.334		1,075	1,075
8″ lifts, no compaction	270.000	C.Y.	1.801		305	305
Compaction, air tamped, add	270.000	C.Y.	18.001		2,625	2,625
30′ x 66′, 2″ lifts, no compaction	60.000	C.Y.	.400		67.50	67.50
Compaction, air tamped, add	60.000	C.Y.	4.000		585	585
4″ lifts, no compaction	130.000	C.Y.	.867		146	146
Compaction, air tamped, add	130.000	C.Y.	8.667		1,275	1,275
8″ lifts, no compaction	320.000	C.Y.	2.134		360	360
Compaction, air tamped, add	320.000	C.Y.	21.334		3,125	3,125
Rough grade, 30′ from building, 24′ x 38′	280.000	C.Y.	1.868		315	315
26′ x 46′	340.000	C.Y.	2.268		380	380
26′ x 60′	420.000	C.Y.	2.801		475	475
30′ x 66′	500.000	C.Y.	3.335		560	560

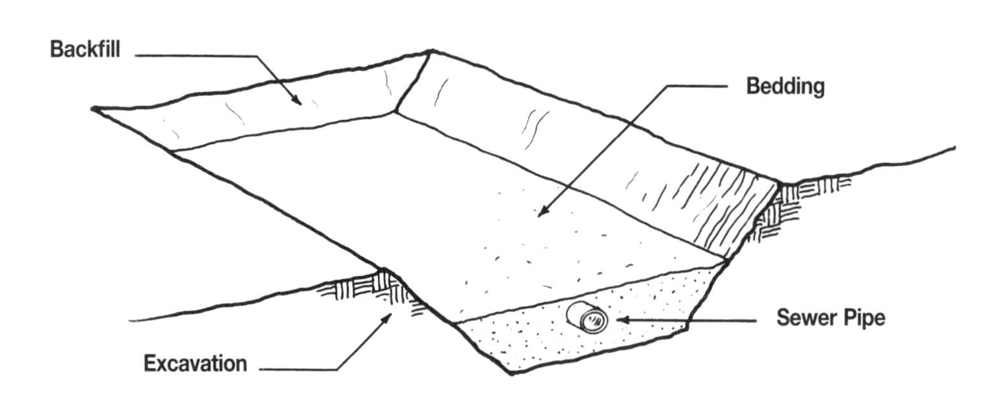

System Description		QUAN.	UNIT	LABOR HOURS	COST PER L.F.		
					MAT.	INST.	TOTAL
2′ DEEP							
Excavation, backhoe		.296	C.Y.	.032		1.62	1.62
Bedding, sand		.111	C.Y.	.044	1.47	1.56	3.03
Utility, sewer, 6″ cast iron		1.000	L.F.	.283	14.23	11.35	25.58
Backfill, incl. compaction		.185	C.Y.	.044		1.31	1.31
	TOTAL		L.F.	.403	15.70	15.84	31.54
4′ DEEP							
Excavation, backhoe		.889	C.Y.	.095		4.88	4.88
Bedding, sand		.111	C.Y.	.044	1.47	1.56	3.03
Utility, sewer, 6″ cast iron		1.000	L.F.	.283	14.23	11.35	25.58
Backfill, incl. compaction		.778	C.Y.	.183		5.52	5.52
	TOTAL		L.F.	.605	15.70	23.31	39.01
6′ DEEP							
Excavation, backhoe		1.770	C.Y.	.189		9.72	9.72
Bedding, sand		.111	C.Y.	.044	1.47	1.56	3.03
Utility, sewer, 6″ cast iron		1.000	L.F.	.283	14.23	11.35	25.58
Backfill, incl. compaction		1.660	C.Y.	.391		11.79	11.79
	TOTAL		L.F.	.907	15.70	34.42	50.12
8′ DEEP							
Excavation, backhoe		2.960	C.Y.	.316		16.25	16.25
Bedding, sand		.111	C.Y.	.044	1.47	1.56	3.03
Utility, sewer, 6″ cast iron		1.000	L.F.	.283	14.23	11.35	25.58
Backfill, incl. compaction		2.850	C.Y.	.671		20.24	20.24
	TOTAL		L.F.	1.314	15.70	49.40	65.10

The costs in this system are based on a cost per linear foot of trench, and based on 2′ wide at bottom of trench up to 6′ deep.

Description	QUAN.	UNIT	LABOR HOURS	COST PER L.F.		
				MAT.	INST.	TOTAL

 Important: See the Reference Section for critical supporting data - Reference Nos., Crews & Location Factors

Utility Trenching Price Sheet	QUAN.	UNIT	LABOR HOURS	MAT.	INST.	TOTAL
Excavation, bottom of trench 2′ wide, 2′ deep	.296	C.Y.	.032		1.62	1.62
4′ deep	.889	C.Y.	.095		4.88	4.88
6′ deep	1.770	C.Y.	.142		7.80	7.80
8′ deep	2.960	C.Y.	.105		14.70	14.70
Bedding, sand, bottom of trench 2′ wide, no compaction, pipe, 2″ diameter	.070	C.Y.	.028	.93	.98	1.91
4″ diameter	.084	C.Y.	.034	1.11	1.18	2.29
6″ diameter	.105	C.Y.	.042	1.39	1.47	2.86
8″ diameter	.122	C.Y.	.049	1.62	1.71	3.33
Compacted, pipe, 2″ diameter	.074	C.Y.	.030	.98	1.04	2.02
4″ diameter	.092	C.Y.	.037	1.22	1.29	2.51
6″ diameter	.111	C.Y.	.044	1.47	1.56	3.03
8″ diameter	.129	C.Y.	.052	1.71	1.81	3.52
3/4″ stone, bottom of trench 2′ wide, pipe, 4″ diameter	.082	C.Y.	.033	1.09	1.15	2.24
6″ diameter	.099	C.Y.	.040	1.31	1.38	2.69
3/8″ stone, bottom of trench 2′ wide, pipe, 4″ diameter	.084	C.Y.	.034	1.11	1.18	2.29
6″ diameter	.102	C.Y.	.041	1.35	1.43	2.78
Utilities, drainage & sewerage, corrugated plastic, 6″ diameter	1.000	L.F.	.069	5.85	2.15	8
8″ diameter	1.000	L.F.	.072	8.85	2.24	11.09
Bituminous fiber, 4″ diameter	1.000	L.F.	.064	2.78	2.01	4.79
6″ diameter	1.000	L.F.	.069	5.85	2.15	8
8″ diameter	1.000	L.F.	.072	8.85	2.24	11.09
Concrete, non-reinforced, 6″ diameter	1.000	L.F.	.181	4.91	6.80	11.71
8″ diameter	1.000	L.F.	.214	5.40	8.05	13.45
PVC, SDR 35, 4″ diameter	1.000	L.F.	.064	2.78	2.01	4.79
6″ diameter	1.000	L.F.	.069	5.85	2.15	8
8″ diameter	1.000	L.F.	.072	8.85	2.24	11.09
Vitrified clay, 4″ diameter	1.000	L.F.	.091	1.87	2.84	4.71
6″ diameter	1.000	L.F.	.120	3.12	3.76	6.88
8″ diameter	1.000	L.F.	.140	4.43	5.45	9.88
Gas & service, polyethylene, 1-1/4″ diameter	1.000	L.F.	.059	1.04	2.17	3.21
Steel sched.40, 1″ diameter	1.000	L.F.	.107	3.14	4.95	8.09
2″ diameter	1.000	L.F.	.114	4.92	5.30	10.22
Sub-drainage, PVC, perforated, 3″ diameter	1.000	L.F.	.064	2.78	2.01	4.79
4″ diameter	1.000	L.F.	.064	2.78	2.01	4.79
5″ diameter	1.000	L.F.	.069	5.85	2.15	8
6″ diameter	1.000	L.F.	.069	5.85	2.15	8
Porous wall concrete, 4″ diameter	1.000	L.F.	.072	2.26	2.24	4.50
Vitrified clay, perforated, 4″ diameter	1.000	L.F.	.120	2.38	4.51	6.89
6″ diameter	1.000	L.F.	.152	3.93	5.75	9.68
Water service, copper, type K, 3/4″	1.000	L.F.	.083	3.04	3.85	6.89
1″ diameter	1.000	L.F.	.093	4.27	4.31	8.58
PVC, 3/4″	1.000	L.F.	.121	1.04	5.60	6.64
1″ diameter	1.000	L.F.	.134	1.17	6.20	7.37
Backfill, bottom of trench 2′ wide no compact, 2′ deep, pipe, 2″ diameter	.226	L.F.	.053		1.60	1.60
4″ diameter	.212	L.F.	.050		1.51	1.51
6″ diameter	.185	L.F.	.044		1.31	1.31
4′ deep, pipe, 2″ diameter	.819	C.Y.	.193		5.80	5.80
4″ diameter	.805	C.Y.	.189		5.70	5.70
6″ diameter	.778	C.Y.	.183		5.50	5.50
6′ deep, pipe, 2″ diameter	1.700	C.Y.	.400		12.05	12.05
4″ diameter	1.690	C.Y.	.398		12	12
6″ diameter	1.660	C.Y.	.391		11.80	11.80
8′ deep, pipe, 2″ diameter	2.890	C.Y.	.680		20.50	20.50
4″ diameter	2.870	C.Y.	.675		20.50	20.50
6″ diameter	2.850	C.Y.	.671		20	20

16 Sidewalk Systems

Asphalt
Brick Edge
Gravel Fill

System Description	QUAN.	UNIT	LABOR HOURS	COST PER S.F.		
				MAT.	INST.	TOTAL
ASPHALT SIDEWALK SYSTEM, 3′ WIDE WALK						
Gravel fill, 4″ deep	1.000	S.F.	.001	.29	.03	.32
Compact fill	.012	C.Y.			.01	.01
Handgrade	1.000	S.F.	.004		.13	.13
Walking surface, bituminous paving, 2″ thick	1.000	S.F.	.007	.49	.26	.75
Edging, brick, laid on edge	.670	L.F.	.079	1.52	2.97	4.49
TOTAL		S.F.	.091	2.30	3.40	5.70
CONCRETE SIDEWALK SYSTEM, 3′ WIDE WALK						
Gravel fill, 4″ deep	1.000	S.F.	.001	.29	.03	.32
Compact fill	.012	C.Y.			.01	.01
Handgrade	1.000	S.F.	.004		.13	.13
Walking surface, concrete, 4″ thick	1.000	S.F.	.040	1.60	1.49	3.09
Edging, brick, laid on edge	.670	L.F.	.079	1.52	2.97	4.49
TOTAL		S.F.	.124	3.41	4.63	8.04
PAVERS, BRICK SIDEWALK SYSTEM, 3′ WIDE WALK						
Sand base fill, 4″ deep	1.000	S.F.	.001	.35	.06	.41
Compact fill	.012	C.Y.			.01	.01
Handgrade	1.000	S.F.	.004		.13	.13
Walking surface, brick pavers	1.000	S.F.	.160	2.78	6	8.78
Edging, redwood, untreated, 1″ x 4″	.670	L.F.	.032	1.61	1.37	2.98
TOTAL		S.F.	.197	4.74	7.57	12.31

The costs in this system are based on a cost per square foot of sidewalk area. Concrete used is 3000 p.s.i.

Description	QUAN.	UNIT	LABOR HOURS	COST PER S.F.		
				MAT.	INST.	TOTAL

Important: See the Reference Section for critical supporting data - Reference Nos., Crews & Location Factors

Sidewalk Price Sheet	QUAN.	UNIT	LABOR HOURS	COST PER S.F.		
				MAT.	INST.	TOTAL
Base, crushed stone, 3″ deep	1.000	S.F.	.001	.33	.08	.41
6″ deep	1.000	S.F.	.001	.66	.08	.74
9″ deep	1.000	S.F.	.002	.96	.11	1.07
12″ deep	1.000	S.F.	.002	1.56	.13	1.69
Bank run gravel, 6″ deep	1.000	S.F.	.001	.43	.05	.48
9″ deep	1.000	S.F.	.001	.63	.07	.70
12″ deep	1.000	S.F.	.001	.86	.09	.95
Compact base, 3″ deep	.009	C.Y.	.001		.01	.01
6″ deep	.019	C.Y.	.001		.02	.02
9″ deep	.028	C.Y.	.001		.03	.03
Handgrade	1.000	S.F.	.004		.13	.13
Surface, brick, pavers dry joints, laid flat, running bond	1.000	S.F.	.160	2.78	6	8.78
Basket weave	1.000	S.F.	.168	3.16	6.30	9.46
Herringbone	1.000	S.F.	.174	3.16	6.50	9.66
Laid on edge, running bond	1.000	S.F.	.229	2.72	8.55	11.27
Mortar jts. laid flat, running bond	1.000	S.F.	.192	3.34	7.20	10.54
Basket weave	1.000	S.F.	.202	3.79	7.55	11.34
Herringbone	1.000	S.F.	.209	3.79	7.80	11.59
Laid on edge, running bond	1.000	S.F.	.274	3.26	10.25	13.51
Bituminous paving, 1-1/2″ thick	1.000	S.F.	.006	.37	.19	.56
2″ thick	1.000	S.F.	.007	.49	.26	.75
2-1/2″ thick	1.000	S.F.	.008	.62	.28	.90
Sand finish, 3/4″ thick	1.000	S.F.	.001	.20	.09	.29
1″ thick	1.000	S.F.	.001	.24	.11	.35
Concrete, reinforced, broom finish, 4″ thick	1.000	S.F.	.040	1.60	1.49	3.09
5″ thick	1.000	S.F.	.044	2.13	1.64	3.77
6″ thick	1.000	S.F.	.047	2.48	1.75	4.23
Crushed stone, white marble, 3″ thick	1.000	S.F.	.009	.22	.28	.50
Bluestone, 3″ thick	1.000	S.F.	.009	.24	.28	.52
Flagging, bluestone, 1″	1.000	S.F.	.198	5.20	7.40	12.60
1-1/2″	1.000	S.F.	.188	9.45	7.05	16.50
Slate, natural cleft, 3/4″	1.000	S.F.	.174	5.80	6.50	12.30
Random rect., 1/2″	1.000	S.F.	.152	12.55	5.70	18.25
Granite blocks	1.000	S.F.	.174	7	6.50	13.50
Edging, corrugated aluminum, 4″, 3′ wide walk	.666	L.F.	.008	.27	.35	.62
4′ wide walk	.500	L.F.	.006	.21	.26	.47
6″, 3′ wide walk	.666	L.F.	.010	.34	.41	.75
4′ wide walk	.500	L.F.	.007	.26	.31	.57
Redwood-cedar-cypress, 1″ x 4″, 3′ wide walk	.666	L.F.	.021	.81	.90	1.71
4′ wide walk	.500	L.F.	.016	.61	.68	1.29
2″ x 4″, 3′ wide walk	.666	L.F.	.032	1.61	1.37	2.98
4′ wide walk	.500	L.F.	.024	1.21	1.03	2.24
Brick, dry joints, 3′ wide walk	.666	L.F.	.079	1.52	2.97	4.49
4′ wide walk	.500	L.F.	.059	1.14	2.22	3.36
Mortar joints, 3′ wide walk	.666	L.F.	.095	1.83	3.57	5.40
4′ wide walk	.500	L.F.	.071	1.36	2.66	4.02

20 Driveway Systems

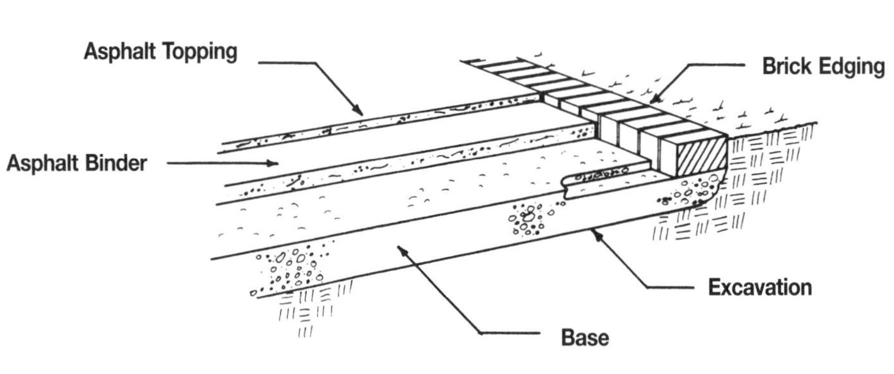

System Description	QUAN.	UNIT	LABOR HOURS	COST PER S.F.		
				MAT.	INST.	TOTAL
ASPHALT DRIVEWAY TO 10′ WIDE						
Excavation, driveway to 10′ wide, 6″ deep	.019	C.Y.			.03	.03
Base, 6″ crushed stone	1.000	S.F.	.001	.66	.08	.74
Handgrade base	1.000	S.F.	.004		.13	.13
2″ thick base	1.000	S.F.	.002	.49	.15	.64
1″ topping	1.000	S.F.	.001	.24	.11	.35
Edging, brick pavers	.200	L.F.	.024	.45	.89	1.34
TOTAL		S.F.	.032	1.84	1.39	3.23
CONCRETE DRIVEWAY TO 10′ WIDE						
Excavation, driveway to 10′ wide, 6″ deep	.019	C.Y.			.03	.03
Base, 6″ crushed stone	1.000	S.F.	.001	.66	.08	.74
Handgrade base	1.000	S.F.	.004		.13	.13
Surface, concrete, 4″ thick	1.000	S.F.	.040	1.60	1.49	3.09
Edging, brick pavers	.200	L.F.	.024	.45	.89	1.34
TOTAL		S.F.	.069	2.71	2.62	5.33
PAVERS, BRICK DRIVEWAY TO 10′ WIDE						
Excavation, driveway to 10′ wide, 6″ deep	.019	C.Y.			.03	.03
Base, 6″ sand	1.000	S.F.	.001	.56	.10	.66
Handgrade base	1.000	S.F.	.004		.13	.13
Surface, pavers, brick laid flat, running bond	1.000	S.F.	.160	2.78	6	8.78
Edging, redwood, untreated, 2″ x 4″	.200	L.F.	.010	.48	.41	.89
TOTAL		S.F.	.175	3.82	6.67	10.49

Description	QUAN.	UNIT	LABOR HOURS	COST PER S.F.		
				MAT.	INST.	TOTAL

Important: See the Reference Section for critical supporting data - Reference Nos., Crews & Location Factors

Driveway Price Sheet	QUAN.	UNIT	LABOR HOURS	COST PER S.F.		
				MAT.	INST.	TOTAL
Excavation, by machine, 10′ wide, 6″ deep	.019	C.Y.	.001		.03	.03
12″ deep	.037	C.Y.	.001		.05	.05
18″ deep	.055	C.Y.	.001		.08	.08
20′ wide, 6″ deep	.019	C.Y.	.001		.03	.03
12″ deep	.037	C.Y.	.001		.05	.05
18″ deep	.055	C.Y.	.001		.08	.08
Base, crushed stone, 10′ wide, 3″ deep	1.000	S.F.	.001	.33	.04	.37
6″ deep	1.000	S.F.	.001	.66	.08	.74
9″ deep	1.000	S.F.	.002	.96	.11	1.07
20′ wide, 3″ deep	1.000	S.F.	.001	.33	.04	.37
6″ deep	1.000	S.F.	.001	.66	.08	.74
9″ deep	1.000	S.F.	.002	.96	.11	1.07
Bank run gravel, 10′ wide, 3″ deep	1.000	S.F.	.001	.22	.03	.25
6″ deep	1.000	S.F.	.001	.43	.05	.48
9″ deep	1.000	S.F.	.001	.63	.07	.70
20′ wide, 3″ deep	1.000	S.F.	.001	.22	.03	.25
6″ deep	1.000	S.F.	.001	.43	.05	.48
9″ deep	1.000	S.F.	.001	.63	.07	.70
Handgrade, 10′ wide	1.000	S.F.	.004		.13	.13
20′ wide	1.000	S.F.	.004		.13	.13
Surface, asphalt, 10′ wide, 3/4″ topping, 1″ base	1.000	S.F.	.002	.58	.20	.78
2″ base	1.000	S.F.	.003	.69	.24	.93
1″ topping, 1″ base	1.000	S.F.	.002	.62	.22	.84
2″ base	1.000	S.F.	.003	.73	.26	.99
20′ wide, 3/4″ topping, 1″ base	1.000	S.F.	.002	.58	.20	.78
2″ base	1.000	S.F.	.003	.69	.24	.93
1″ topping, 1″ base	1.000	S.F.	.002	.62	.22	.84
2″ base	1.000	S.F.	.003	.73	.26	.99
Concrete, 10′ wide, 4″ thick	1.000	S.F.	.040	1.60	1.49	3.09
6″ thick	1.000	S.F.	.047	2.48	1.75	4.23
20′ wide, 4″ thick	1.000	S.F.	.040	1.60	1.49	3.09
6″ thick	1.000	S.F.	.047	2.48	1.75	4.23
Paver, brick 10′ wide dry joints, running bond, laid flat	1.000	S.F.	.160	2.78	6	8.78
Laid on edge	1.000	S.F.	.229	2.72	8.55	11.27
Mortar joints, laid flat	1.000	S.F.	.192	3.34	7.20	10.54
Laid on edge	1.000	S.F.	.274	3.26	10.25	13.51
20′ wide, running bond, dry jts., laid flat	1.000	S.F.	.160	2.78	6	8.78
Laid on edge	1.000	S.F.	.229	2.72	8.55	11.27
Mortar joints, laid flat	1.000	S.F.	.192	3.34	7.20	10.54
Laid on edge	1.000	S.F.	.274	3.26	10.25	13.51
Crushed stone, 10′ wide, white marble, 3″	1.000	S.F.	.009	.22	.28	.50
Bluestone, 3″	1.000	S.F.	.009	.24	.28	.52
20′ wide, white marble, 3″	1.000	S.F.	.009	.22	.28	.50
Bluestone, 3″	1.000	S.F.	.009	.24	.28	.52
Soil cement, 10′ wide	1.000	S.F.	.007	.23	.69	.92
20′ wide	1.000	S.F.	.007	.23	.69	.92
Granite blocks, 10′ wide	1.000	S.F.	.174	7	6.50	13.50
20′ wide	1.000	S.F.	.174	7	6.50	13.50
Asphalt block, solid 1-1/4″ thick	1.000	S.F.	.119	5.10	4.44	9.54
Solid 3″ thick	1.000	S.F.	.123	7.15	4.61	11.76
Edging, brick, 10′ wide	.200	L.F.	.024	.45	.89	1.34
20′ wide	.100	L.F.	.012	.23	.44	.67
Redwood, untreated 2″ x 4″, 10′ wide	.200	L.F.	.010	.48	.41	.89
20′ wide	.100	L.F.	.005	.24	.21	.45
Granite, 4 1/2″ x 12″ straight, 10′ wide	.200	L.F.	.032	1.12	1.67	2.79
20′ wide	.100	L.F.	.016	.56	.84	1.40
Finishes, asphalt sealer, 10′ wide	1.000	S.F.	.023	.58	.72	1.30
20′ wide	1.000	S.F.	.023	.58	.72	1.30
Concrete, exposed aggregate 10′ wide	1.000	S.F.	.013	.18	.50	.68
20′ wide	1.000	S.F.	.013	.18	.50	.68

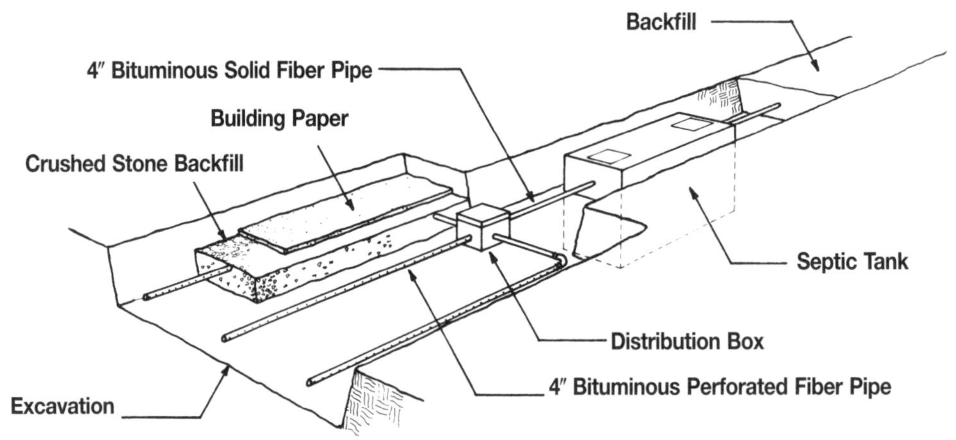

24 Septic Systems

Backfill

4″ Bituminous Solid Fiber Pipe

Building Paper

Crushed Stone Backfill

Septic Tank

Distribution Box

4″ Bituminous Perforated Fiber Pipe

Excavation

System Description	QUAN.	UNIT	LABOR HOURS	COST EACH		
				MAT.	INST.	TOTAL
SEPTIC SYSTEM WITH 1000 S.F. LEACHING FIELD, 1000 GALLON TANK						
Tank, 1000 gallon, concrete	1.000	Ea.	3.500	645	137	782
Distribution box, concrete	1.000	Ea.	1.000	120	30	150
4″ PVC pipe	25.000	L.F.	1.600	69.50	50.25	119.75
Tank and field excavation	119.000	C.Y.	13.130		913.92	913.92
Crushed stone backfill	76.000	C.Y.	12.160	1,900	532.76	2,432.76
Backfill with excavated material	36.000	C.Y.	.240		40.32	40.32
Building paper	125.000	S.Y.	2.430	45	101.25	146.25
4″ PVC perforated pipe	145.000	L.F.	9.280	403.10	291.45	694.55
4″ pipe fittings	2.000	Ea.	1.939	27.50	81	108.50
TOTAL		Ea.	45.279	3,210.10	2,177.95	5,388.05
SEPTIC SYSTEM WITH 2 LEACHING PITS, 1000 GALLON TANK						
Tank, 1000 gallon, concrete	1.000	Ea.	3.500	645	137	782
Distribution box, concrete	1.000	Ea.	1.000	120	30	150
4″ PVC pipe	75.000	L.F.	4.800	208.50	150.75	359.25
Excavation for tank only	20.000	C.Y.	2.207		153.60	153.60
Crushed stone backfill	10.000	C.Y.	1.600	250	70.10	320.10
Backfill with excavated material	55.000	C.Y.	.367		61.60	61.60
Pits, 6′ diameter, including excavation and stone backfill	2.000	Ea.		1,490		1,490
TOTAL		Ea.	13.474	2,713.50	603.05	3,316.55

The costs in this system include all necessary piping and excavation.

Description	QUAN.	UNIT	LABOR HOURS	COST EACH		
				MAT.	INST.	TOTAL

Important: See the Reference Section for critical supporting data - Reference Nos., Crews & Location Factors

Septic Systems Price Sheet	QUAN.	UNIT	LABOR HOURS	COST EACH		
				MAT.	INST.	TOTAL
Tank, precast concrete, 1000 gallon	1.000	Ea.	3.500	645	137	782
2000 gallon	1.000	Ea.	5.600	1,275	219	1,494
Distribution box, concrete, 5 outlets	1.000	Ea.	1.000	120	30	150
12 outlets	1.000	Ea.	2.000	325	60.50	385.50
4″ pipe, PVC, solid	25.000	L.F.	1.600	69.50	50.50	120
Tank and field excavation, 1000 S.F. field	119.000	C.Y.	6.565		915	915
2000 S.F. field	190.000	C.Y.	10.482		1,450	1,450
Tank excavation only, 1000 gallon tank	20.000	C.Y.	1.103		154	154
2000 gallon tank	32.000	C.Y.	1.765		246	246
Backfill, crushed stone 1000 S.F. field	76.000	C.Y.	12.160	1,900	530	2,430
2000 S.F. field	140.000	C.Y.	22.400	3,500	980	4,480
Backfill with excavated material, 1000 S.F. field	36.000	C.Y.	.240		40	40
2000 S.F. field	60.000	C.Y.	.400		67.50	67.50
6′ diameter pits	55.000	C.Y.	.367		61.50	61.50
3′ diameter pits	42.000	C.Y.	.280		47.50	47.50
Building paper, 1000 S.F. field	125.000	S.Y.	2.376	44	99	143
2000 S.F. field	250.000	S.Y.	4.860	90	203	293
4″ pipe, PVC, perforated, 1000 S.F. field	145.000	L.F.	9.280	405	291	696
2000 S.F. field	265.000	L.F.	16.960	735	535	1,270
Pipe fittings, bituminous fiber, 1000 S.F. field	2.000	Ea.	1.939	27.50	81	108.50
2000 S.F. field	4.000	Ea.	3.879	55	162	217
Leaching pit, including excavation and stone backfill, 3′ diameter	1.000	Ea.		560		560
6′ diameter	1.000	Ea.		745		745

60 Chain Link Fence Price Sheet

System Description	QUAN.	UNIT	LABOR HOURS	COST PER UNIT		
				MAT.	INST.	TOTAL
Chain link fence						
Galv.9ga. wire, 1-5/8"post 10'O.C., 1-3/8"top rail, 2"corner post, 3'hi	1.000	L.F.	.130	6.85	4.06	10.91
4' high	1.000	L.F.	.141	10.20	4.42	14.62
6' high	1.000	L.F.	.209	11.55	6.55	18.10
Add for gate 3' wide 1-3/8" frame 3' high	1.000	Ea.	2.000	61.50	62.50	124
4' high	1.000	Ea.	2.400	76	75	151
6' high	1.000	Ea.	2.400	137	75	212
Add for gate 4' wide 1-3/8" frame 3' high	1.000	Ea.	2.667	72	83.50	155.50
4' high	1.000	Ea.	2.667	94	83.50	177.50
6' high	1.000	Ea.	3.000	173	94	267
Alum.9ga. wire, 1-5/8"post, 10'O.C., 1-3/8"top rail, 2"corner post,3'hi	1.000	L.F.	.130	8.40	4.06	12.46
4' high	1.000	L.F.	.141	9.55	4.42	13.97
6' high	1.000	L.F.	.209	12.25	6.55	18.80
Add for gate 3' wide 1-3/8" frame 3' high	1.000	Ea.	2.000	82	62.50	144.50
4' high	1.000	Ea.	2.400	112	75	187
6' high	1.000	Ea.	2.400	167	75	242
Add for gate 4' wide 1-3/8" frame 3' high	1.000	Ea.	2.400	112	75	187
4' high	1.000	Ea.	2.667	149	83.50	232.50
6' high	1.000	Ea.	3.000	232	94	326
Vinyl 9ga. wire, 1-5/8"post 10'O.C., 1-3/8"top rail, 2"corner post,3'hi	1.000	L.F.	.130	7.45	4.06	11.51
4' high	1.000	L.F.	.141	12.20	4.42	16.62
6' high	1.000	L.F.	.209	13.95	6.55	20.50
Add for gate 3' wide 1-3/8" frame 3' high	1.000	Ea.	2.000	91.50	62.50	154
4' high	1.000	Ea.	2.400	119	75	194
6' high	1.000	Ea.	2.400	183	75	258
Add for gate 4' wide 1-3/8" frame 3' high	1.000	Ea.	2.400	124	75	199
4' high	1.000	Ea.	2.667	164	83.50	247.50
6' high	1.000	Ea.	3.000	237	94	331
Tennis court, chain link fence, 10' high						
Galv.11ga.wire, 2"post 10'O.C., 1-3/8"top rail, 2-1/2"corner post	1.000	L.F.	.253	18.25	7.90	26.15
Add for gate 3' wide 1-3/8" frame	1.000	Ea.	2.400	228	75	303
Alum.11ga.wire, 2"post 10'O.C., 1-3/8"top rail, 2-1/2"corner post	1.000	L.F.	.253	26	7.90	33.90
Add for gate 3' wide 1-3/8" frame	1.000	Ea.	2.400	297	75	372
Vinyl 11ga.wire,2"post 10' O.C.,1-3/8"top rail,2-1/2"corner post	1.000	L.F.	.253	22	7.90	29.90
Add for gate 3' wide 1-3/8" frame	1.000	Ea.	2.400	330	75	405
Railings, commercial						
Aluminum balcony rail, 1-1/2" posts with pickets	1.000	L.F.	.164	53	9.15	62.15
With expanded metal panels	1.000	L.F.	.164	68	9.15	77.15
With porcelain enamel panel inserts	1.000	L.F.	.164	60.50	9.15	69.65
Mild steel, ornamental rounded top rail	1.000	L.F.	.164	59	9.15	68.15
As above, but pitch down stairs	1.000	L.F.	.183	64.50	10.20	74.70
Steel pipe, welded, 1-1/2" round, painted	1.000	L.F.	.160	19.55	8.95	28.50
Galvanized	1.000	L.F.	.160	27.50	8.95	36.45
Residential, stock units, mild steel, deluxe	1.000	L.F.	.102	13.55	5.65	19.20
Economy	1.000	L.F.	.102	10.15	5.65	15.80

64 | Wood Fence Price Sheet

System Description	QUAN.	UNIT	LABOR HOURS	COST PER UNIT		
				MAT.	INST.	TOTAL
Basketweave, 3/8"x4" boards, 2"x4" stringers on spreaders, 4"x4" posts						
No. 1 cedar, 6' high	1.000	L.F.	.150	9.25	5.55	14.80
Treated pine, 6' high	1.000	L.F.	.160	11.30	5.95	17.25
Board fence, 1"x4" boards, 2"x4" rails, 4"x4" posts						
Preservative treated, 2 rail, 3' high	1.000	L.F.	.166	6.90	6.15	13.05
4' high	1.000	L.F.	.178	7.55	6.60	14.15
3 rail, 5' high	1.000	L.F.	.185	8.55	6.90	15.45
6' high	1.000	L.F.	.192	9.80	7.15	16.95
Western cedar, No. 1, 2 rail, 3' high	1.000	L.F.	.166	7.50	6.15	13.65
3 rail, 4' high	1.000	L.F.	.178	8.90	6.60	15.50
5' high	1.000	L.F.	.185	10.25	6.90	17.15
6' high	1.000	L.F.	.192	11.25	7.15	18.40
No. 1 cedar, 2 rail, 3' high	1.000	L.F.	.166	11.30	6.15	17.45
4' high	1.000	L.F.	.178	12.85	6.60	19.45
3 rail, 5' high	1.000	L.F.	.185	14.85	6.90	21.75
6' high	1.000	L.F.	.192	16.55	7.15	23.70
Shadow box, 1"x6" boards, 2"x4" rails, 4"x4" posts						
Fir, pine or spruce, treated, 3 rail, 6' high	1.000	L.F.	.160	12.65	5.95	18.60
No. 1 cedar, 3 rail, 4' high	1.000	L.F.	.185	15.55	6.90	22.45
6' high	1.000	L.F.	.192	19.20	7.15	26.35
Open rail, split rails, No. 1 cedar, 2 rail, 3' high	1.000	L.F.	.150	6.25	5.55	11.80
3 rail, 4' high	1.000	L.F.	.160	8.40	5.95	14.35
No. 2 cedar, 2 rail, 3' high	1.000	L.F.	.150	4.86	5.55	10.41
3 rail, 4' high	1.000	L.F.	.160	5.55	5.95	11.50
Open rail, rustic rails, No. 1 cedar, 2 rail, 3' high	1.000	L.F.	.150	3.89	5.55	9.44
3 rail, 4' high	1.000	L.F.	.160	5.25	5.95	11.20
No. 2 cedar, 2 rail, 3' high	1.000	L.F.	.150	3.74	5.55	9.29
3 rail, 4' high	1.000	L.F.	.160	3.95	5.95	9.90
Rustic picket, molded pine pickets, 2 rail, 3' high	1.000	L.F.	.171	5.50	6.35	11.85
3 rail, 4' high	1.000	L.F.	.197	6.35	7.30	13.65
No. 1 cedar, 2 rail, 3' high	1.000	L.F.	.171	7.50	6.35	13.85
3 rail, 4' high	1.000	L.F.	.197	8.65	7.30	15.95
Picket fence, fir, pine or spruce, preserved, treated						
2 rail, 3' high	1.000	L.F.	.171	4.81	6.35	11.16
3 rail, 4' high	1.000	L.F.	.185	5.65	6.90	12.55
Western cedar, 2 rail, 3' high	1.000	L.F.	.171	6	6.35	12.35
3 rail, 4' high	1.000	L.F.	.185	6.15	6.90	13.05
No. 1 cedar, 2 rail, 3' high	1.000	L.F.	.171	12	6.35	18.35
3 rail, 4' high	1.000	L.F.	.185	14	6.90	20.90
Stockade, No. 1 cedar, 3-1/4" rails, 6' high	1.000	L.F.	.150	11.35	5.55	16.90
8' high	1.000	L.F.	.155	14.65	5.75	20.40
No. 2 cedar, treated rails, 6' high	1.000	L.F.	.150	11.35	5.55	16.90
Treated pine, treated rails, 6' high	1.000	L.F.	.150	11.10	5.55	16.65
Gates, No. 2 cedar, picket, 3'-6" wide 4' high	1.000	Ea.	2.667	60	99	159
No. 2 cedar, rustic round, 3' wide, 3' high	1.000	Ea.	2.667	76.50	99	175.50
No. 2 cedar, stockade screen, 3'-6" wide, 6' high	1.000	Ea.	3.000	66.50	111	177.50
General, wood, 3'-6" wide, 4' high	1.000	Ea.	2.400	58	89.50	147.50
6' high	1.000	Ea.	3.000	72.50	111	183.50

Division 2 Foundations

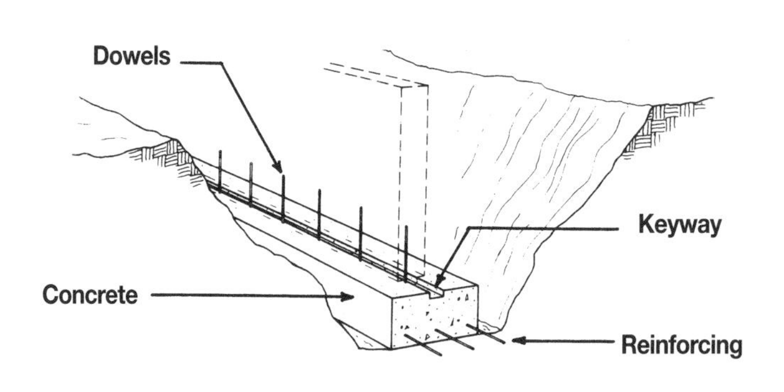

System Description	QUAN.	UNIT	LABOR HOURS	COST PER L.F.		
				MAT.	INST.	TOTAL
8″ THICK BY 18″ WIDE FOOTING						
Concrete, 3000 psi	.040	C.Y.		3.82		3.82
Place concrete, direct chute	.040	C.Y.	.016		.54	.54
Forms, footing, 4 uses	1.330	SFCA	.103	.86	3.74	4.60
Reinforcing, 1/2″ diameter bars, 2 each	1.380	Lb.	.011	.68	.50	1.18
Keyway, 2″ x 4″, beveled, 4 uses	1.000	L.F.	.015	.22	.64	.86
Dowels, 1/2″ diameter bars, 2′ long, 6′ O.C.	.166	Ea.	.006	.11	.26	.37
TOTAL		L.F.	.151	5.69	5.68	11.37
12″ THICK BY 24″ WIDE FOOTING						
Concrete, 3000 psi	.070	C.Y.		6.69		6.69
Place concrete, direct chute	.070	C.Y.	.028		.93	.93
Forms, footing, 4 uses	2.000	SFCA	.155	1.30	5.62	6.92
Reinforcing, 1/2″ diameter bars, 2 each	1.380	Lb.	.011	.68	.50	1.18
Keyway, 2″ x 4″, beveled, 4 uses	1.000	L.F.	.015	.22	.64	.86
Dowels, 1/2″ diameter bars, 2′ long, 6′ O.C.	.166	Ea.	.006	.11	.26	.37
TOTAL		L.F.	.215	9	7.95	16.95
12″ THICK BY 36″ WIDE FOOTING						
Concrete, 3000 psi	.110	C.Y.		10.51		10.51
Place concrete, direct chute	.110	C.Y.	.044		1.46	1.46
Forms, footing, 4 uses	2.000	SFCA	.155	1.30	5.62	6.92
Reinforcing, 1/2″ diameter bars, 2 each	1.380	Lb.	.011	.68	.50	1.18
Keyway, 2″ x 4″, beveled, 4 uses	1.000	L.F.	.015	.22	.64	.86
Dowels, 1/2″ diameter bars, 2′ long, 6′ O.C.	.166	Ea.	.006	.11	.26	.37
TOTAL		L.F.	.231	12.82	8.48	21.30

The footing costs in this system are on a cost per linear foot basis

Description	QUAN.	UNIT	LABOR HOURS	COST PER S.F.		
				MAT.	INST.	TOTAL

Footing Price Sheet	QUAN.	UNIT	LABOR HOURS	COST PER L.F. MAT.	INST.	TOTAL
Concrete, 8″ thick by 18″ wide footing						
2000 psi concrete	.040	C.Y.		3.68		3.68
2500 psi concrete	.040	C.Y.		3.74		3.74
3000 psi concrete	.040	C.Y.		3.82		3.82
3500 psi concrete	.040	C.Y.		3.92		3.92
4000 psi concrete	.040	C.Y.		4		4
12″ thick by 24″ wide footing						
2000 psi concrete	.070	C.Y.		6.45		6.45
2500 psi concrete	.070	C.Y.		6.55		6.55
3000 psi concrete	.070	C.Y.		6.70		6.70
3500 psi concrete	.070	C.Y.		6.85		6.85
4000 psi concrete	.070	C.Y.		7		7
12″ thick by 36″ wide footing						
2000 psi concrete	.110	C.Y.		10.10		10.10
2500 psi concrete	.110	C.Y.		10.30		10.30
3000 psi concrete	.110	C.Y.		10.50		10.50
3500 psi concrete	.110	C.Y.		10.80		10.80
4000 psi concrete	.110	C.Y.		11		11
Place concrete, 8″ thick by 18″ wide footing, direct chute	.040	C.Y.	.016		.54	.54
Pumped concrete	.040	C.Y.	.017		.79	.79
Crane & bucket	.040	C.Y.	.032		1.57	1.57
12″ thick by 24″ wide footing, direct chute	.070	C.Y.	.028		.93	.93
Pumped concrete	.070	C.Y.	.030		1.37	1.37
Crane & bucket	.070	C.Y.	.056		2.75	2.75
12″ thick by 36″ wide footing, direct chute	.110	C.Y.	.044		1.46	1.46
Pumped concrete	.110	C.Y.	.047		2.16	2.16
Crane & bucket	.110	C.Y.	.088		4.33	4.33
Forms, 8″ thick footing, 1 use	1.330	SFCA	.140	.25	5.15	5.40
4 uses	1.330	SFCA	.103	.86	3.74	4.60
12″ thick footing, 1 use	2.000	SFCA	.211	.38	7.75	8.13
4 uses	2.000	SFCA	.155	1.30	5.60	6.90
Reinforcing, 3/8″ diameter bar, 1 each	.400	Lb.	.003	.20	.14	.34
2 each	.800	Lb.	.006	.39	.29	.68
3 each	1.200	Lb.	.009	.59	.43	1.02
1/2″ diameter bar, 1 each	.700	Lb.	.005	.34	.25	.59
2 each	1.380	Lb.	.011	.68	.50	1.18
3 each	2.100	Lb.	.016	1.03	.76	1.79
5/8″ diameter bar, 1 each	1.040	Lb.	.008	.51	.37	.88
2 each	2.080	Lb.	.016	1.02	.75	1.77
Keyway, beveled, 2″ x 4″, 1 use	1.000	L.F.	.030	.44	1.28	1.72
2 uses	1.000	L.F.	.023	.33	.96	1.29
2″ x 6″, 1 use	1.000	L.F.	.032	.66	1.36	2.02
2 uses	1.000	L.F.	.024	.50	1.02	1.52
Dowels, 2 feet long, 6′ O.C., 3/8″ bar	.166	Ea.	.005	.06	.24	.30
1/2″ bar	.166	Ea.	.006	.11	.26	.37
5/8″ bar	.166	Ea.	.006	.18	.29	.47
3/4″ bar	.166	Ea.	.006	.18	.29	.47

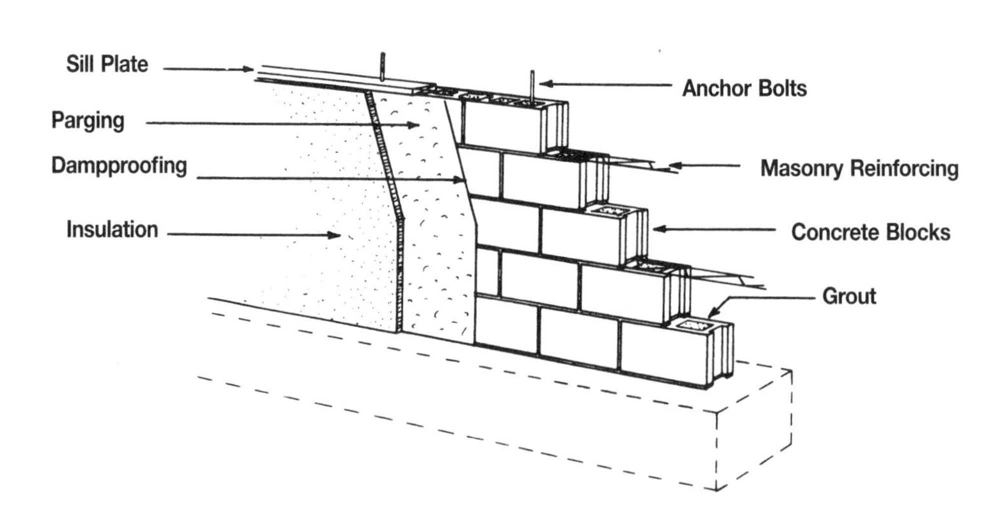

System Description	QUAN.	UNIT	LABOR HOURS	COST PER S.F. MAT.	COST PER S.F. INST.	COST PER S.F. TOTAL
8" WALL, GROUTED, FULL HEIGHT						
Concrete block, 8" x 16" x 8"	1.000	S.F.	.094	2.54	3.62	6.16
Masonry reinforcing, every second course	.750	L.F.	.002	.14	.08	.22
Parging, plastering with portland cement plaster, 1 coat	1.000	S.F.	.014	.24	.55	.79
Dampproofing, bituminous coating, 1 coat	1.000	S.F.	.012	.10	.47	.57
Insulation, 1" rigid polystyrene	1.000	S.F.	.010	.52	.42	.94
Grout, solid, pumped	1.000	S.F.	.059	1.03	2.23	3.26
Anchor bolts, 1/2" diameter, 8" long, 4' O.C.	.060	Ea.	.002	.05	.10	.15
Sill plate, 2" x 4", treated	.250	L.F.	.007	.13	.31	.44
TOTAL		S.F.	.200	4.75	7.78	12.53
12" WALL, GROUTED, FULL HEIGHT						
Concrete block, 8" x 16" x 12"	1.000	S.F.	.160	3.50	6	9.50
Masonry reinforcing, every second course	.750	L.F.	.003	.17	.13	.30
Parging, plastering with portland cement plaster, 1 coat	1.000	S.F.	.014	.24	.55	.79
Dampproofing, bituminous coating, 1 coat	1.000	S.F.	.012	.10	.47	.57
Insulation, 1" rigid polystyrene	1.000	S.F.	.010	.52	.42	.94
Grout, solid, pumped	1.000	S.F.	.063	1.69	2.37	4.06
Anchor bolts, 1/2" diameter, 8" long, 4' O.C.	.060	Ea.	.002	.05	.10	.15
Sill plate, 2" x 4", treated	.250	L.F.	.007	.13	.31	.44
TOTAL		S.F.	.271	6.40	10.35	16.75

The costs in this system are based on a square foot of wall. Do not subtract for window or door openings.

Description	QUAN.	UNIT	LABOR HOURS	COST PER S.F. MAT.	COST PER S.F. INST.	COST PER S.F. TOTAL

Block Wall Systems	QUAN.	UNIT	LABOR HOURS	COST PER S.F.		
				MAT.	INST.	TOTAL
Concrete, block, 8" x 16" x, 6" thick	1.000	S.F.	.089	2.35	3.38	5.73
8" thick	1.000	S.F.	.093	2.54	3.62	6.16
10" thick	1.000	S.F.	.095	3.09	4.40	7.49
12" thick	1.000	S.F.	.122	3.50	6	9.50
Solid block, 8" x 16" x, 6" thick	1.000	S.F.	.091	2.42	3.50	5.92
8" thick	1.000	S.F.	.096	3.23	3.71	6.94
10" thick	1.000	S.F.	.096	3.23	3.71	6.94
12" thick	1.000	S.F.	.126	4.70	5.15	9.85
Masonry reinforcing, wire strips, to 8" wide, every course	1.500	L.F.	.004	.29	.17	.46
Every 2nd course	.750	L.F.	.002	.14	.08	.22
Every 3rd course	.500	L.F.	.001	.10	.06	.16
Every 4th course	.400	L.F.	.001	.08	.04	.12
Wire strips to 12" wide, every course	1.500	L.F.	.006	.33	.26	.59
Every 2nd course	.750	L.F.	.003	.17	.13	.30
Every 3rd course	.500	L.F.	.002	.11	.09	.20
Every 4th course	.400	L.F.	.002	.09	.07	.16
Parging, plastering with portland cement plaster, 1 coat	1.000	S.F.	.014	.24	.55	.79
2 coats	1.000	S.F.	.022	.37	.85	1.22
Dampproofing, bituminous, brushed on, 1 coat	1.000	S.F.	.012	.10	.47	.57
2 coats	1.000	S.F.	.016	.19	.62	.81
Sprayed on, 1 coat	1.000	S.F.	.010	.10	.37	.47
2 coats	1.000	S.F.	.016	.19	.62	.81
Troweled on, 1/16" thick	1.000	S.F.	.016	.20	.62	.82
1/8" thick	1.000	S.F.	.020	.35	.78	1.13
1/2" thick	1.000	S.F.	.023	1.15	.89	2.04
Insulation, rigid, fiberglass, 1.5#/C.F., unfaced						
1-1/2" thick R 6.2	1.000	S.F.	.008	.54	.34	.88
2" thick R 8.5	1.000	S.F.	.008	.65	.34	.99
3" thick R 13	1.000	S.F.	.010	.75	.42	1.17
Foamglass, 1-1/2" thick R 2.64	1.000	S.F.	.010	1.38	.42	1.80
2" thick R 5.26	1.000	S.F.	.011	3.31	.46	3.77
Perlite, 1" thick R 2.77	1.000	S.F.	.010	.31	.42	.73
2" thick R 5.55	1.000	S.F.	.011	.59	.46	1.05
Polystyrene, extruded, 1" thick R 5.4	1.000	S.F.	.010	.52	.42	.94
2" thick R 10.8	1.000	S.F.	.011	1.43	.46	1.89
Molded 1" thick R 3.85	1.000	S.F.	.010	.24	.42	.66
2" thick R 7.7	1.000	S.F.	.011	.75	.46	1.21
Grout, concrete block cores, 6" thick	1.000	S.F.	.044	.77	1.67	2.44
8" thick	1.000	S.F.	.059	1.03	2.23	3.26
10" thick	1.000	S.F.	.061	1.36	2.30	3.66
12" thick	1.000	S.F.	.063	1.69	2.37	4.06
Anchor bolts, 2' on center, 1/2" diameter, 8" long	.120	Ea.	.005	.09	.20	.29
12" long	.120	Ea.	.005	.17	.21	.38
3/4" diameter, 8" long	.120	Ea.	.006	.20	.26	.46
12" long	.120	Ea.	.006	.25	.27	.52
4' on center, 1/2" diameter, 8" long	.060	Ea.	.002	.05	.10	.15
12" long	.060	Ea.	.003	.09	.11	.20
3/4" diameter, 8" long	.060	Ea.	.003	.10	.13	.23
12" long	.060	Ea.	.003	.13	.14	.27
Sill plates, treated, 2" x 4"	.250	L.F.	.007	.13	.31	.44
4" x 4"	.250	L.F.	.007	.38	.29	.67

12 Concrete Wall Systems

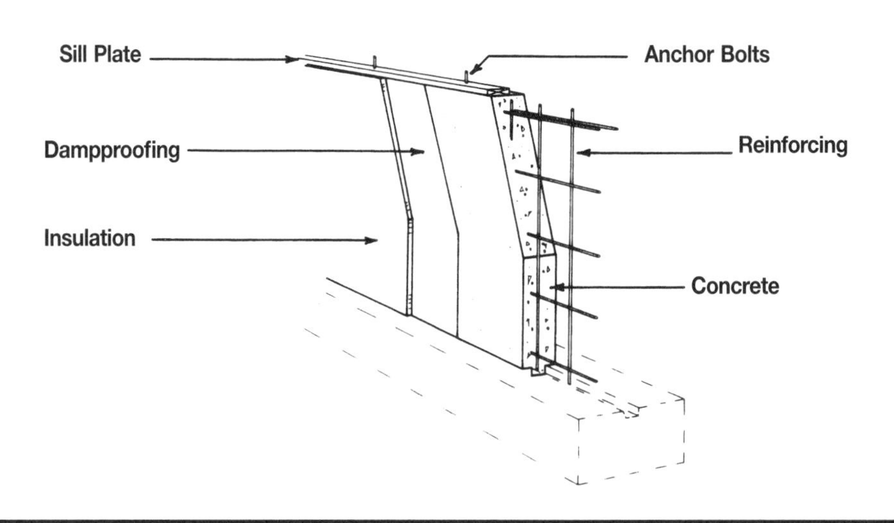

System Description	QUAN.	UNIT	LABOR HOURS	COST PER S.F.		
				MAT.	INST.	TOTAL
8″ THICK, POURED CONCRETE WALL						
Concrete, 8″ thick , 3000 psi	.025	C.Y.		2.39		2.39
Forms, prefabricated plywood, 4 uses per month	2.000	SFCA	.076	1.38	2.82	4.20
Reinforcing, light	.670	Lb.	.004	.33	.17	.50
Placing concrete, direct chute	.025	C.Y.	.013		.44	.44
Dampproofing, brushed on, 2 coats	1.000	S.F.	.016	.19	.62	.81
Rigid insulation, 1″ polystyrene	1.000	S.F.	.010	.52	.42	.94
Anchor bolts, 1/2″ diameter, 12″ long, 4′ O.C.	.060	Ea.	.003	.09	.11	.20
Sill plates, 2″ x 4″, treated	.250	L.F.	.007	.13	.31	.44
TOTAL		S.F.	.129	5.03	4.89	9.92
12″ THICK, POURED CONCRETE WALL						
Concrete, 12″ thick, 3000 psi	.040	C.Y.		3.82		3.82
Forms, prefabricated plywood, 4 uses per month	2.000	SFCA	.076	1.38	2.82	4.20
Reinforcing, light	1.000	Lb.	.005	.49	.25	.74
Placing concrete, direct chute	.040	C.Y.	.019		.64	.64
Dampproofing, brushed on, 2 coats	1.000	S.F.	.016	.19	.62	.81
Rigid insulation, 1″ polystyrene	1.000	S.F.	.010	.52	.42	.94
Anchor bolts, 1/2″ diameter, 12″ long, 4′ O.C.	.060	Ea.	.003	.09	.11	.20
Sill plates, 2″ x 4″ treated	.250	L.F.	.007	.13	.31	.44
TOTAL		S.F.	.136	6.62	5.17	11.79

The costs in this system are based on sq. ft. of wall. Do not subtract for window and door openings. The costs assume a 4′ high wall.

Description	QUAN.	UNIT	LABOR HOURS	COST PER S.F.		
				MAT.	INST.	TOTAL

Important: See the Reference Section for critical supporting data - Reference Nos., Crews & Location Factors

Concrete Wall Price Sheet	QUAN.	UNIT	LABOR HOURS	COST PER S.F.		
				MAT.	INST.	TOTAL
Formwork, prefabricated plywood, 1 use per month	2.000	SFCA	.081	4.20	3	7.20
4 uses per month	2.000	SFCA	.076	1.38	2.82	4.20
Job built forms, 1 use per month	2.000	SFCA	.320	5.45	11.80	17.25
4 uses per month	2.000	SFCA	.221	2.04	8.15	10.19
Reinforcing, 8″ wall, light reinforcing	.670	Lb.	.004	.33	.17	.50
Heavy reinforcing	1.500	Lb.	.008	.74	.38	1.12
10″ wall, light reinforcing	.850	Lb.	.005	.42	.21	.63
Heavy reinforcing	2.000	Lb.	.011	.98	.50	1.48
12″ wall light reinforcing	1.000	Lb.	.005	.49	.25	.74
Heavy reinforcing	2.250	Lb.	.012	1.10	.56	1.66
Placing concrete, 8″ wall, direct chute	.025	C.Y.	.013		.44	.44
Pumped concrete	.025	C.Y.	.016		.74	.74
Crane & bucket	.025	C.Y.	.023		1.10	1.10
10″ wall, direct chute	.030	C.Y.	.016		.54	.54
Pumped concrete	.030	C.Y.	.019		.88	.88
Crane & bucket	.030	C.Y.	.027		1.32	1.32
12″ wall, direct chute	.040	C.Y.	.019		.64	.64
Pumped concrete	.040	C.Y.	.023		1.07	1.07
Crane & bucket	.040	C.Y.	.032		1.57	1.57
Dampproofing, bituminous, brushed on, 1 coat	1.000	S.F.	.012	.10	.47	.57
2 coats	1.000	S.F.	.016	.19	.62	.81
Sprayed on, 1 coat	1.000	S.F.	.010	.10	.37	.47
2 coats	1.000	S.F.	.016	.19	.62	.81
Troweled on, 1/16″ thick	1.000	S.F.	.016	.20	.62	.82
1/8″ thick	1.000	S.F.	.020	.35	.78	1.13
1/2″ thick	1.000	S.F.	.023	1.15	.89	2.04
Insulation rigid, fiberglass, 1.5#/C.F., unfaced						
1-1/2″ thick, R 6.2	1.000	S.F.	.008	.54	.34	.88
2″ thick, R 8.3	1.000	S.F.	.008	.65	.34	.99
3″ thick, R 12.4	1.000	S.F.	.010	.75	.42	1.17
Foamglass, 1-1/2″ thick R 2.64	1.000	S.F.	.010	1.38	.42	1.80
2″ thick R 5.26	1.000	S.F.	.011	3.31	.46	3.77
Perlite, 1″ thick R 2.77	1.000	S.F.	.010	.31	.42	.73
2″ thick R 5.55	1.000	S.F.	.011	.59	.46	1.05
Polystyrene, extruded, 1″ thick R 5.40	1.000	S.F.	.010	.52	.42	.94
2″ thick R 10.8	1.000	S.F.	.011	1.43	.46	1.89
Molded, 1″ thick R 3.85	1.000	S.F.	.010	.24	.42	.66
2″ thick R 7.70	1.000	S.F.	.011	.75	.46	1.21
Anchor bolts, 2′ on center, 1/2″ diameter, 8″ long	.120	Ea.	.005	.09	.20	.29
12″ long	.120	Ea.	.005	.17	.21	.38
3/4″ diameter, 8″ long	.120	Ea.	.006	.20	.26	.46
12″ long	.120	Ea.	.006	.25	.27	.52
Sill plates, treated lumber, 2″ x 4″	.250	L.F.	.007	.13	.31	.44
4″ x 4″	.250	L.F.	.007	.38	.29	.67

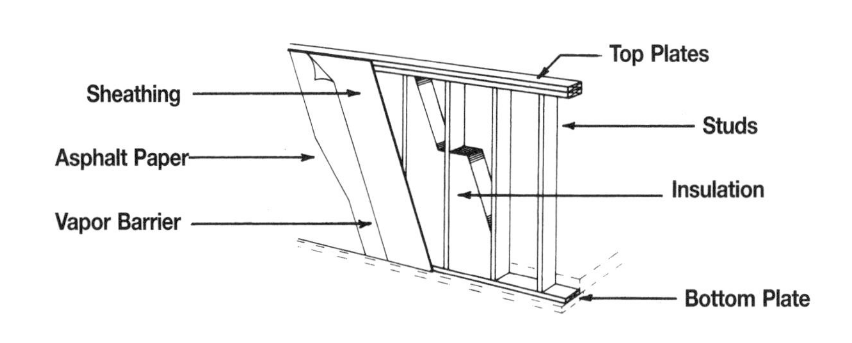

System Description	QUAN.	UNIT	LABOR HOURS	COST PER S.F.		
				MAT.	INST.	TOTAL
2" X 4" STUDS, 16" O.C., WALL						
Studs, 2" x 4", 16" O.C., treated	1.000	L.F.	.015	.53	.61	1.14
Plates, double top plate, single bottom plate, treated, 2" x 4"	.750	L.F.	.011	.40	.46	.86
Sheathing, 1/2", exterior grade, CDX, treated	1.000	S.F.	.014	.88	.60	1.48
Asphalt paper, 15# roll	1.100	S.F.	.002	.04	.10	.14
Vapor barrier, 4 mil polyethylene	1.000	S.F.	.002	.04	.09	.13
Insulation, batts, fiberglass, 3-1/2" thick, R 11	1.000	S.F.	.005	.32	.21	.53
TOTAL		S.F.	.049	2.21	2.07	4.28
2" X 6" STUDS, 16" O.C., WALL						
Studs, 2" x 6", 16" O.C., treated	1.000	L.F.	.016	.82	.68	1.50
Plates, double top plate, single bottom plate, treated, 2" x 6"	.750	L.F.	.012	.62	.51	1.13
Sheathing, 5/8" exterior grade, CDX, treated	1.000	S.F.	.015	1.36	.64	2
Asphalt paper, 15# roll	1.100	S.F.	.002	.04	.10	.14
Vapor barrier, 4 mil polyethylene	1.000	S.F.	.002	.04	.09	.13
Insulation, batts, fiberglass, 6" thick, R 19	1.000	S.F.	.006	.46	.25	.71
TOTAL		S.F.	.053	3.34	2.27	5.61
2" X 8" STUDS, 16" O.C., WALL						
Studs, 2" x 8", 16" O.C. treated	1.000	L.F.	.018	1.07	.75	1.82
Plates, double top plate, single bottom plate, treated, 2" x 8"	.750	L.F.	.013	.80	.56	1.36
Sheathing, 3/4" exterior grade, CDX, treated	1.000	S.F.	.016	1.72	.69	2.41
Asphalt paper, 15# roll	1.100	S.F.	.002	.04	.10	.14
Vapor barrier, 4 mil polyethylene	1.000	S.F.	.002	.04	.09	.13
Insulation, batts, fiberglass, 9" thick, R 30	1.000	S.F.	.006	.78	.25	1.03
TOTAL		S.F.	.057	4.45	2.44	6.89

The costs in this system are based on a sq. ft. of wall area. Do not Subtract for window or door openings. The costs assume a 4' high wall.

Description	QUAN.	UNIT	LABOR HOURS	COST PER S.F.		
				MAT.	INST.	TOTAL

Important: See the Reference Section for critical supporting data - Reference Nos., Crews & Location Factors

Wood Wall Foundation Price Sheet	QUAN.	UNIT	LABOR HOURS	COST PER S.F.		
				MAT.	INST.	TOTAL
Studs, treated, 2″ x 4″, 12″ O.C.	1.250	L.F.	.018	.66	.76	1.42
16″ O.C.	1.000	L.F.	.015	.53	.61	1.14
2″ x 6″, 12″ O.C.	1.250	L.F.	.020	1.03	.85	1.88
16″ O.C.	1.000	L.F.	.016	.82	.68	1.50
2″ x 8″, 12″ O.C.	1.250	L.F.	.022	1.34	.94	2.28
16″ O.C.	1.000	L.F.	.018	1.07	.75	1.82
Plates, treated double top single bottom, 2″ x 4″	.750	L.F.	.011	.40	.46	.86
2″ x 6″	.750	L.F.	.012	.62	.51	1.13
2″ x 8″	.750	L.F.	.013	.80	.56	1.36
Sheathing, treated exterior grade CDX, 1/2″ thick	1.000	S.F.	.014	.88	.60	1.48
5/8″ thick	1.000	S.F.	.015	1.36	.64	2
3/4″ thick	1.000	S.F.	.016	1.72	.69	2.41
Asphalt paper, 15# roll	1.100	S.F.	.002	.04	.10	.14
Vapor barrier, polyethylene, 4 mil	1.000	S.F.	.002	.03	.09	.12
10 mil	1.000	S.F.	.002	.06	.09	.15
Insulation, rigid, fiberglass, 1.5#/C.F., unfaced	1.000	S.F.	.008	.42	.34	.76
1-1/2″ thick, R 6.2	1.000	S.F.	.008	.54	.34	.88
2″ thick, R 8.3	1.000	S.F.	.008	.65	.34	.99
3″ thick, R 12.4	1.000	S.F.	.010	.77	.43	1.20
Foamglass 1 1/2″ thick, R 2.64	1.000	S.F.	.010	1.38	.42	1.80
2″ thick, R 5.26	1.000	S.F.	.011	3.31	.46	3.77
Perlite 1″ thick, R 2.77	1.000	S.F.	.010	.31	.42	.73
2″ thick, R 5.55	1.000	S.F.	.011	.59	.46	1.05
Polystyrene, extruded, 1″ thick, R 5.40	1.000	S.F.	.010	.52	.42	.94
2″ thick, R 10.8	1.000	S.F.	.011	1.43	.46	1.89
Molded 1″ thick, R 3.85	1.000	S.F.	.010	.24	.42	.66
2″ thick, R 7.7	1.000	S.F.	.011	.75	.46	1.21
Non rigid, batts, fiberglass, paper backed, 3-1/2″ thick roll, R 11	1.000	S.F.	.005	.32	.21	.53
6″, R 19	1.000	S.F.	.006	.46	.25	.71
9″, R 30	1.000	S.F.	.006	.78	.25	1.03
12″, R 38	1.000	S.F.	.006	.94	.25	1.19
Mineral fiber, paper backed, 3-1/2″, R 13	1.000	S.F.	.005	.33	.21	.54
6″, R 19	1.000	S.F.	.005	.44	.21	.65
10″, R 30	1.000	S.F.	.006	.65	.25	.90

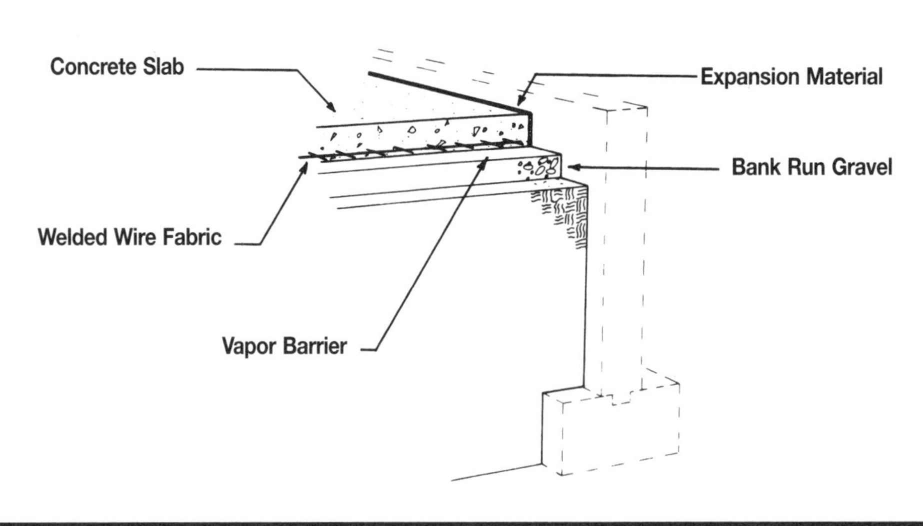

System Description	QUAN.	UNIT	LABOR HOURS	COST PER S.F. MAT.	COST PER S.F. INST.	COST PER S.F. TOTAL
4″ THICK SLAB						
Concrete, 4″ thick, 3000 psi concrete	.012	C.Y.		1.15		1.15
Place concrete, direct chute	.012	C.Y.	.005		.18	.18
Bank run gravel, 4″ deep	1.000	S.F.	.001	.32	.04	.36
Polyethylene vapor barrier, .006″ thick	1.000	S.F.	.002	.04	.09	.13
Edge forms, expansion material	.100	L.F.	.005	.03	.19	.22
Welded wire fabric, 6 x 6, 10/10 (W1.4/W1.4)	1.100	S.F.	.005	.14	.24	.38
Steel trowel finish	1.000	S.F.	.015		.57	.57
TOTAL		S.F.	.033	1.68	1.31	2.99
6″ THICK SLAB						
Concrete, 6″ thick, 3000 psi concrete	.019	C.Y.		1.81		1.81
Place concrete, direct chute	.019	C.Y.	.008		.28	.28
Bank run gravel, 4″ deep	1.000	S.F.	.001	.32	.04	.36
Polyethylene vapor barrier, .006″ thick	1.000	S.F.	.002	.04	.09	.13
Edge forms, expansion material	.100	L.F.	.005	.03	.19	.22
Welded wire fabric, 6 x 6, 10/10 (W1.4/W1.4)	1.100	S.F.	.005	.14	.24	.38
Steel trowel finish	1.000	S.F.	.015		.57	.57
TOTAL		S.F.	.036	2.34	1.41	3.75

The slab costs in this section are based on a cost per square foot of floor area.

Description	QUAN.	UNIT	LABOR HOURS	COST PER S.F. MAT.	COST PER S.F. INST.	COST PER S.F. TOTAL

Floor Slab Price Sheet	QUAN.	UNIT	LABOR HOURS	COST PER S.F.		
				MAT.	INST.	TOTAL
Concrete, 4" thick slab, 2000 psi concrete	.012	C.Y.		1.10		1.10
2500 psi concrete	.012	C.Y.		1.12		1.12
3000 psi concrete	.012	C.Y.		1.15		1.15
3500 psi concrete	.012	C.Y.		1.18		1.18
4000 psi concrete	.012	C.Y.		1.20		1.20
4500 psi concrete	.012	C.Y.		1.22		1.22
5" thick slab, 2000 psi concrete	.015	C.Y.		1.38		1.38
2500 psi concrete	.015	C.Y.		1.40		1.40
3000 psi concrete	.015	C.Y.		1.43		1.43
3500 psi concrete	.015	C.Y.		1.47		1.47
4000 psi concrete	.015	C.Y.		1.50		1.50
4500 psi concrete	.015	C.Y.		1.53		1.53
6" thick slab, 2000 psi concrete	.019	C.Y.		1.75		1.75
2500 psi concrete	.019	C.Y.		1.78		1.78
3000 psi concrete	.019	C.Y.		1.81		1.81
3500 psi concrete	.019	C.Y.		1.86		1.86
4000 psi concrete	.019	C.Y.		1.90		1.90
4500 psi concrete	.019	C.Y.		1.94		1.94
Place concrete, 4" slab, direct chute	.012	C.Y.	.005		.18	.18
Pumped concrete	.012	C.Y.	.006		.28	.28
Crane & bucket	.012	C.Y.	.008		.38	.38
5" slab, direct chute	.015	C.Y.	.007		.22	.22
Pumped concrete	.015	C.Y.	.007		.34	.34
Crane & bucket	.015	C.Y.	.010		.48	.48
6" slab, direct chute	.019	C.Y.	.008		.28	.28
Pumped concrete	.019	C.Y.	.009		.43	.43
Crane & bucket	.019	C.Y.	.012		.61	.61
Gravel, bank run, 4" deep	1.000	S.F.	.001	.32	.04	.36
6" deep	1.000	S.F.	.001	.43	.05	.48
9" deep	1.000	S.F.	.001	.63	.07	.70
12" deep	1.000	S.F.	.001	.86	.09	.95
3/4" crushed stone, 3" deep	1.000	S.F.	.001	.33	.04	.37
6" deep	1.000	S.F.	.001	.66	.08	.74
9" deep	1.000	S.F.	.002	.96	.11	1.07
12" deep	1.000	S.F.	.002	1.56	.13	1.69
Vapor barrier polyethylene, .004" thick	1.000	S.F.	.002	.03	.09	.12
.006" thick	1.000	S.F.	.002	.04	.09	.13
Edge forms, expansion material, 4" thick slab	.100	L.F.	.004	.02	.13	.15
6" thick slab	.100	L.F.	.005	.03	.19	.22
Welded wire fabric 6 x 6, 10/10 (W1.4/W1.4)	1.100	S.F.	.005	.14	.24	.38
6 x 6, 6/6 (W2.9/W2.9)	1.100	S.F.	.006	.21	.29	.50
4 x 4, 10/10 (W1.4/W1.4)	1.100	S.F.	.006	.21	.26	.47
Finish concrete, screed finish	1.000	S.F.	.009		.35	.35
Float finish	1.000	S.F.	.011		.43	.43
Steel trowel, for resilient floor	1.000	S.F.	.013		.52	.52
For finished floor	1.000	S.F.	.015		.57	.57

Division 3
Framing

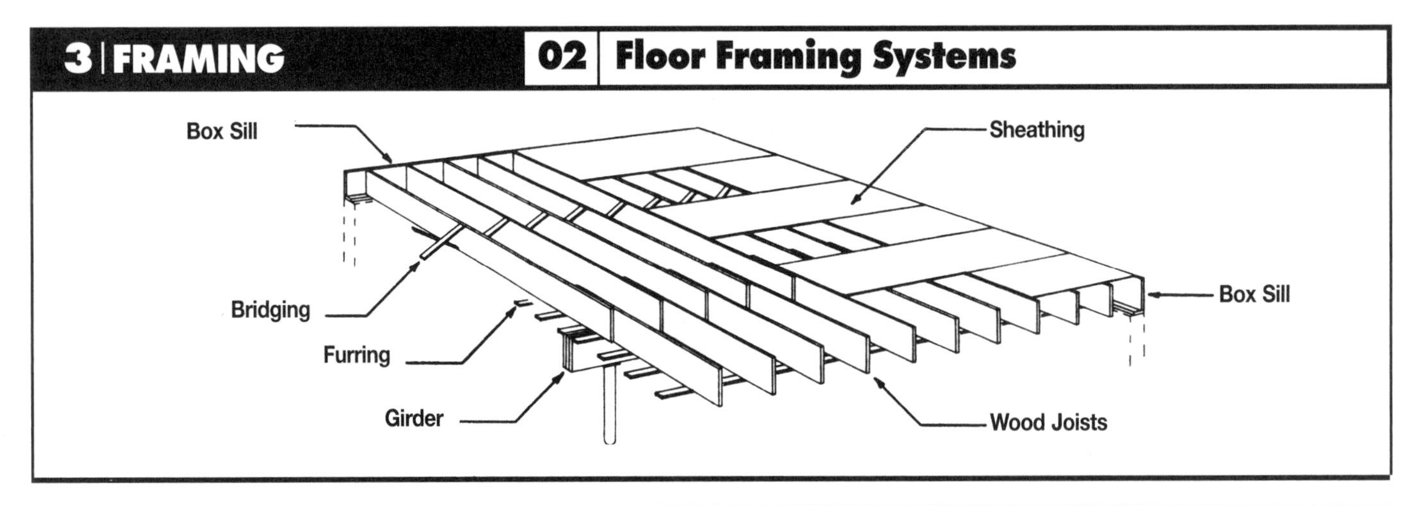

System Description	QUAN.	UNIT	LABOR HOURS	COST PER S.F. MAT.	INST.	TOTAL
2″ X 8″, 16″ O.C.						
Wood joists, 2″ x 8″, 16″ O.C.	1.000	L.F.	.015	.94	.61	1.55
Bridging, 1″ x 3″, 6′ O.C.	.080	Pr.	.005	.04	.21	.25
Box sills, 2″ x 8″	.150	L.F.	.002	.14	.09	.23
Concrete filled steel column, 4″ diameter	.125	L.F.	.002	.12	.10	.22
Girder, built up from three 2″ x 8″	.125	L.F.	.013	.35	.56	.91
Sheathing, plywood, subfloor, 5/8″ CDX	1.000	S.F.	.012	.73	.50	1.23
Furring, 1″ x 3″, 16″ O.C.	1.000	L.F.	.023	.31	.97	1.28
TOTAL		S.F.	.072	2.63	3.04	5.67
2″ X 10″, 16″ O.C.						
Wood joists, 2″ x 10″, 16″ OC	1.000	L.F.	.018	1.42	.75	2.17
Bridging, 1″ x 3″, 6′ OC	.080	Pr.	.005	.04	.21	.25
Box sills, 2″ x 10″	.150	L.F.	.003	.21	.11	.32
Girder, built up from three 2″ x 10″	.125	L.F.	.002	.12	.10	.22
Sheathing, plywood, subfloor, 5/8″ CDX	1.000	S.F.	.012	.73	.50	1.23
Furring, 1″ x 3″,16″ OC	1.000	L.F.	.023	.31	.97	1.28
TOTAL		S.F.	.077	3.36	3.24	6.60
2″ X 12″, 16″ O.C.						
Wood joists, 2″ x 12″, 16″ O.C.	1.000	L.F.	.018	1.76	.77	2.53
Bridging, 1″ x 3″, 6′ O.C.	.080	Pr.	.005	.04	.21	.25
Box sills, 2″ x 12″	.150	L.F.	.003	.26	.12	.38
Concrete filled steel column, 4″ diameter	.125	L.F.	.002	.12	.10	.22
Girder, built up from three 2″ x 12″	.125	L.F.	.015	.66	.64	1.30
Sheathing, plywood, subfloor, 5/8″ CDX	1.000	S.F.	.012	.73	.50	1.23
Furring, 1″ x 3″, 16″ O.C.	1.000	L.F.	.023	.31	.97	1.28
TOTAL		S.F.	.078	3.88	3.31	7.19

Floor costs on this page are given on a cost per square foot basis.

Description	QUAN.	UNIT	LABOR HOURS	COST PER S.F. MAT.	INST.	TOTAL

Floor Framing Price Sheet (Wood)	QUAN.	UNIT	LABOR HOURS	COST PER S.F.		
				MAT.	INST.	TOTAL
Joists, #2 or better, pine, 2″ x 4″, 12″ O.C.	1.250	L.F.	.016	.51	.68	1.19
16″ O.C.	1.000	L.F.	.013	.41	.54	.95
2″ x 6″, 12″ O.C.	1.250	L.F.	.016	.81	.68	1.49
16″ O.C.	1.000	L.F.	.013	.65	.54	1.19
2″ x 8″, 12″ O.C.	1.250	L.F.	.018	1.18	.76	1.94
16″ O.C.	1.000	L.F.	.015	.94	.61	1.55
2″ x 10″, 12″ O.C.	1.250	L.F.	.022	1.78	.94	2.72
16″ O.C.	1.000	L.F.	.018	1.42	.75	2.17
2″x 12″, 12″ O.C.	1.250	L.F.	.023	2.20	.96	3.16
16″ O.C.	1.000	L.F.	.018	1.76	.77	2.53
Bridging, wood 1″ x 3″, joists 12″ O.C.	.100	Pr.	.006	.05	.26	.31
16″ O.C.	.080	Pr.	.005	.04	.21	.25
Metal, galvanized, joists 12″ O.C.	.100	Pr.	.006	.13	.26	.39
16″ O.C.	.080	Pr.	.005	.10	.21	.31
Compression type, joists 12″ O.C.	.100	Pr.	.004	.15	.17	.32
16″ O.C.	.080	Pr.	.003	.12	.14	.26
Box sills, #2 or better pine, 2″ x 4″	.150	L.F.	.002	.06	.08	.14
2″ x 6″	.150	L.F.	.002	.10	.08	.18
2″ x 8″	.150	L.F.	.002	.14	.09	.23
2″ x 10″	.150	L.F.	.003	.21	.11	.32
2″ x 12″	.150	L.F.	.003	.26	.12	.38
Girders, including lally columns, 3 pieces spiked together, 2″ x 8″	.125	L.F.	.015	.47	.66	1.13
2″ x 10″	.125	L.F.	.016	.65	.70	1.35
2″ x 12″	.125	L.F.	.017	.78	.74	1.52
Solid girders, 3″ x 8″	.040	L.F.	.004	.24	.17	.41
3″ x 10″	.040	L.F.	.004	.27	.18	.45
3″ x 12″	.040	L.F.	.004	.30	.19	.49
4″ x 8″	.040	L.F.	.004	.24	.18	.42
4″ x 10″	.040	L.F.	.004	.27	.19	.46
4″ x 12″	.040	L.F.	.004	.30	.20	.50
Steel girders, bolted & including fabrication, wide flange shapes						
12″ deep, 14#/l.f.	.040	L.F.	.003	.75	.23	.98
10″ deep, 15#/l.f.	.040	L.F.	.003	.75	.23	.98
8″ deep, 10#/l.f.	.040	L.F.	.003	.50	.23	.73
6″ deep, 9#/l.f.	.040	L.F.	.003	.45	.23	.68
5″ deep, 16#/l.f.	.040	L.F.	.003	.75	.23	.98
Sheathing, plywood exterior grade CDX, 1/2″ thick	1.000	S.F.	.011	.68	.48	1.16
5/8″ thick	1.000	S.F.	.012	.73	.50	1.23
3/4″ thick	1.000	S.F.	.013	.90	.54	1.44
Boards, 1″ x 8″ laid regular	1.000	S.F.	.016	1.23	.68	1.91
Laid diagonal	1.000	S.F.	.019	1.23	.80	2.03
1″ x 10″ laid regular	1.000	S.F.	.015	1.50	.61	2.11
Laid diagonal	1.000	S.F.	.018	1.50	.75	2.25
Furring, 1″ x 3″, 12″ O.C.	1.250	L.F.	.029	.39	1.21	1.60
16″ O.C.	1.000	L.F.	.023	.31	.97	1.28
24″ O.C.	.750	L.F.	.017	.23	.73	.96

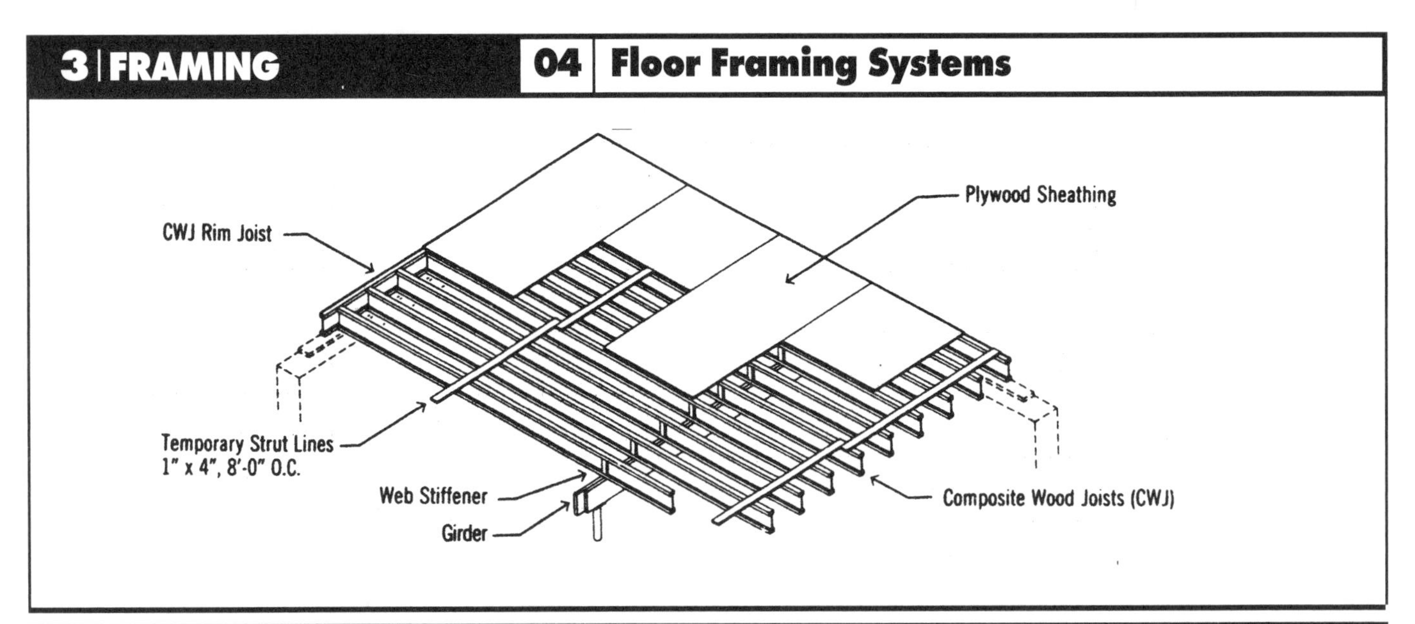

System Description	QUAN.	UNIT	LABOR HOURS	COST PER S.F.		
				MAT.	INST.	TOTAL
9-1/2" COMPOSITE WOOD JOISTS, 16" O.C.						
CWJ, 9-1/2", 16" O.C., 15' span	1.000	L.F.	.018	2	.75	2.75
Temp. strut line, 1" x 4", 8' O.C.	.160	L.F.	.003	.06	.14	.20
CWJ rim joist, 9-1/2"	.150	L.F.	.003	.30	.11	.41
Concrete filled steel column, 4" diameter	.125	L.F.	.002	.12	.10	.22
Girder, built up from three 2" x 8"	.125	L.F.	.013	.35	.56	.91
Sheathing, plywood, subfloor, 5/8" CDX	1.000	S.F.	.012	.73	.50	1.23
TOTAL		S.F.	.051	3.56	2.16	5.72
11-1/2" COMPOSITE WOOD JOISTS, 16" O.C.						
CWJ, 11-1/2", 16" O.C., 18' span	1.000	L.F.	.018	2.15	.77	2.92
Temp. strut line, 1" x 4", 8' O.C.	.160	L.F.	.003	.06	.14	.20
CWJ rim joist, 11-1/2"	.150	L.F.	.003	.32	.12	.44
Concrete filled steel column, 4" diameter	.125	L.F.	.002	.12	.10	.22
Girder, built up from three 2" x 10"	.125	L.F.	.014	.53	.60	1.13
Sheathing, plywood, subfloor, 5/8" CDX	1.000	S.F.	.012	.73	.50	1.23
TOTAL		S.F.	.052	3.91	2.23	6.14
14" COMPOSITE WOOD JOISTS, 16" O.C.						
CWJ, 14", 16" O.C., 22' span	1.000	L.F.	.020	2.48	.83	3.31
Temp. strut line, 1" x 4", 8' O.C.	.160	L.F.	.003	.06	.14	.20
CWJ rim joist, 14"	.150	L.F.	.003	.37	.12	.49
Concrete filled steel column, 4" diameter	.600	L.F.	.002	.12	.10	.22
Girder, built up from three 2" x 12"	.600	L.F.	.015	.66	.64	1.30
Sheathing, plywood, subfloor, 5/8" CDX	1.000	S.F.	.012	.73	.50	1.23
TOTAL		S.F.	.055	4.42	2.33	6.75

Floor costs on this page are given on a cost per square foot basis.

Description	QUAN.	UNIT	LABOR HOURS	COST PER S.F.		
				MAT.	INST.	TOTAL

 Important: See the Reference Section for critical supporting data - Reference Nos., Crews & Location Factors

Floor Framing Price Sheet (Wood)	QUAN.	UNIT	LABOR HOURS	COST PER S.F.		
				MAT.	INST.	TOTAL
Composite wood joist 9-1/2" deep, 12" O.C.	1.250	L.F.	.022	2.50	.94	3.44
16" O.C.	1.000	L.F.	.018	2	.75	2.75
11-1/2" deep, 12" O.C.	1.250	L.F.	.023	2.69	.96	3.65
16" O.C.	1.000	L.F.	.018	2.15	.77	2.92
14" deep, 12" O.C.	1.250	L.F.	.024	3.09	1.03	4.12
16" O.C.	1.000	L.F.	.020	2.48	.83	3.31
16 " deep, 12" O.C.	1.250	L.F.	.026	3.50	1.08	4.58
16" O.C.	1.000	L.F.	.021	2.80	.87	3.67
CWJ rim joist, 9-1/2"	.150	L.F.	.003	.30	.11	.41
11-1/2"	.150	L.F.	.003	.32	.12	.44
14"	.150	L.F.	.003	.37	.12	.49
16"	.150	L.F.	.003	.42	.13	.55
Girders, including lally columns, 3 pieces spiked together, 2" x 8"	.125	L.F.	.015	.47	.66	1.13
2" x 10"	.125	L.F.	.016	.65	.70	1.35
2" x 12"	.125	L.F.	.017	.78	.74	1.52
Solid girders, 3" x 8"	.040	L.F.	.004	.24	.17	.41
3" x 10"	.040	L.F.	.004	.27	.18	.45
3" x 12"	.040	L.F.	.004	.30	.19	.49
4" x 8"	.040	L.F.	.004	.24	.18	.42
4" x 10"	.040	L.F.	.004	.27	.19	.46
4" x 12"	.040	L.F.	.004	.30	.20	.50
Steel girders, bolted & including fabrication, wide flange shapes						
12" deep, 14#/l.f.	.040	L.F.	.061	17.20	5.10	22.30
10" deep, 15#/l.f.	.040	L.F.	.067	18.80	5.60	24.40
8" deep, 10#/l.f.	.040	L.F.	.067	12.55	5.60	18.15
6" deep, 9#/l.f.	.040	L.F.	.067	11.30	5.60	16.90
5" deep, 16#/l.f.	.040	L.F.	.064	18.20	5.40	23.60
Sheathing, plywood exterior grade CDX, 1/2" thick	1.000	S.F.	.011	.68	.48	1.16
5/8" thick	1.000	S.F.	.012	.73	.50	1.23
3/4" thick	1.000	S.F.	.013	.90	.54	1.44
Boards, 1" x 8" laid regular	1.000	S.F.	.016	1.23	.68	1.91
Laid diagonal	1.000	S.F.	.019	1.23	.80	2.03
1" x 10" laid regular	1.000	S.F.	.015	1.50	.61	2.11
Laid diagonal	1.000	S.F.	.018	1.50	.75	2.25
Furring, 1" x 3", 12" O.C.	1.250	L.F.	.029	.39	1.21	1.60
16" O.C.	1.000	L.F.	.023	.31	.97	1.28
24" O.C.	.750	L.F.	.017	.23	.73	.96

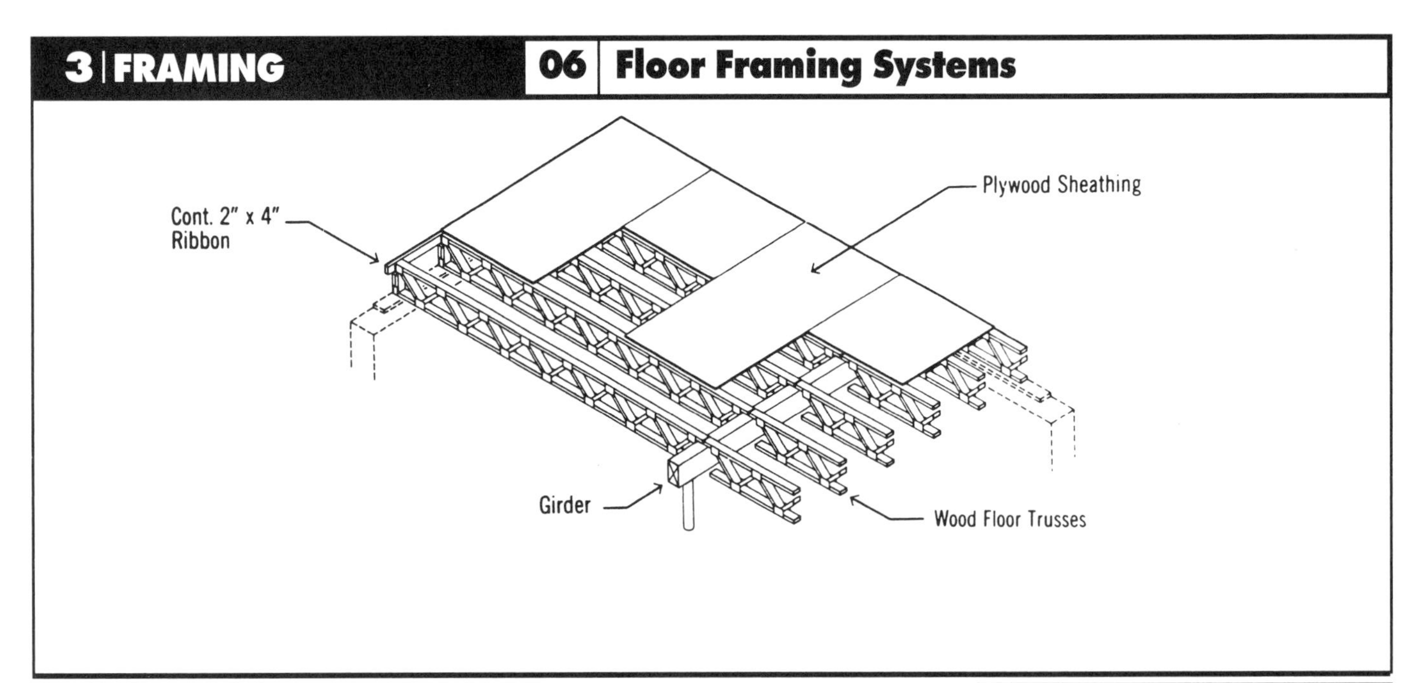

System Description	QUAN.	UNIT	LABOR HOURS	COST PER S.F.		
				MAT.	INST.	TOTAL
12" OPEN WEB JOISTS, 16" O.C.						
OWJ 12", 16" O.C., 21' span	1.000	L.F.	.018	2	.77	2.77
Continuous ribbing, 2" x 4"	.150	L.F.	.002	.06	.08	.14
Concrete filled steel column, 4" diameter	.125	L.F.	.002	.12	.10	.22
Girder, built up from three 2" x 8"	.125	L.F.	.013	.35	.56	.91
Sheathing, plywood, subfloor, 5/8" CDX	1.000	S.F.	.012	.73	.50	1.23
Furring, 1" x 3", 16" O.C.	1.000	L.F.	.023	.31	.97	1.28
TOTAL		S.F.	.070	3.57	2.98	6.55
14" OPEN WEB WOOD JOISTS, 16" O.C.						
OWJ 14", 16" O.C., 22' span	1.000	L.F.	.020	2.33	.83	3.16
Continuous ribbing, 2" x 4"	.150	L.F.	.002	.06	.08	.14
Concrete filled steel column, 4" diameter	.125	L.F.	.002	.12	.10	.22
Girder, built up from three 2" x 10"	.125	L.F.	.014	.53	.60	1.13
Sheathing, plywood, subfloor, 5/8" CDX	1.000	S.F.	.012	.73	.50	1.23
Furring, 1" x 3",16" O.C.	1.000	L.F.	.023	.31	.97	1.28
TOTAL		S.F.	.073	4.08	3.08	7.16
16" OPEN WEB WOOD JOISTS, 16" O.C.						
OWJ 16", 16" O.C., 24' span	1.000	L.F.	.021	2.43	.87	3.30
Continuous ribbing, 2" x 4"	.150	L.F.	.002	.06	.08	.14
Concrete filled steel column, 4" diameter	.125	L.F.	.002	.12	.10	.22
Girder, built up from three 2" x 12"	.125	L.F.	.015	.66	.64	1.30
Sheathing, plywood, subfloor, 5/8" CDX	1.000	S.F.	.012	.73	.50	1.23
Furring, 1" x 3", 16" O.C.	1.000	L.F.	.023	.31	.97	1.28
TOTAL		S.F.	.075	4.31	3.16	7.47

Floor costs on this page are given on a cost per square foot basis.

Description	QUAN.	UNIT	LABOR HOURS	COST PER S.F.		
				MAT.	INST.	TOTAL

Important: See the Reference Section for critical supporting data - Reference Nos., Crews & Location Factors

Floor Framing Price Sheet (Wood)

Floor Framing Price Sheet (Wood)	QUAN.	UNIT	LABOR HOURS	COST PER S.F.		
				MAT.	INST.	TOTAL
Open web joists, 12″ deep, 12″ O.C.	1.250	L.F.	.023	2.50	.96	3.46
16″ O.C.	1.000	L.F.	.018	2	.77	2.77
14″ deep, 12″ O.C.	1.250	L.F.	.024	2.91	1.03	3.94
16″ O.C.	1.000	L.F.	.020	2.33	.83	3.16
16 ″ deep, 12″ O.C.	1.250	L.F.	.026	3.03	1.08	4.11
16″ O.C.	1.000	L.F.	.021	2.43	.87	3.30
18″ deep, 12″ O.C.	1.250	L.F.	.027	3.09	1.14	4.23
16″ O.C.	1.000	L.F.	.022	2.48	.92	3.40
Continuous ribbing, 2″ x 4″	.150	L.F.	.002	.06	.08	.14
2″ x 6″	.150	L.F.	.002	.10	.08	.18
2″ x 8″	.150	L.F.	.002	.14	.09	.23
2″ x 10″	.150	L.F.	.003	.21	.11	.32
2″ x 12″	.150	L.F.	.003	.26	.12	.38
Girders, including lally columns, 3 pieces spiked together, 2″ x 8″	.125	L.F.	.015	.47	.66	1.13
2″ x 10″	.125	L.F.	.016	.65	.70	1.35
2″ x 12″	.125	L.F.	.017	.78	.74	1.52
Solid girders, 3″ x 8″	.040	L.F.	.004	.24	.17	.41
3″ x 10″	.040	L.F.	.004	.27	.18	.45
3″ x 12″	.040	L.F.	.004	.30	.19	.49
4″ x 8″	.040	L.F.	.004	.24	.18	.42
4″ x 10″	.040	L.F.	.004	.27	.19	.46
4″ x 12″	.040	L.F.	.004	.30	.20	.50
Steel girders, bolted & including fabrication, wide flange shapes						
12″ deep, 14#/l.f.	.040	L.F.	.061	17.20	5.10	22.30
10″ deep, 15#/l.f.	.040	L.F.	.067	18.80	5.60	24.40
8″ deep, 10#/l.f.	.040	L.F.	.067	12.55	5.60	18.15
6″ deep, 9#/l.f.	.040	L.F.	.067	11.30	5.60	16.90
5″ deep, 16#/l.f.	.040	L.F.	.064	18.20	5.40	23.60
Sheathing, plywood exterior grade CDX, 1/2″ thick	1.000	S.F.	.011	.68	.48	1.16
5/8″ thick	1.000	S.F.	.012	.73	.50	1.23
3/4″ thick	1.000	S.F.	.013	.90	.54	1.44
Boards, 1″ x 8″ laid regular	1.000	S.F.	.016	1.23	.68	1.91
Laid diagonal	1.000	S.F.	.019	1.23	.80	2.03
1″ x 10″ laid regular	1.000	S.F.	.015	1.50	.61	2.11
Laid diagonal	1.000	S.F.	.018	1.50	.75	2.25
Furring, 1″ x 3″, 12″ O.C.	1.250	L.F.	.029	.39	1.21	1.60
16″ O.C.	1.000	L.F.	.023	.31	.97	1.28
24″ O.C.	.750	L.F.	.017	.23	.73	.96

08 Exterior Wall Framing Systems

Sheathing
Top Plates
Studs
Bottom Plate
Corner Bracing

System Description	QUAN.	UNIT	LABOR HOURS	COST PER S.F. MAT.	COST PER S.F. INST.	COST PER S.F. TOTAL
2″ X 4″, 16″ O.C.						
2″ x 4″ studs, 16″ O.C.	1.000	L.F.	.015	.41	.61	1.02
Plates, 2″ x 4″, double top, single bottom	.375	L.F.	.005	.15	.23	.38
Corner bracing, let-in, 1″ x 6″	.063	L.F.	.003	.04	.14	.18
Sheathing, 1/2″ plywood, CDX	1.000	S.F.	.011	.68	.48	1.16
TOTAL		S.F.	.034	1.28	1.46	2.74
2″ X 4″, 24″ O.C.						
2″ x 4″ studs, 24″ O.C.	.750	L.F.	.011	.31	.46	.77
Plates, 2″ x 4″, double top, single bottom	.375	L.F.	.005	.15	.23	.38
Corner bracing, let-in, 1″ x 6″	.063	L.F.	.002	.04	.09	.13
Sheathing, 1/2″ plywood, CDX	1.000	S.F.	.011	.68	.48	1.16
TOTAL		S.F.	.029	1.18	1.26	2.44
2″ X 6″, 16″ O.C.						
2″ x 6″ studs, 16″ O.C.	1.000	L.F.	.016	.65	.68	1.33
Plates, 2″ x 6″, double top, single bottom	.375	L.F.	.006	.24	.26	.50
Corner bracing, let-in, 1″ x 6″	.063	L.F.	.003	.04	.14	.18
Sheathing, 1/2″ plywood, CDX	1.000	S.F.	.014	.68	.60	1.28
TOTAL		S.F.	.039	1.61	1.68	3.29
2″ X 6″, 24″ O.C.						
2″ x 6″ studs, 24″ O.C.	.750	L.F.	.012	.49	.51	1
Plates, 2″ x 6″, double top, single bottom	.375	L.F.	.006	.24	.26	.50
Corner bracing, let-in, 1″ x 6″	.063	L.F.	.002	.04	.09	.13
Sheathing, 1/2″ plywood, CDX	1.000	S.F.	.011	.68	.48	1.16
TOTAL		S.F.	.031	1.45	1.34	2.79

The wall costs on this page are given in cost per square foot of wall.
For window and door openings see below.

Description	QUAN.	UNIT	LABOR HOURS	COST PER S.F. MAT.	COST PER S.F. INST.	COST PER S.F. TOTAL

Exterior Wall Framing Price Sheet	QUAN.	UNIT	LABOR HOURS	COST PER S.F.		
				MAT.	INST.	TOTAL
Studs, #2 or better, 2" x 4", 12" O.C.	1.250	L.F.	.018	.51	.76	1.27
16" O.C.	1.000	L.F.	.015	.41	.61	1.02
24" O.C.	.750	L.F.	.011	.31	.46	.77
32" O.C.	.600	L.F.	.009	.25	.37	.62
2" x 6", 12" O.C.	1.250	L.F.	.020	.81	.85	1.66
16" O.C.	1.000	L.F.	.016	.65	.68	1.33
24" O.C.	.750	L.F.	.012	.49	.51	1
32" O.C.	.600	L.F.	.010	.39	.41	.80
2" x 8", 12" O.C.	1.250	L.F.	.025	1.50	1.06	2.56
16" O.C.	1.000	L.F.	.020	1.20	.85	2.05
24" O.C.	.750	L.F.	.015	.90	.64	1.54
32" O.C.	.600	L.F.	.012	.72	.51	1.23
Plates, #2 or better, double top, single bottom, 2" x 4"	.375	L.F.	.005	.15	.23	.38
2" x 6"	.375	L.F.	.006	.24	.26	.50
2" x 8"	.375	L.F.	.008	.45	.32	.77
Corner bracing, let-in 1" x 6" boards, studs, 12" O.C.	.070	L.F.	.004	.04	.16	.20
16" O.C.	.063	L.F.	.003	.04	.14	.18
24" O.C.	.063	L.F.	.002	.04	.09	.13
32" O.C.	.057	L.F.	.002	.03	.08	.11
Let-in steel ("T" shape), studs, 12" O.C.	.070	L.F.	.001	.04	.04	.08
16" O.C.	.063	L.F.	.001	.04	.04	.08
24" O.C.	.063	L.F.	.001	.04	.04	.08
32" O.C.	.057	L.F.	.001	.03	.03	.06
Sheathing, plywood CDX, 3/8" thick	1.000	S.F.	.010	.50	.44	.94
1/2" thick	1.000	S.F.	.011	.68	.48	1.16
5/8" thick	1.000	S.F.	.012	.73	.52	1.25
3/4" thick	1.000	S.F.	.013	.90	.56	1.46
Boards, 1" x 6", laid regular	1.000	S.F.	.025	1.32	1.04	2.36
Laid diagonal	1.000	S.F.	.027	1.32	1.16	2.48
1" x 8", laid regular	1.000	S.F.	.021	1.23	.88	2.11
Laid diagonal	1.000	S.F.	.025	1.23	1.04	2.27
Wood fiber, regular, no vapor barrier, 1/2" thick	1.000	S.F.	.013	.61	.56	1.17
5/8" thick	1.000	S.F.	.013	.79	.56	1.35
Asphalt impregnated 25/32" thick	1.000	S.F.	.013	.31	.56	.87
1/2" thick	1.000	S.F.	.013	.20	.56	.76
Polystyrene, regular, 3/4" thick	1.000	S.F.	.010	.52	.42	.94
2" thick	1.000	S.F.	.011	1.43	.46	1.89
Fiberglass, foil faced, 1" thick	1.000	S.F.	.008	1.05	.34	1.39
2" thick	1.000	S.F.	.009	1.77	.38	2.15

Window & Door Openings	QUAN.	UNIT	LABOR HOURS	COST EACH		
				MAT.	INST.	TOTAL
The following costs are to be added to the total costs of the wall for each opening. Do not subtract the area of the openings.						
Headers, 2" x 6" double, 2' long	4.000	L.F.	.178	2.60	7.50	10.10
3' long	6.000	L.F.	.267	3.90	11.30	15.20
4' long	8.000	L.F.	.356	5.20	15.05	20.25
5' long	10.000	L.F.	.444	6.50	18.80	25.30
2" x 8" double, 4' long	8.000	L.F.	.376	7.50	15.90	23.40
5' long	10.000	L.F.	.471	9.40	19.90	29.30
6' long	12.000	L.F.	.565	11.30	24	35.30
8' long	16.000	L.F.	.753	15.05	32	47.05
2" x 10" double, 4' long	8.000	L.F.	.400	11.35	16.90	28.25
6' long	12.000	L.F.	.600	17.05	25.50	42.55
8' long	16.000	L.F.	.800	22.50	34	56.50
10' long	20.000	L.F.	1.000	28.50	42	70.50
2" x 12" double, 8' long	16.000	L.F.	.853	28	36	64
12' long	24.000	L.F.	1.280	42	54	96

12 | Gable End Roof Framing Systems

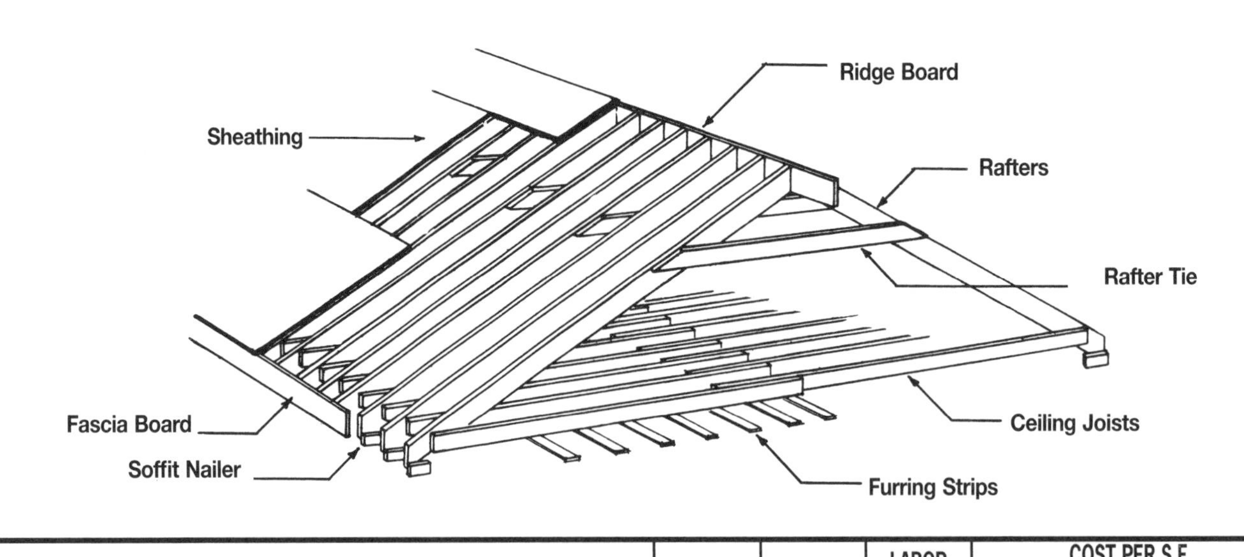

System Description	QUAN.	UNIT	LABOR HOURS	COST PER S.F.		
				MAT.	INST.	TOTAL
2" X 6" RAFTERS, 16" O.C., 4/12 PITCH						
Rafters, 2" x 6", 16" O.C., 4/12 pitch	1.170	L.F.	.019	.76	.80	1.56
Ceiling joists, 2" x 4", 16" O.C.	1.000	L.F.	.013	.41	.54	.95
Ridge board, 2" x 6"	.050	L.F.	.002	.03	.07	.10
Fascia board, 2" x 6"	.100	L.F.	.005	.07	.23	.30
Rafter tie, 1" x 4", 4' O.C.	.060	L.F.	.001	.02	.05	.07
Soffit nailer (outrigger), 2" x 4", 24" O.C.	.170	L.F.	.004	.07	.19	.26
Sheathing, exterior, plywood, CDX, 1/2" thick	1.170	S.F.	.013	.80	.56	1.36
Furring strips, 1" x 3", 16" O.C.	1.000	L.F.	.023	.31	.97	1.28
TOTAL		S.F.	.080	2.47	3.41	5.88
2" X 8" RAFTERS, 16" O.C., 4/12 PITCH						
Rafters, 2" x 8", 16" O.C., 4/12 pitch	1.170	L.F.	.020	1.10	.83	1.93
Ceiling joists, 2" x 6", 16" O.C.	1.000	L.F.	.013	.65	.54	1.19
Ridge board, 2" x 8"	.050	L.F.	.002	.05	.08	.13
Fascia board, 2" x 8"	.100	L.F.	.007	.09	.30	.39
Rafter tie, 1" x 4", 4' O.C.	.060	L.F.	.001	.02	.05	.07
Soffit nailer (outrigger), 2" x 4", 24" O.C.	.170	L.F.	.004	.07	.19	.26
Sheathing, exterior, plywood, CDX, 1/2" thick	1.170	S.F.	.013	.80	.56	1.36
Furring strips, 1" x 3", 16" O.C.	1.000	L.F.	.023	.31	.97	1.28
TOTAL		S.F.	.083	3.09	3.52	6.61

The cost of this system is based on the square foot of plan area.
All quantities have been adjusted accordingly.

Description	QUAN.	UNIT	LABOR HOURS	COST PER S.F.		
				MAT.	INST.	TOTAL

Important: See the Reference Section for critical supporting data - Reference Nos., Crews & Location Factors

Gable End Roof Framing Price Sheet	QUAN.	UNIT	LABOR HOURS	COST PER S.F.		
				MAT.	INST.	TOTAL
Rafters, #2 or better, 16″ O.C., 2″ x 6″, 4/12 pitch	1.170	L.F.	.019	.76	.80	1.56
8/12 pitch	1.330	L.F.	.027	.86	1.13	1.99
2″ x 8″, 4/12 pitch	1.170	L.F.	.020	1.10	.83	1.93
8/12 pitch	1.330	L.F.	.028	1.25	1.20	2.45
2″ x 10″, 4/12 pitch	1.170	L.F.	.030	1.66	1.25	2.91
8/12 pitch	1.330	L.F.	.043	1.89	1.82	3.71
24″ O.C., 2″ x 6″, 4/12 pitch	.940	L.F.	.015	.61	.64	1.25
8/12 pitch	1.060	L.F.	.021	.69	.90	1.59
2″ x 8″, 4/12 pitch	.940	L.F.	.016	.88	.67	1.55
8/12 pitch	1.060	L.F.	.023	1	.95	1.95
2″ x 10″, 4/12 pitch	.940	L.F.	.024	1.33	1.01	2.34
8/12 pitch	1.060	L.F.	.034	1.51	1.45	2.96
Ceiling joist, #2 or better, 2″ x 4″, 16″ O.C.	1.000	L.F.	.013	.41	.54	.95
24″ O.C.	.750	L.F.	.010	.31	.41	.72
2″ x 6″, 16″ O.C.	1.000	L.F.	.013	.65	.54	1.19
24″ O.C.	.750	L.F.	.010	.49	.41	.90
2″ x 8″, 16″ O.C.	1.000	L.F.	.015	.94	.61	1.55
24″ O.C.	.750	L.F.	.011	.71	.46	1.17
2″ x 10″, 16″ O.C.	1.000	L.F.	.018	1.42	.75	2.17
24″ O.C.	.750	L.F.	.013	1.07	.56	1.63
Ridge board, #2 or better, 1″ x 6″	.050	L.F.	.001	.04	.06	.10
1″ x 8″	.050	L.F.	.001	.05	.06	.11
1″ x 10″	.050	L.F.	.002	.07	.07	.14
2″ x 6″	.050	L.F.	.002	.03	.07	.10
2″ x 8″	.050	L.F.	.002	.05	.08	.13
2″ x 10″	.050	L.F.	.002	.07	.08	.15
Fascia board, #2 or better, 1″ x 6″	.100	L.F.	.004	.05	.17	.22
1″ x 8″	.100	L.F.	.005	.06	.19	.25
1″ x 10″	.100	L.F.	.005	.07	.22	.29
2″ x 6″	.100	L.F.	.006	.08	.24	.32
2″ x 8″	.100	L.F.	.007	.09	.30	.39
2″ x 10″	.100	L.F.	.004	.28	.15	.43
Rafter tie, #2 or better, 4′ O.C., 1″ x 4″	.060	L.F.	.001	.02	.05	.07
1″ x 6″	.060	L.F.	.001	.03	.06	.09
2″ x 4″	.060	L.F.	.002	.03	.07	.10
2″ x 6″	.060	L.F.	.002	.04	.09	.13
Soffit nailer (outrigger), 2″ x 4″, 16″ O.C.	.220	L.F.	.006	.09	.24	.33
24″ O.C.	.170	L.F.	.004	.07	.19	.26
2″ x 6″, 16″ O.C.	.220	L.F.	.006	.10	.27	.37
24″ O.C.	.170	L.F.	.005	.08	.22	.30
Sheathing, plywood CDX, 4/12 pitch, 3/8″ thick.	1.170	S.F.	.012	.59	.51	1.10
1/2″ thick	1.170	S.F.	.013	.80	.56	1.36
5/8″ thick	1.170	S.F.	.014	.85	.61	1.46
8/12 pitch, 3/8″	1.330	S.F.	.014	.67	.59	1.26
1/2″ thick	1.330	S.F.	.015	.90	.64	1.54
5/8″ thick	1.330	S.F.	.016	.97	.69	1.66
Boards, 4/12 pitch roof, 1″ x 6″	1.170	S.F.	.026	1.54	1.09	2.63
1″ x 8″	1.170	S.F.	.021	1.44	.90	2.34
8/12 pitch roof, 1″ x 6″	1.330	S.F.	.029	1.76	1.24	3
1″ x 8″	1.330	S.F.	.024	1.64	1.02	2.66
Furring, 1″ x 3″, 12″ O.C.	1.200	L.F.	.027	.37	1.16	1.53
16″ O.C.	1.000	L.F.	.023	.31	.97	1.28
24″ O.C.	.800	L.F.	.018	.25	.78	1.03

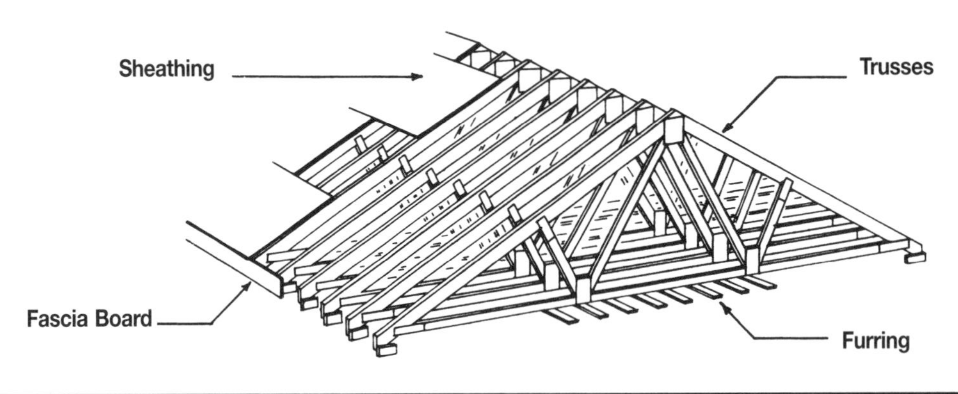

System Description	QUAN.	UNIT	LABOR HOURS	COST PER S.F.		
				MAT.	INST.	TOTAL
TRUSS, 16" O.C., 4/12 PITCH, 1' OVERHANG, 26' SPAN						
Truss, 40# loading, 16" O.C., 4/12 pitch, 26' span	.030	Ea.	.021	2.70	1.17	3.87
Fascia board, 2" x 6"	.100	L.F.	.005	.07	.23	.30
Sheathing, exterior, plywood, CDX, 1/2" thick	1.170	S.F.	.013	.80	.56	1.36
Furring, 1" x 3", 16" O.C.	1.000	L.F.	.023	.31	.97	1.28
TOTAL		S.F.	.062	3.88	2.93	6.81
TRUSS, 16" O.C., 8/12 PITCH, 1' OVERHANG, 26' SPAN						
Truss, 40# loading, 16" O.C., 8/12 pitch, 26' span	.030	Ea.	.023	3.18	1.27	4.45
Fascia board, 2" x 6"	.100	L.F.	.005	.07	.23	.30
Sheathing, exterior, plywood, CDX, 1/2" thick	1.330	S.F.	.015	.90	.64	1.54
Furring, 1" x 3", 16" O.C.	1.000	L.F.	.023	.31	.97	1.28
TOTAL		S.F.	.066	4.46	3.11	7.57
TRUSS, 24" O.C., 4/12 PITCH, 1' OVERHANG, 26' SPAN						
Truss, 40# loading, 24" O.C., 4/12 pitch, 26' span	.020	Ea.	.014	1.80	.78	2.58
Fascia board, 2" x 6"	.100	L.F.	.005	.07	.23	.30
Sheathing, exterior, plywood, CDX, 1/2" thick	1.170	S.F.	.013	.80	.56	1.36
Furring, 1" x 3", 16" O.C.	1.000	L.F.	.023	.31	.97	1.28
TOTAL		S.F.	.055	2.98	2.54	5.52
TRUSS, 24" O.C., 8/12 PITCH, 1' OVERHANG, 26' SPAN						
Truss, 40# loading, 24" O.C., 8/12 pitch, 26' span	.020	Ea.	.015	2.12	.85	2.97
Fascia board, 2" x 6"	.100	L.F.	.005	.07	.23	.30
Sheathing, exterior, plywood, CDX, 1/2" thick	1.330	S.F.	.015	.90	.64	1.54
Furring, 1" x 3", 16" O.C.	1.000	L.F.	.023	.31	.97	1.28
TOTAL		S.F.	.058	3.40	2.69	6.09

The cost of this system is based on the square foot of plan area.
A one foot overhang is included.

Description	QUAN.	UNIT	LABOR HOURS	COST PER S.F.		
				MAT.	INST.	TOTAL

Truss Roof Framing Price Sheet	QUAN.	UNIT	LABOR HOURS	COST PER S.F.		
				MAT.	INST.	TOTAL
Truss, 40# loading, including 1′ overhang, 4/12 pitch, 24′ span, 16″ O.C.	.033	Ea.	.022	1.98	1.21	3.19
24″ O.C.	.022	Ea.	.015	1.32	.81	2.13
26′ span, 16″ O.C.	.030	Ea.	.021	2.70	1.17	3.87
24″ O.C.	.020	Ea.	.014	1.80	.78	2.58
28′ span, 16″ O.C.	.027	Ea.	.020	1.89	1.13	3.02
24″ O.C.	.019	Ea.	.014	1.33	.79	2.12
32′ span, 16″ O.C.	.024	Ea.	.019	2.64	1.06	3.70
24″ O.C.	.016	Ea.	.013	1.76	.70	2.46
36′ span, 16″ O.C.	.022	Ea.	.019	3.26	1.06	4.32
24″ O.C.	.015	Ea.	.013	2.22	.72	2.94
8/12 pitch, 24′ span, 16″ O.C.	.033	Ea.	.024	3.23	1.33	4.56
24″ O.C.	.022	Ea.	.016	2.16	.89	3.05
26′ span, 16″ O.C.	.030	Ea.	.023	3.18	1.27	4.45
24″ O.C.	.020	Ea.	.015	2.12	.85	2.97
28′ span, 16″ O.C.	.027	Ea.	.022	3.11	1.21	4.32
24″ O.C.	.019	Ea.	.016	2.19	.85	3.04
32′ span, 16″ O.C.	.024	Ea.	.021	3.24	1.17	4.41
24″ O.C.	.016	Ea.	.014	2.16	.79	2.95
36′ span, 16″ O.C.	.022	Ea.	.021	3.52	1.18	4.70
24″ O.C.	.015	Ea.	.015	2.40	.81	3.21
Fascia board, #2 or better, 1″ x 6″	.100	L.F.	.004	.05	.17	.22
1″ x 8″	.100	L.F.	.005	.06	.19	.25
1″ x 10″	.100	L.F.	.005	.07	.22	.29
2″ x 6″	.100	L.F.	.006	.08	.24	.32
2″ x 8″	.100	L.F.	.007	.09	.30	.39
2″ x 10″	.100	L.F.	.009	.14	.38	.52
Sheathing, plywood CDX, 4/12 pitch, 3/8″ thick	1.170	S.F.	.012	.59	.51	1.10
1/2″ thick	1.170	S.F.	.013	.80	.56	1.36
5/8″ thick	1.170	S.F.	.014	.85	.61	1.46
8/12 pitch, 3/8″ thick	1.330	S.F.	.014	.67	.59	1.26
1/2″ thick	1.330	S.F.	.015	.90	.64	1.54
5/8″ thick	1.330	S.F.	.016	.97	.69	1.66
Boards, 4/12 pitch, 1″ x 6″	1.170	S.F.	.026	1.54	1.09	2.63
1″ x 8″	1.170	S.F.	.021	1.44	.90	2.34
8/12 pitch, 1″ x 6″	1.330	S.F.	.029	1.76	1.24	3
1″ x 8″	1.330	S.F.	.024	1.64	1.02	2.66
Furring, 1″ x 3″, 12″ O.C.	1.200	L.F.	.027	.37	1.16	1.53
16″ O.C.	1.000	L.F.	.023	.31	.97	1.28
24″ O.C.	.800	L.F.	.018	.25	.78	1.03

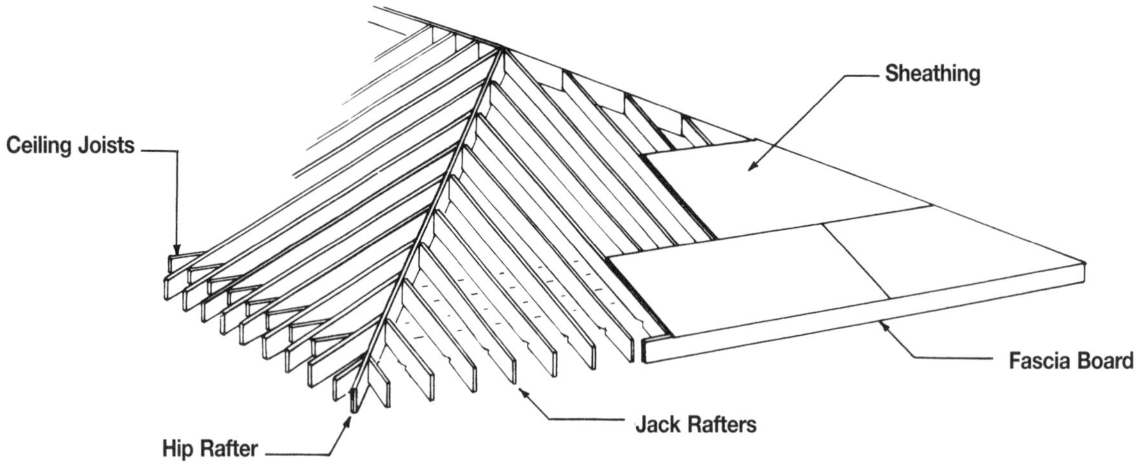

Sheathing
Ceiling Joists
Fascia Board
Jack Rafters
Hip Rafter

System Description	QUAN.	UNIT	LABOR HOURS	COST PER S.F.		
				MAT.	INST.	TOTAL
2" X 6", 16" O.C., 4/12 PITCH						
Hip rafters, 2" x 8", 4/12 pitch	.160	L.F.	.004	.15	.15	.30
Jack rafters, 2" x 6", 16" O.C., 4/12 pitch	1.430	L.F.	.038	.93	1.62	2.55
Ceiling joists, 2" x 6", 16" O.C.	1.000	L.F.	.013	.65	.54	1.19
Fascia board, 2" x 8"	.220	L.F.	.016	.21	.66	.87
Soffit nailer (outrigger), 2" x 4", 24" O.C.	.220	L.F.	.006	.09	.24	.33
Sheathing, 1/2" exterior plywood, CDX	1.570	S.F.	.018	1.07	.75	1.82
Furring strips, 1" x 3", 16" O.C.	1.000	L.F.	.023	.31	.97	1.28
TOTAL		S.F.	.118	3.41	4.93	8.34
2" X 8", 16" O.C., 4/12 PITCH						
Hip rafters, 2" x 10", 4/12 pitch	.160	L.F.	.004	.23	.19	.42
Jack rafters, 2" x 8", 16" O.C., 4/12 pitch	1.430	L.F.	.047	1.34	1.97	3.31
Ceiling joists, 2" x 6", 16" O.C.	1.000	L.F.	.013	.65	.54	1.19
Fascia board, 2" x 8"	.220	L.F.	.012	.16	.51	.67
Soffit nailer (outrigger), 2" x 4", 24" O.C.	.220	L.F.	.006	.09	.24	.33
Sheathing, 1/2" exterior plywood, CDX	1.570	S.F.	.018	1.07	.75	1.82
Furring strips, 1" x 3", 16" O.C.	1.000	L.F.	.023	.31	.97	1.28
TOTAL		S.F.	.123	3.85	5.17	9.02

The cost of this system is based on S.F. of plan area. Measurement is area under the hip roof only. See gable roof system for added costs.

Description	QUAN.	UNIT	LABOR HOURS	COST PER S.F.		
				MAT.	INST.	TOTAL

Hip Roof Framing Price Sheet	QUAN.	UNIT	LABOR HOURS	COST PER S.F.		
				MAT.	INST.	TOTAL
Hip rafters, #2 or better, 2″ x 6″, 4/12 pitch	.160	L.F.	.003	.10	.14	.24
8/12 pitch	.210	L.F.	.006	.14	.24	.38
2″ x 8″, 4/12 pitch	.160	L.F.	.004	.15	.15	.30
8/12 pitch	.210	L.F.	.006	.20	.26	.46
2″ x 10″, 4/12 pitch	.160	L.F.	.004	.23	.19	.42
8/12 pitch roof	.210	L.F.	.008	.30	.32	.62
Jack rafters, #2 or better, 16″ O.C., 2″ x 6″, 4/12 pitch	1.430	L.F.	.038	.93	1.62	2.55
8/12 pitch	1.800	L.F.	.061	1.17	2.56	3.73
2″ x 8″, 4/12 pitch	1.430	L.F.	.047	1.34	1.97	3.31
8/12 pitch	1.800	L.F.	.075	1.69	3.17	4.86
2″ x 10″, 4/12 pitch	1.430	L.F.	.051	2.03	2.15	4.18
8/12 pitch	1.800	L.F.	.082	2.56	3.47	6.03
24″ O.C., 2″ x 6″, 4/12 pitch	1.150	L.F.	.031	.75	1.30	2.05
8/12 pitch	1.440	L.F.	.048	.94	2.04	2.98
2″ x 8″, 4/12 pitch	1.150	L.F.	.038	1.08	1.59	2.67
8/12 pitch	1.440	L.F.	.060	1.35	2.53	3.88
2″ x 10″, 4/12 pitch	1.150	L.F.	.041	1.63	1.73	3.36
8/12 pitch	1.440	L.F.	.066	2.04	2.78	4.82
Ceiling joists, #2 or better, 2″ x 4″, 16″ O.C.	1.000	L.F.	.013	.41	.54	.95
24″ O.C.	.750	L.F.	.010	.31	.41	.72
2″ x 6″, 16″ O.C.	1.000	L.F.	.013	.65	.54	1.19
24″ O.C.	.750	L.F.	.010	.49	.41	.90
2″ x 8″, 16″ O.C.	1.000	L.F.	.015	.94	.61	1.55
24″ O.C.	.750	L.F.	.011	.71	.46	1.17
2″ x 10″, 16″ O.C.	1.000	L.F.	.018	1.42	.75	2.17
24″ O.C.	.750	L.F.	.013	1.07	.56	1.63
Fascia board, #2 or better, 1″ x 6″	.220	L.F.	.009	.11	.36	.47
1″ x 8″	.220	L.F.	.010	.13	.42	.55
1″ x 10″	.220	L.F.	.011	.15	.47	.62
2″ x 6″	.220	L.F.	.013	.17	.53	.70
2″ x 8″	.220	L.F.	.016	.21	.66	.87
2″ x 10″	.220	L.F.	.020	.31	.83	1.14
Soffit nailer (outrigger), 2″ x 4″, 16″ O.C.	.280	L.F.	.007	.11	.31	.42
24″ O.C.	.220	L.F.	.006	.09	.24	.33
2″ x 8″, 16″ O.C.	.280	L.F.	.007	.20	.28	.48
24″ O.C.	.220	L.F.	.005	.16	.23	.39
Sheathing, plywood CDX, 4/12 pitch, 3/8″ thick	1.570	S.F.	.016	.79	.69	1.48
1/2″ thick	1.570	S.F.	.018	1.07	.75	1.82
5/8″ thick	1.570	S.F.	.019	1.15	.82	1.97
8/12 pitch, 3/8″ thick	1.900	S.F.	.020	.95	.84	1.79
1/2″ thick	1.900	S.F.	.022	1.29	.91	2.20
5/8″ thick	1.900	S.F.	.023	1.39	.99	2.38
Boards, 4/12 pitch, 1″ x 6″ boards	1.450	S.F.	.032	1.91	1.35	3.26
1″ x 8″ boards	1.450	S.F.	.027	1.78	1.12	2.90
8/12 pitch, 1″ x 6″ boards	1.750	S.F.	.039	2.31	1.63	3.94
1″ x 8″ boards	1.750	S.F.	.032	2.15	1.35	3.50
Furring, 1″ x 3″, 12″ O.C.	1.200	L.F.	.027	.37	1.16	1.53
16″ O.C.	1.000	L.F.	.023	.31	.97	1.28
24″ O.C.	.800	L.F.	.018	.25	.78	1.03

24 Gambrel Roof Framing Systems

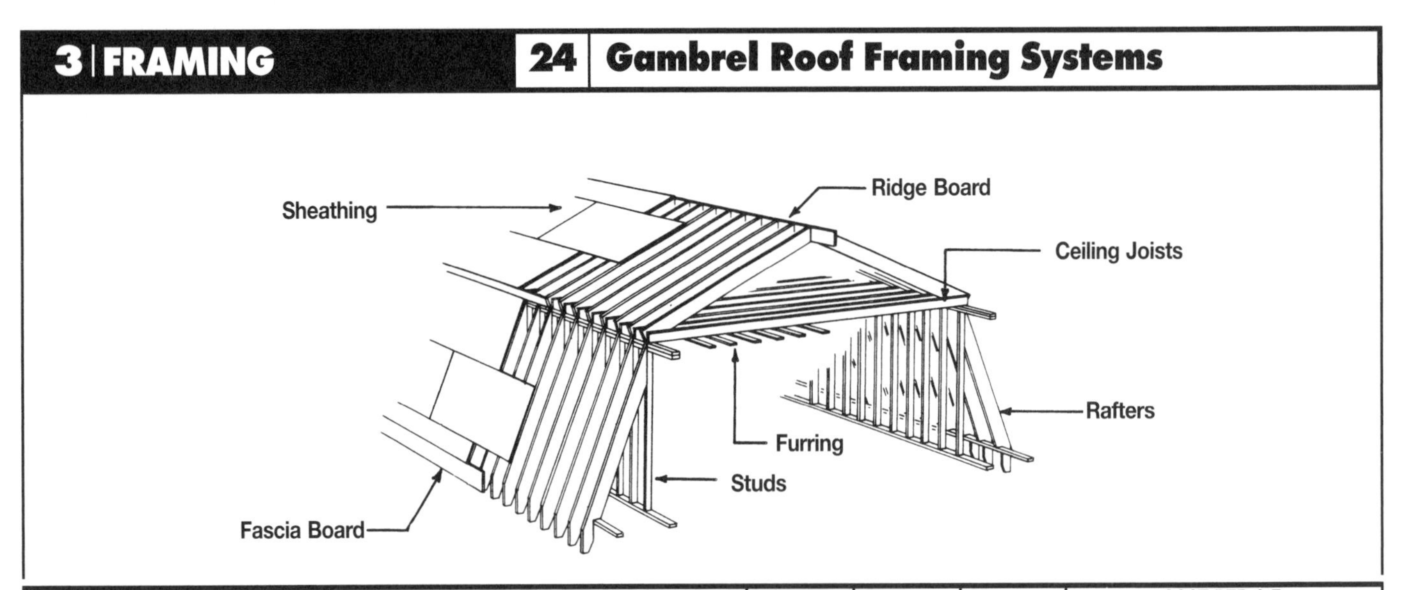

System Description	QUAN.	UNIT	LABOR HOURS	COST PER S.F.		
				MAT.	INST.	TOTAL
2″ X 6″ RAFTERS, 16″ O.C.						
Roof rafters, 2″ x 6″, 16″ O.C.	1.430	L.F.	.029	.93	1.22	2.15
Ceiling joists, 2″ x 6″, 16″ O.C.	.710	L.F.	.009	.46	.38	.84
Stud wall, 2″ x 4″, 16″ O.C., including plates	.790	L.F.	.012	.32	.53	.85
Furring strips, 1″ x 3″, 16″ O.C.	.710	L.F.	.016	.22	.69	.91
Ridge board, 2″ x 8″	.050	L.F.	.002	.05	.08	.13
Fascia board, 2″ x 6″	.100	L.F.	.006	.08	.24	.32
Sheathing, exterior grade plywood, 1/2″ thick	1.450	S.F.	.017	.99	.70	1.69
TOTAL		S.F.	.091	3.05	3.84	6.89
2″ X 8″ RAFTERS, 16″ O.C.						
Roof rafters, 2″ x 8″, 16″ O.C.	1.430	L.F.	.031	1.34	1.29	2.63
Ceiling joists, 2″ x 6″, 16″ O.C.	.710	L.F.	.009	.46	.38	.84
Stud wall, 2″ x 4″, 16″ O.C., including plates	.790	L.F.	.012	.32	.53	.85
Furring strips, 1″ x 3″, 16″ O.C.	.710	L.F.	.016	.22	.69	.91
Ridge board, 2″ x 8″	.050	L.F.	.002	.05	.08	.13
Fascia board, 2″ x 8″	.100	L.F.	.007	.09	.30	.39
Sheathing, exterior grade plywood, 1/2″ thick	1.450	S.F.	.017	.99	.70	1.69
TOTAL		S.F.	.094	3.47	3.97	7.44

The cost of this system is based on the square foot of plan area on the first floor.

Description	QUAN.	UNIT	LABOR HOURS	COST PER S.F.		
				MAT.	INST.	TOTAL

Gambrel Roof Framing Price Sheet	QUAN.	UNIT	LABOR HOURS	COST PER S.F.		
				MAT.	INST.	TOTAL
Roof rafters, #2 or better, 2″ x 6″, 16″ O.C.	1.430	L.F.	.029	.93	1.22	2.15
24″ O.C.	1.140	L.F.	.023	.74	.97	1.71
2″ x 8″, 16″ O.C.	1.430	L.F.	.031	1.34	1.29	2.63
24″ O.C.	1.140	L.F.	.024	1.07	1.03	2.10
2″ x 10″, 16″ O.C.	1.430	L.F.	.046	2.03	1.96	3.99
24″ O.C.	1.140	L.F.	.037	1.62	1.56	3.18
Ceiling joist, #2 or better, 2″ x 4″, 16″ O.C.	.710	L.F.	.009	.29	.38	.67
24″ O.C.	.570	L.F.	.007	.23	.31	.54
2″ x 6″, 16″ O.C.	.710	L.F.	.009	.46	.38	.84
24″ O.C.	.570	L.F.	.007	.37	.31	.68
2″ x 8″, 16″ O.C.	.710	L.F.	.010	.67	.43	1.10
24″ O.C.	.570	L.F.	.008	.54	.35	.89
Stud wall, #2 or better, 2″ x 4″, 16″ O.C.	.790	L.F.	.012	.32	.53	.85
24″ O.C.	.630	L.F.	.010	.26	.42	.68
2″ x 6″, 16″ O.C.	.790	L.F.	.014	.51	.60	1.11
24″ O.C.	.630	L.F.	.011	.41	.48	.89
Furring, 1″ x 3″, 16″ O.C.	.710	L.F.	.016	.22	.69	.91
24″ O.C.	.590	L.F.	.013	.18	.57	.75
Ridge board, #2 or better, 1″ x 6″	.050	L.F.	.001	.04	.06	.10
1″ x 8″	.050	L.F.	.001	.05	.06	.11
1″ x 10″	.050	L.F.	.002	.07	.07	.14
2″ x 6″	.050	L.F.	.002	.03	.07	.10
2″ x 8″	.050	L.F.	.002	.05	.08	.13
2″ x 10″	.050	L.F.	.002	.07	.08	.15
Fascia board, #2 or better, 1″ x 6″	.100	L.F.	.004	.05	.17	.22
1″ x 8″	.100	L.F.	.005	.06	.19	.25
1″ x 10″	.100	L.F.	.005	.07	.22	.29
2″ x 6″	.100	L.F.	.006	.08	.24	.32
2″ x 8″	.100	L.F.	.007	.09	.30	.39
2″ x 10″	.100	L.F.	.009	.14	.38	.52
Sheathing, plywood, exterior grade CDX, 3/8″ thick	1.450	S.F.	.015	.73	.64	1.37
1/2″ thick	1.450	S.F.	.017	.99	.70	1.69
5/8″ thick	1.450	S.F.	.018	1.06	.75	1.81
3/4″ thick	1.450	S.F.	.019	1.31	.81	2.12
Boards, 1″ x 6″, laid regular	1.450	S.F.	.032	1.91	1.35	3.26
Laid diagonal	1.450	S.F.	.036	1.91	1.51	3.42
1″ x 8″, laid regular	1.450	S.F.	.027	1.78	1.12	2.90
Laid diagonal	1.450	S.F.	.032	1.78	1.35	3.13

28 Mansard Roof Framing Systems

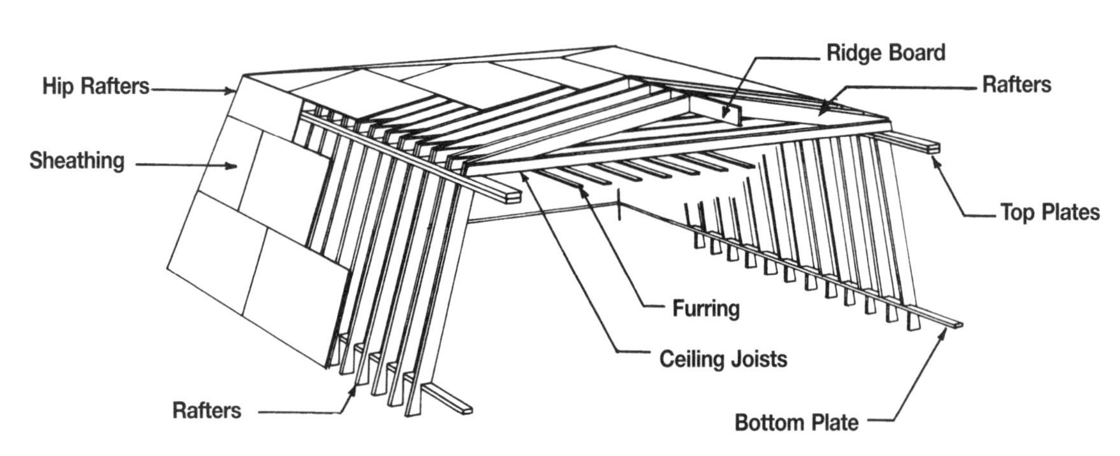

System Description	QUAN.	UNIT	LABOR HOURS	COST PER S.F.		
				MAT.	INST.	TOTAL
2" X 6" RAFTERS, 16" O.C.						
Roof rafters, 2" x 6", 16" O.C.	1.210	L.F.	.033	.79	1.39	2.18
Rafter plates, 2" x 6", double top, single bottom	.364	L.F.	.010	.24	.42	.66
Ceiling joists, 2" x 4", 16" O.C.	.920	L.F.	.012	.38	.50	.88
Hip rafter, 2" x 6"	.070	L.F.	.002	.05	.09	.14
Jack rafter, 2" x 6", 16" O.C.	1.000	L.F.	.039	.65	1.65	2.30
Ridge board, 2" x 6"	.018	L.F.	.001	.01	.02	.03
Sheathing, exterior grade plywood, 1/2" thick	2.210	S.F.	.025	1.50	1.06	2.56
Furring strips, 1" x 3", 16" O.C.	.920	L.F.	.021	.29	.89	1.18
TOTAL		S.F.	.143	3.91	6.02	9.93
2" X 8" RAFTERS, 16" O.C.						
Roof rafters, 2" x 8", 16" O.C.	1.210	L.F.	.036	1.14	1.51	2.65
Rafter plates, 2" x 8", double top, single bottom	.364	L.F.	.011	.34	.46	.80
Ceiling joists, 2" x 6", 16" O.C.	.920	L.F.	.012	.60	.50	1.10
Hip rafter, 2" x 8"	.070	L.F.	.002	.07	.10	.17
Jack rafter, 2" x 8", 16" O.C.	1.000	L.F.	.048	.94	2.02	2.96
Ridge board, 2" x 8"	.018	L.F.	.001	.02	.03	.05
Sheathing, exterior grade plywood, 1/2" thick	2.210	S.F.	.025	1.50	1.06	2.56
Furring strips, 1" x 3", 16" O.C.	.920	L.F.	.021	.29	.89	1.18
TOTAL		S.F.	.156	4.90	6.57	11.47

The cost of this system is based on the square foot of plan area.

Description	QUAN.	UNIT	LABOR HOURS	COST PER S.F.		
				MAT.	INST.	TOTAL

Mansard Roof Framing Price Sheet

Mansard Roof Framing Price Sheet	QUAN.	UNIT	LABOR HOURS	COST PER S.F.		
				MAT.	INST.	TOTAL
Roof rafters, #2 or better, 2″ x 6″, 16″ O.C.	1.210	L.F.	.033	.79	1.39	2.18
24″ O.C.	.970	L.F.	.026	.63	1.12	1.75
2″ x 8″, 16″ O.C.	1.210	L.F.	.036	1.14	1.51	2.65
24″ O.C.	.970	L.F.	.029	.91	1.21	2.12
2″ x 10″, 16″ O.C.	1.210	L.F.	.046	1.72	1.92	3.64
24″ O.C.	.970	L.F.	.037	1.38	1.54	2.92
Rafter plates, #2 or better double top single bottom, 2″ x 6″	.364	L.F.	.010	.24	.42	.66
2″ x 8″	.364	L.F.	.011	.34	.46	.80
2″ x 10″	.364	L.F.	.014	.52	.58	1.10
Ceiling joist, #2 or better, 2″ x 4″, 16″ O.C.	.920	L.F.	.012	.38	.50	.88
24″ O.C.	.740	L.F.	.009	.30	.40	.70
2″ x 6″, 16″ O.C.	.920	L.F.	.012	.60	.50	1.10
24″ O.C.	.740	L.F.	.009	.48	.40	.88
2″ x 8″, 16″ O.C.	.920	L.F.	.013	.86	.56	1.42
24″ O.C.	.740	L.F.	.011	.70	.45	1.15
Hip rafter, #2 or better, 2″ x 6″	.070	L.F.	.002	.05	.09	.14
2″ x 8″	.070	L.F.	.002	.07	.10	.17
2″ x 10″	.070	L.F.	.003	.10	.12	.22
Jack rafter, #2 or better, 2″ x 6″, 16″ O.C.	1.000	L.F.	.039	.65	1.65	2.30
24″ O.C.	.800	L.F.	.031	.52	1.32	1.84
2″ x 8″, 16″ O.C.	1.000	L.F.	.048	.94	2.02	2.96
24″ O.C.	.800	L.F.	.038	.75	1.62	2.37
Ridge board, #2 or better, 1″ x 6″	.018	L.F.	.001	.01	.02	.03
1″ x 8″	.018	L.F.	.001	.02	.02	.04
1″ x 10″	.018	L.F.	.001	.03	.02	.05
2″ x 6″	.018	L.F.	.001	.01	.02	.03
2″ x 8″	.018	L.F.	.001	.02	.03	.05
2″ x 10″	.018	L.F.	.001	.03	.03	.06
Sheathing, plywood exterior grade CDX, 3/8″ thick	2.210	S.F.	.023	1.11	.97	2.08
1/2″ thick	2.210	S.F.	.025	1.50	1.06	2.56
5/8″ thick	2.210	S.F.	.027	1.61	1.15	2.76
3/4″ thick	2.210	S.F.	.029	1.99	1.24	3.23
Boards, 1″ x 6″, laid regular	2.210	S.F.	.049	2.92	2.06	4.98
Laid diagonal	2.210	S.F.	.054	2.92	2.30	5.22
1″ x 8″, laid regular	2.210	S.F.	.040	2.72	1.70	4.42
Laid diagonal	2.210	S.F.	.049	2.72	2.06	4.78
Furring, 1″ x 3″, 12″ O.C.	1.150	L.F.	.026	.36	1.12	1.48
24″ O.C.	.740	L.F.	.017	.23	.72	.95

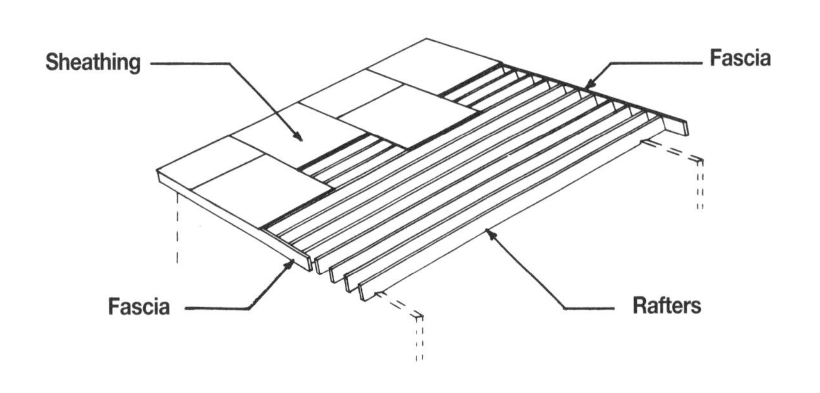

System Description	QUAN.	UNIT	LABOR HOURS	COST PER S.F.		
				MAT.	INST.	TOTAL
2″ X 6″,16″ O.C., 4/12 PITCH						
Rafters, 2″ x 6″, 16″ O.C., 4/12 pitch	1.170	L.F.	.019	.76	.80	1.56
Fascia, 2″ x 6″	.100	L.F.	.006	.08	.24	.32
Bridging, 1″ x 3″, 6′ O.C.	.080	Pr.	.005	.04	.21	.25
Sheathing, exterior grade plywood, 1/2″ thick	1.230	S.F.	.014	.84	.59	1.43
TOTAL		S.F.	.044	1.72	1.84	3.56
2″ X 6″, 24″ O.C., 4/12 PITCH						
Rafters, 2″ x 6″, 24″ O.C., 4/12 pitch	.940	L.F.	.015	.61	.64	1.25
Fascia, 2″ x 6″	.100	L.F.	.006	.08	.24	.32
Bridging, 1″ x 3″, 6′ O.C.	.060	Pr.	.004	.03	.16	.19
Sheathing, exterior grade plywood, 1/2″ thick	1.230	S.F.	.014	.84	.59	1.43
TOTAL		S.F.	.039	1.56	1.63	3.19
2″ X 8″, 16″ O.C., 4/12 PITCH						
Rafters, 2″ x 8″, 16″ O.C., 4/12 pitch	1.170	L.F.	.020	1.10	.83	1.93
Fascia, 2″ x 8″	.100	L.F.	.007	.09	.30	.39
Bridging, 1″ x 3″, 6′ O.C.	.080	Pr.	.005	.04	.21	.25
Sheathing, exterior grade plywood, 1/2″ thick	1.230	S.F.	.014	.84	.59	1.43
TOTAL		S.F.	.046	2.07	1.93	4
2″ X 8″, 24″ O.C., 4/12 PITCH						
Rafters, 2″ x 8″, 24″ O.C., 4/12 pitch	.940	L.F.	.016	.88	.67	1.55
Fascia, 2″ x 8″	.100	L.F.	.007	.09	.30	.39
Bridging, 1″ x 3″, 6′ O.C.	.060	Pr.	.004	.03	.16	.19
Sheathing, exterior grade plywood, 1/2″ thick	1.230	S.F.	.014	.84	.59	1.43
TOTAL		S.F.	.041	1.84	1.72	3.56

The cost of this system is based on the square foot of plan area.
A 1′ overhang is assumed. No ceiling joists or furring are included.

Description	QUAN.	UNIT	LABOR HOURS	COST PER S.F.		
				MAT.	INST.	TOTAL

 Important: See the Reference Section for critical supporting data - Reference Nos., Crews & Location Factors

Shed/Flat Roof Framing Price Sheet	QUAN.	UNIT	LABOR HOURS	COST PER S.F.		
				MAT.	INST.	TOTAL
Rafters, #2 or better, 16" O.C., 2" x 4", 0 - 4/12 pitch	1.170	L.F.	.014	.57	.59	1.16
5/12 - 8/12 pitch	1.330	L.F.	.020	.65	.85	1.50
2" x 6", 0 - 4/12 pitch	1.170	L.F.	.019	.76	.80	1.56
5/12 - 8/12 pitch	1.330	L.F.	.027	.86	1.13	1.99
2" x 8", 0 - 4/12 pitch	1.170	L.F.	.020	1.10	.83	1.93
5/12 - 8/12 pitch	1.330	L.F.	.028	1.25	1.20	2.45
2" x 10", 0 - 4/12 pitch	1.170	L.F.	.030	1.66	1.25	2.91
5/12 - 8/12 pitch	1.330	L.F.	.043	1.89	1.82	3.71
24" O.C., 2" x 4", 0 - 4/12 pitch	.940	L.F.	.011	.46	.48	.94
5/12 - 8/12 pitch	1.060	L.F.	.021	.69	.90	1.59
2" x 6", 0 - 4/12 pitch	.940	L.F.	.015	.61	.64	1.25
5/12 - 8/12 pitch	1.060	L.F.	.021	.69	.90	1.59
2" x 8", 0 - 4/12 pitch	.940	L.F.	.016	.88	.67	1.55
5/12 - 8/12 pitch	1.060	L.F.	.023	1	.95	1.95
2" x 10", 0 - 4/12 pitch	.940	L.F.	.024	1.33	1.01	2.34
5/12 - 8/12 pitch	1.060	L.F.	.034	1.51	1.45	2.96
Fascia, #2 or better,, 1" x 4"	.100	L.F.	.003	.04	.12	.16
1" x 6"	.100	L.F.	.004	.05	.17	.22
1" x 8"	.100	L.F.	.005	.06	.19	.25
1" x 10"	.100	L.F.	.005	.07	.22	.29
2" x 4"	.100	L.F.	.005	.06	.20	.26
2" x 6"	.100	L.F.	.006	.08	.24	.32
2" x 8"	.100	L.F.	.007	.09	.30	.39
2" x 10"	.100	L.F.	.009	.14	.38	.52
Bridging, wood 6' O.C., 1" x 3", rafters, 16" O.C.	.080	Pr.	.005	.04	.21	.25
24" O.C.	.060	Pr.	.004	.03	.16	.19
Metal, galvanized, rafters, 16" O.C.	.080	Pr.	.005	.10	.21	.31
24" O.C.	.060	Pr.	.003	.09	.14	.23
Compression type, rafters, 16" O.C.	.080	Pr.	.003	.12	.14	.26
24" O.C.	.060	Pr.	.002	.09	.10	.19
Sheathing, plywood, exterior grade, 3/8" thick, flat 0 - 4/12 pitch	1.230	S.F.	.013	.62	.54	1.16
5/12 - 8/12 pitch	1.330	S.F.	.014	.67	.59	1.26
1/2" thick, flat 0 - 4/12 pitch	1.230	S.F.	.014	.84	.59	1.43
5/12 - 8/12 pitch	1.330	S.F.	.015	.90	.64	1.54
5/8" thick, flat 0 - 4/12 pitch	1.230	S.F.	.015	.90	.64	1.54
5/12 - 8/12 pitch	1.330	S.F.	.016	.97	.69	1.66
3/4" thick, flat 0 - 4/12 pitch	1.230	S.F.	.016	1.11	.69	1.80
5/12 - 8/12 pitch	1.330	S.F.	.018	1.20	.74	1.94
Boards, 1" x 6", laid regular, flat 0 - 4/12 pitch	1.230	S.F.	.027	1.62	1.14	2.76
5/12 - 8/12 pitch	1.330	S.F.	.041	1.76	1.73	3.49
Laid diagonal, flat 0 - 4/12 pitch	1.230	S.F.	.030	1.62	1.28	2.90
5/12 - 8/12 pitch	1.330	S.F.	.044	1.76	1.88	3.64
1" x 8", laid regular, flat 0 - 4/12 pitch	1.230	S.F.	.022	1.51	.95	2.46
5/12 - 8/12 pitch	1.330	S.F.	.034	1.64	1.41	3.05
Laid diagonal, flat 0 - 4/12 pitch	1.230	S.F.	.027	1.51	1.14	2.65
5/12 - 8/12 pitch	1.330	S.F.	.044	1.76	1.88	3.64

40 Gable Dormer Framing Systems

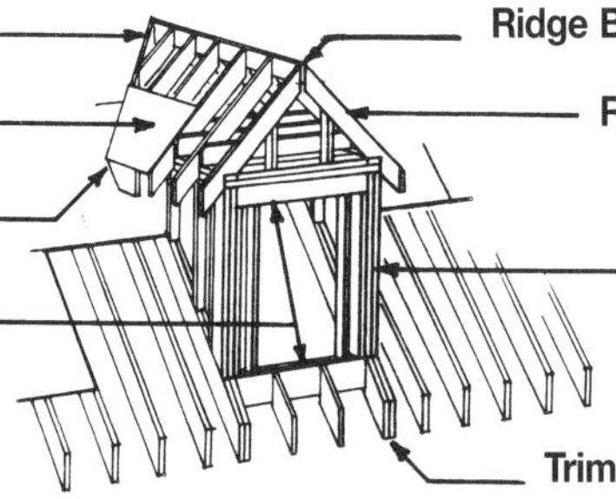

Valley Rafter
Sheathing
Fascia Board
Headers
Ridge Board
Rafters
Studs & Plates
Trimmer Rafters

System Description	QUAN.	UNIT	LABOR HOURS	COST PER S.F. MAT.	COST PER S.F. INST.	COST PER S.F. TOTAL
2" X 6", 16" O.C.						
Dormer rafter, 2" x 6", 16" O.C.	1.330	L.F.	.036	.86	1.53	2.39
Ridge board, 2" x 6"	.280	L.F.	.009	.18	.38	.56
Trimmer rafters, 2" x 6"	.880	L.F.	.014	.57	.60	1.17
Wall studs & plates, 2" x 4", 16" O.C.	3.160	L.F.	.056	1.30	2.37	3.67
Fascia, 2" x 6"	.220	L.F.	.012	.16	.51	.67
Valley rafter, 2" x 6", 16" O.C.	.280	L.F.	.009	.18	.37	.55
Cripple rafter, 2" x 6", 16" O.C.	.560	L.F.	.022	.36	.92	1.28
Headers, 2" x 6", doubled	.670	L.F.	.030	.44	1.26	1.70
Ceiling joist, 2" x 4", 16" O.C.	1.000	L.F.	.013	.41	.54	.95
Sheathing, exterior grade plywood, 1/2" thick	3.610	S.F.	.041	2.45	1.73	4.18
TOTAL		S.F.	.242	6.91	10.21	17.12
2" X 8", 16" O.C.						
Dormer rafter, 2" x 8", 16" O.C.	1.330	L.F.	.039	1.25	1.66	2.91
Ridge board, 2" x 8"	.280	L.F.	.010	.26	.42	.68
Trimmer rafter, 2" x 8"	.880	L.F.	.015	.83	.62	1.45
Wall studs & plates, 2" x 4", 16" O.C.	3.160	L.F.	.056	1.30	2.37	3.67
Fascia, 2" x 8"	.220	L.F.	.016	.21	.66	.87
Valley rafter, 2" x 8", 16" O.C.	.280	L.F.	.010	.26	.40	.66
Cripple rafter, 2" x 8", 16" O.C.	.560	L.F.	.027	.53	1.13	1.66
Headers, 2" x 8", doubled	.670	L.F.	.032	.63	1.33	1.96
Ceiling joist, 2" x 4", 16" O.C.	1.000	L.F.	.013	.41	.54	.95
Sheathing,, exterior grade plywood, 1/2" thick	3.610	S.F.	.041	2.45	1.73	4.18
TOTAL		S.F.	.259	8.13	10.86	18.99

The cost in this system is based on the square foot of plan area.
The measurement being the plan area of the dormer only.

Description	QUAN.	UNIT	LABOR HOURS	COST PER S.F. MAT.	COST PER S.F. INST.	COST PER S.F. TOTAL

 Important: See the Reference Section for critical supporting data - Reference Nos., Crews & Location Factors

Gable Dormer Framing Price Sheet	QUAN.	UNIT	LABOR HOURS	COST PER S.F.		
				MAT.	INST.	TOTAL
Dormer rafters, #2 or better, 2″ x 4″, 16″ O.C.	1.330	L.F.	.029	.69	1.22	1.91
24″ O.C.	1.060	L.F.	.023	.55	.98	1.53
2″ x 6″, 16″ O.C.	1.330	L.F.	.036	.86	1.53	2.39
24″ O.C.	1.060	L.F.	.029	.69	1.22	1.91
2″ x 8″, 16″ O.C.	1.330	L.F.	.039	1.25	1.66	2.91
24″ O.C.	1.060	L.F.	.031	1	1.33	2.33
Ridge board, #2 or better, 1″ x 4″	.280	L.F.	.006	.17	.25	.42
1″ x 6″	.280	L.F.	.007	.22	.32	.54
1″ x 8″	.280	L.F.	.008	.27	.34	.61
2″ x 4″	.280	L.F.	.007	.15	.30	.45
2″ x 6″	.280	L.F.	.009	.18	.38	.56
2″ x 8″	.280	L.F.	.010	.26	.42	.68
Trimmer rafters, #2 or better, 2″ x 4″	.880	L.F.	.011	.46	.48	.94
2″ x 6″	.880	L.F.	.014	.57	.60	1.17
2″ x 8″	.880	L.F.	.015	.83	.62	1.45
2″ x 10″	.880	L.F.	.022	1.25	.94	2.19
Wall studs & plates, #2 or better, 2″ x 4″ studs, 16″ O.C.	3.160	L.F.	.056	1.30	2.37	3.67
24″ O.C.	2.800	L.F.	.050	1.15	2.10	3.25
2″ x 6″ studs, 16″ O.C.	3.160	L.F.	.063	2.05	2.69	4.74
24″ O.C.	2.800	L.F.	.056	1.82	2.38	4.20
Fascia, #2 or better, 1″ x 4″	.220	L.F.	.006	.08	.27	.35
1″ x 6″	.220	L.F.	.008	.10	.33	.43
1″ x 8″	.220	L.F.	.009	.12	.38	.50
2″ x 4″	.220	L.F.	.011	.14	.45	.59
2″ x 6″	.220	L.F.	.014	.18	.57	.75
2″ x 8″	.220	L.F.	.016	.21	.66	.87
Valley rafter, #2 or better, 2″ x 4″	.280	L.F.	.007	.15	.30	.45
2″ x 6″	.280	L.F.	.009	.18	.37	.55
2″ x 8″	.280	L.F.	.010	.26	.40	.66
2″ x 10″	.280	L.F.	.012	.40	.50	.90
Cripple rafter, #2 or better, 2″ x 4″, 16″ O.C.	.560	L.F.	.018	.29	.74	1.03
24″ O.C.	.450	L.F.	.014	.23	.59	.82
2″ x 6″, 16″ O.C.	.560	L.F.	.022	.36	.92	1.28
24″ O.C.	.450	L.F.	.018	.29	.74	1.03
2″ x 8″, 16″ O.C.	.560	L.F.	.027	.53	1.13	1.66
24″ O.C.	.450	L.F.	.021	.42	.91	1.33
Headers, #2 or better double header, 2″ x 4″	.670	L.F.	.024	.35	1.02	1.37
2″ x 6″	.670	L.F.	.030	.44	1.26	1.70
2″ x 8″	.670	L.F.	.032	.63	1.33	1.96
2″ x 10″	.670	L.F.	.034	.95	1.41	2.36
Ceiling joist, #2 or better, 2″ x 4″, 16″ O.C.	1.000	L.F.	.013	.41	.54	.95
24″ O.C.	.800	L.F.	.010	.33	.43	.76
2″ x 6″, 16″ O.C.	1.000	L.F.	.013	.65	.54	1.19
24″ O.C.	.800	L.F.	.010	.52	.43	.95
Sheathing, plywood exterior grade, 3/8″ thick	3.610	S.F.	.038	1.81	1.59	3.40
1/2″ thick	3.610	S.F.	.041	2.45	1.73	4.18
5/8″ thick	3.610	S.F.	.044	2.64	1.88	4.52
3/4″ thick	3.610	S.F.	.048	3.25	2.02	5.27
Boards, 1″ x 6″, laid regular	3.610	S.F.	.089	4.77	3.75	8.52
Laid diagonal	3.610	S.F.	.099	4.77	4.19	8.96
1″ x 8″, laid regular	3.610	S.F.	.076	4.44	3.18	7.62
Laid diagonal	3.610	S.F.	.089	4.44	3.75	8.19

44 Shed Dormer Framing Systems

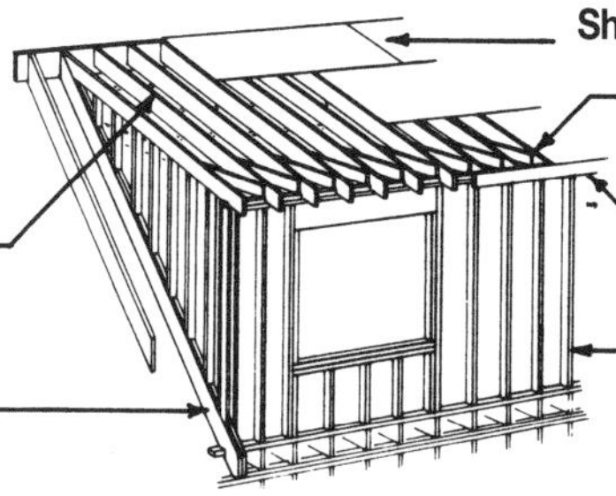

Sheathing
Ceiling Joists
Fascia Board
Rafters
Studs & Plates
Trimmer Rafters

System Description	QUAN.	UNIT	LABOR HOURS	COST PER S.F.		
				MAT.	INST.	TOTAL
2″ X 6″ RAFTERS, 16″ O.C.						
Dormer rafter, 2″ x 6″, 16″ O.C.	1.080	L.F.	.029	.70	1.24	1.94
Trimmer rafter, 2″ x 6″	.400	L.F.	.006	.26	.27	.53
Studs & plates, 2″ x 4″, 16″ O.C.	2.750	L.F.	.049	1.13	2.06	3.19
Fascia, 2″ x 6″	.250	L.F.	.014	.18	.57	.75
Ceiling joist, 2″ x 4″, 16″ O.C.	1.000	L.F.	.013	.41	.54	.95
Sheathing, exterior grade plywood, CDX, 1/2″ thick	2.940	S.F.	.034	2	1.41	3.41
TOTAL		S.F.	.145	4.68	6.09	10.77
2″ X 8″ RAFTERS, 16″ O.C.						
Dormer rafter, 2″ x 8″, 16″ O.C.	1.080	L.F.	.032	1.02	1.35	2.37
Trimmer rafter, 2″ x 8″	.400	L.F.	.007	.38	.28	.66
Studs & plates, 2″ x 4″, 16″ O.C.	2.750	L.F.	.049	1.13	2.06	3.19
Fascia, 2″ x 8″	.250	L.F.	.018	.24	.75	.99
Ceiling joist, 2″ x 6″, 16″ O.C.	1.000	L.F.	.013	.65	.54	1.19
Sheathing, exterior grade plywood, CDX, 1/2″ thick	2.940	S.F.	.034	2	1.41	3.41
TOTAL		S.F.	.153	5.42	6.39	11.81
2″ X 10″ RAFTERS, 16″ O.C.						
Dormer rafter, 2″ x 10″, 16″ O.C.	1.080	L.F.	.041	1.53	1.72	3.25
Trimmer rafter, 2″ x 10″	.400	L.F.	.010	.57	.43	1
Studs & plates, 2″ x 4″, 16″ O.C.	2.750	L.F.	.049	1.13	2.06	3.19
Fascia, 2″ x 10″	.250	L.F.	.022	.36	.94	1.30
Ceiling joist, 2″ x 6″, 16″ O.C.	1.000	L.F.	.013	.65	.54	1.19
Sheathing, exterior grade plywood, CDX, 1/2″ thick	2.940	S.F.	.034	2	1.41	3.41
TOTAL		S.F.	.169	6.24	7.10	13.34

The cost in this system is based on the square foot of plan area.
The measurement is the plan area of the dormer only.

Description	QUAN.	UNIT	LABOR HOURS	COST PER S.F.		
				MAT.	INST.	TOTAL

Shed Dormer Framing Price Sheet	QUAN.	UNIT	LABOR HOURS	COST PER S.F.		
				MAT.	INST.	TOTAL
Dormer rafters, #2 or better, 2″ x 4″, 16″ O.C.	1.080	L.F.	.023	.56	.99	1.55
24″ O.C.	.860	L.F.	.019	.45	.79	1.24
2″ x 6″, 16″ O.C.	1.080	L.F.	.029	.70	1.24	1.94
24″ O.C.	.860	L.F.	.023	.56	.99	1.55
2″ x 8″, 16″ O.C.	1.080	L.F.	.032	1.02	1.35	2.37
24″ O.C.	.860	L.F.	.025	.81	1.08	1.89
2″ x 10″, 16″ O.C.	1.080	L.F.	.041	1.53	1.72	3.25
24″ O.C.	.860	L.F.	.032	1.22	1.37	2.59
Trimmer rafter, #2 or better, 2″ x 4″	.400	L.F.	.005	.21	.22	.43
2″ x 6″	.400	L.F.	.006	.26	.27	.53
2″ x 8″	.400	L.F.	.007	.38	.28	.66
2″ x 10″	.400	L.F.	.010	.57	.43	1
Studs & plates, #2 or better, 2″ x 4″, 16″ O.C.	2.750	L.F.	.049	1.13	2.06	3.19
24″ O.C.	2.200	L.F.	.039	.90	1.65	2.55
2″ x 6″, 16″ O.C.	2.750	L.F.	.055	1.79	2.34	4.13
24″ O.C.	2.200	L.F.	.044	1.43	1.87	3.30
Fascia, #2 or better, 1″ x 4″	.250	L.F.	.006	.08	.27	.35
1″ x 6″	.250	L.F.	.008	.10	.33	.43
1″ x 8″	.250	L.F.	.009	.12	.38	.50
2″ x 4″	.250	L.F.	.011	.14	.45	.59
2″ x 6″	.250	L.F.	.014	.18	.57	.75
2″ x 8″	.250	L.F.	.018	.24	.75	.99
Ceiling joist, #2 or better, 2″ x 4″, 16″ O.C.	1.000	L.F.	.013	.41	.54	.95
24″ O.C.	.800	L.F.	.010	.33	.43	.76
2″ x 6″, 16″ O.C.	1.000	L.F.	.013	.65	.54	1.19
24″ O.C.	.800	L.F.	.010	.52	.43	.95
2″ x 8″, 16″ O.C.	1.000	L.F.	.015	.94	.61	1.55
24″ O.C.	.800	L.F.	.012	.75	.49	1.24
Sheathing, plywood exterior grade, 3/8″ thick	2.940	S.F.	.031	1.47	1.29	2.76
1/2″ thick	2.940	S.F.	.034	2	1.41	3.41
5/8″ thick	2.940	S.F.	.036	2.15	1.53	3.68
3/4″ thick	2.940	S.F.	.039	2.65	1.65	4.30
Boards, 1″ x 6″, laid regular	2.940	S.F.	.072	3.88	3.06	6.94
Laid diagonal	2.940	S.F.	.080	3.88	3.41	7.29
1″ x 8″, laid regular	2.940	S.F.	.062	3.62	2.59	6.21
Laid diagonal	2.940	S.F.	.072	3.62	3.06	6.68

Window Openings	QUAN.	UNIT	LABOR HOURS	COST EACH		
				MAT.	INST.	TOTAL
The following are to be added to the total cost of the dormers for window openings. Do not subtract window area from the stud wall quantities.						
Headers, 2″ x 6″ doubled, 2′ long	4.000	L.F.	.178	2.60	7.50	10.10
3′ long	6.000	L.F.	.267	3.90	11.30	15.20
4′ long	8.000	L.F.	.356	5.20	15.05	20.25
5′ long	10.000	L.F.	.444	6.50	18.80	25.30
2″ x 8″ doubled, 4′ long	8.000	L.F.	.376	7.50	15.90	23.40
5′ long	10.000	L.F.	.471	9.40	19.90	29.30
6′ long	12.000	L.F.	.565	11.30	24	35.30
8′ long	16.000	L.F.	.753	15.05	32	47.05
2″ x 10″ doubled, 4′ long	8.000	L.F.	.400	11.35	16.90	28.25
6′ long	12.000	L.F.	.600	17.05	25.50	42.55
8′ long	16.000	L.F.	.800	22.50	34	56.50
10′ long	20.000	L.F.	1.000	28.50	42	70.50

48 | Partition Framing Systems

Bracing
Top Plates
Studs
Bottom Plate

System Description	QUAN.	UNIT	LABOR HOURS	COST PER S.F. MAT.	INST.	TOTAL
2" X 4", 16" O.C.						
2" x 4" studs, #2 or better, 16" O.C.	1.000	L.F.	.015	.41	.61	1.02
Plates, double top, single bottom	.375	L.F.	.005	.15	.23	.38
Cross bracing, let-in, 1" x 6"	.080	L.F.	.004	.05	.18	.23
TOTAL		S.F.	.024	.61	1.02	1.63
2" X 4", 24" O.C.						
2" x 4" studs, #2 or better, 24" O.C.	.800	L.F.	.012	.33	.49	.82
Plates, double top, single bottom	.375	L.F.	.005	.15	.23	.38
Cross bracing, let-in, 1" x 6"	.080	L.F.	.003	.05	.12	.17
TOTAL		S.F.	.020	.53	.84	1.37
2" X 6", 16" O.C.						
2" x 6" studs, #2 or better, 16" O.C.	1.000	L.F.	.016	.65	.68	1.33
Plates, double top, single bottom	.375	L.F.	.006	.24	.26	.50
Cross bracing, let-in, 1" x 6"	.080	L.F.	.004	.05	.18	.23
TOTAL		S.F.	.026	.94	1.12	2.06
2" X 6", 24" O.C.						
2" x 6" studs, #2 or better, 24" O.C.	.800	L.F.	.013	.52	.54	1.06
Plates, double top, single bottom	.375	L.F.	.006	.24	.26	.50
Cross bracing, let-in, 1" x 6"	.080	L.F.	.003	.05	.12	.17
TOTAL		S.F.	.022	.81	.92	1.73

The costs in this system are based on a square foot of wall area. Do not subtract for door or window openings.

Description	QUAN.	UNIT	LABOR HOURS	COST PER S.F. MAT.	INST.	TOTAL

 Important: See the Reference Section for critical supporting data - Reference Nos., Crews & Location Factors

Partition Framing Price Sheet	QUAN.	UNIT	LABOR HOURS	COST PER S.F.		
				MAT.	INST.	TOTAL
Wood studs, #2 or better, 2" x 4", 12" O.C.	1.250	L.F.	.018	.51	.76	1.27
16" O.C.	1.000	L.F.	.015	.41	.61	1.02
24" O.C.	.800	L.F.	.012	.33	.49	.82
32" O.C.	.650	L.F.	.009	.27	.40	.67
2" x 6", 12" O.C.	1.250	L.F.	.020	.81	.85	1.66
16" O.C.	1.000	L.F.	.016	.65	.68	1.33
24" O.C.	.800	L.F.	.013	.52	.54	1.06
32" O.C.	.650	L.F.	.010	.42	.44	.86
Plates, #2 or better double top single bottom, 2" x 4"	.375	L.F.	.005	.15	.23	.38
2" x 6"	.375	L.F.	.006	.24	.26	.50
2" x 8"	.375	L.F.	.005	.35	.23	.58
Cross bracing, let-in, 1" x 6" boards studs, 12" O.C.	.080	L.F.	.005	.06	.23	.29
16" O.C.	.080	L.F.	.004	.05	.18	.23
24" O.C.	.080	L.F.	.003	.05	.12	.17
32" O.C.	.080	L.F.	.002	.04	.09	.13
Let-in steel (T shaped) studs, 12" O.C.	.080	L.F.	.001	.06	.06	.12
16" O.C.	.080	L.F.	.001	.05	.05	.10
24" O.C.	.080	L.F.	.001	.05	.04	.09
32" O.C.	.080	L.F.	.001	.04	.04	.08
Steel straps studs, 12" O.C.	.080	L.F.	.001	.07	.05	.12
16" O.C.	.080	L.F.	.001	.07	.04	.11
24" O.C.	.080	L.F.	.001	.07	.04	.11
32" O.C.	.080	L.F.	.001	.07	.04	.11
Metal studs, load bearing 24" O.C., 20 ga. galv., 2-1/2" wide	1.000	S.F.	.015	.58	.63	1.21
3-5/8" wide	1.000	S.F.	.015	.69	.65	1.34
4" wide	1.000	S.F.	.016	.73	.66	1.39
6" wide	1.000	S.F.	.016	.92	.67	1.59
16 ga., 2-1/2" wide	1.000	S.F.	.017	.67	.72	1.39
3-5/8" wide	1.000	S.F.	.017	.80	.74	1.54
4" wide	1.000	S.F.	.018	.85	.75	1.60
6" wide	1.000	S.F.	.018	1.07	.77	1.84
Non-load bearing 24" O.C., 25 ga. galv., 1-5/8" wide	1.000	S.F.	.011	.18	.44	.62
2-1/2" wide	1.000	S.F.	.011	.22	.45	.67
3-5/8" wide	1.000	S.F.	.011	.25	.46	.71
4" wide	1.000	S.F.	.011	.27	.46	.73
6" wide	1.000	S.F.	.011	.39	.47	.86
20 ga., 2-1/2" wide	1.000	S.F.	.013	.35	.56	.91
3-5/8" wide	1.000	S.F.	.014	.42	.57	.99
4" wide	1.000	S.F.	.014	.46	.57	1.03
6" wide	1.000	S.F.	.014	.59	.58	1.17

Window & Door Openings	QUAN.	UNIT	LABOR HOURS	COST EACH		
				MAT.	INST.	TOTAL
The following costs are to be added to the total costs of the walls. Do not subtract openings from total wall area.						
Headers, 2" x 6" double, 2' long	4.000	L.F.	.178	2.60	7.50	10.10
3' long	6.000	L.F.	.267	3.90	11.30	15.20
4' long	8.000	L.F.	.356	5.20	15.05	20.25
5' long	10.000	L.F.	.444	6.50	18.80	25.30
2" x 8" double, 4' long	8.000	L.F.	.376	7.50	15.90	23.40
5' long	10.000	L.F.	.471	9.40	19.90	29.30
6' long	12.000	L.F.	.565	11.30	24	35.30
8' long	16.000	L.F.	.753	15.05	32	47.05
2" x 10" double, 4' long	8.000	L.F.	.400	11.35	16.90	28.25
6' long	12.000	L.F.	.600	17.05	25.50	42.55
8' long	16.000	L.F.	.800	22.50	34	56.50
10' long	20.000	L.F.	1.000	28.50	42	70.50
2" x 12" double, 8' long	16.000	L.F.	.853	28	36	64
12' long	24.000	L.F.	1.280	42	54	96

Division 4
Exterior Walls

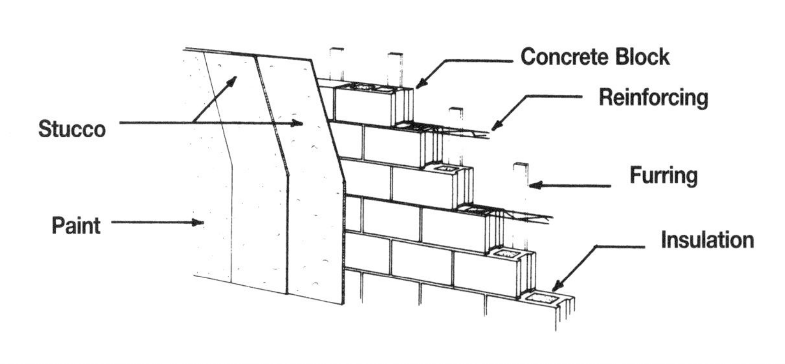

System Description	QUAN.	UNIT	LABOR HOURS	COST PER S.F.		
				MAT.	INST.	TOTAL
6″ THICK CONCRETE BLOCK WALL						
6″ thick concrete block, 6″ x 8″ x 16″	1.000	S.F.	.100	1.86	3.85	5.71
Masonry reinforcing, truss strips every other course	.625	L.F.	.002	.12	.07	.19
Furring, 1″ x 3″, 16″ O.C.	1.000	L.F.	.016	.31	.68	.99
Masonry insulation, poured vermiculite	1.000	S.F.	.013	.62	.56	1.18
Stucco, 2 coats	1.000	S.F.	.069	.20	2.63	2.83
Masonry paint, 2 coats	1.000	S.F.	.016	.21	.59	.80
TOTAL		S.F.	.216	3.32	8.38	11.70
8″ THICK CONCRETE BLOCK WALL						
8″ thick concrete block, 8″ x 8″ x 16″	1.000	S.F.	.107	2.04	4.10	6.14
Masonry reinforcing, truss strips every other course	.625	L.F.	.002	.12	.07	.19
Furring, 1″ x 3″, 16″ O.C.	1.000	L.F.	.016	.31	.68	.99
Masonry insulation, poured vermiculite	1.000	S.F.	.018	.82	.74	1.56
Stucco, 2 coats	1.000	S.F.	.069	.20	2.63	2.83
Masonry paint, 2 coats	1.000	S.F.	.016	.21	.59	.80
TOTAL		S.F.	.228	3.70	8.81	12.51
12″ THICK CONCRETE BLOCK WALL						
12″ thick concrete block, 12″ x 8″ x 16″	1.000	S.F.	.141	2.97	5.30	8.27
Masonry reinforcing, truss strips every other course	.625	L.F.	.003	.14	.11	.25
Furring, 1″ x 3″, 16″ O.C.	1.000	L.F.	.016	.31	.68	.99
Masonry insulation, poured vermiculite	1.000	S.F.	.026	1.22	1.10	2.32
Stucco, 2 coats	1.000	S.F.	.069	.20	2.63	2.83
Masonry paint, 2 coats	1.000	S.F.	.016	.21	.59	.80
TOTAL		S.F.	.271	5.05	10.41	15.46

Costs for this system are based on a square foot of wall area. Do not subtract for window openings.

Description	QUAN.	UNIT	LABOR HOURS	COST PER S.F.		
				MAT.	INST.	TOTAL

Important: See the Reference Section for critical supporting data - Reference Nos., Crews & Location Factors

Masonry Block Price Sheet	QUAN.	UNIT	LABOR HOURS	COST PER S.F.		
				MAT.	INST.	TOTAL
Block concrete, 8″ x 16″ regular, 4″ thick	1.000	S.F.	.093	1.25	3.58	4.83
6″ thick	1.000	S.F.	.100	1.86	3.85	5.71
8″ thick	1.000	S.F.	.107	2.04	4.10	6.14
10″ thick	1.000	S.F.	.111	2.58	4.27	6.85
12″ thick	1.000	S.F.	.141	2.97	5.30	8.27
Solid block, 4″ thick	1.000	S.F.	.096	1.56	3.71	5.27
6″ thick	1.000	S.F.	.104	1.93	4	5.93
8″ thick	1.000	S.F.	.111	2.72	4.27	6.99
10″ thick	1.000	S.F.	.133	3.75	5	8.75
12″ thick	1.000	S.F.	.148	4.17	5.55	9.72
Lightweight, 4″ thick	1.000	S.F.	.093	1.25	3.58	4.83
6″ thick	1.000	S.F.	.100	1.86	3.85	5.71
8″ thick	1.000	S.F.	.107	2.04	4.10	6.14
10″ thick	1.000	S.F.	.111	2.58	4.27	6.85
12″ thick	1.000	S.F.	.141	2.97	5.30	8.27
Split rib profile, 4″ thick	1.000	S.F.	.116	2.72	4.46	7.18
6″ thick	1.000	S.F.	.123	3.15	4.73	7.88
8″ thick	1.000	S.F.	.131	3.64	5.15	8.79
10″ thick	1.000	S.F.	.157	3.90	5.90	9.80
12″ thick	1.000	S.F.	.175	4.33	6.55	10.88
Masonry reinforcing, wire truss strips, every course, 8″ block	1.375	L.F.	.004	.26	.15	.41
12″ block	1.375	L.F.	.006	.30	.23	.53
Every other course, 8″ block	.625	L.F.	.002	.12	.07	.19
12″ block	.625	L.F.	.003	.14	.11	.25
Furring, wood, 1″ x 3″, 12″ O.C.	1.250	L.F.	.020	.39	.85	1.24
16″ O.C.	1.000	L.F.	.016	.31	.68	.99
24″ O.C.	.800	L.F.	.013	.25	.54	.79
32″ O.C.	.640	L.F.	.010	.20	.44	.64
Steel, 3/4″ channels, 12″ O.C.	1.250	L.F.	.034	.25	1.27	1.52
16″ O.C.	1.000	L.F.	.030	.23	1.13	1.36
24″ O.C.	.800	L.F.	.023	.15	.85	1
32″ O.C.	.640	L.F.	.018	.12	.68	.80
Masonry insulation, vermiculite or perlite poured 4″ thick	1.000	S.F.	.009	.40	.36	.76
6″ thick	1.000	S.F.	.013	.62	.56	1.18
8″ thick	1.000	S.F.	.018	.82	.74	1.56
10″ thick	1.000	S.F.	.021	1	.90	1.90
12″ thick	1.000	S.F.	.026	1.22	1.10	2.32
Block inserts polystyrene, 6″ thick	1.000	S.F.		1.12		1.12
8″ thick	1.000	S.F.		1.12		1.12
10″ thick	1.000	S.F.		1.32		1.32
12″ thick	1.000	S.F.		1.39		1.39
Stucco, 1 coat	1.000	S.F.	.057	.16	2.16	2.32
2 coats	1.000	S.F.	.069	.20	2.63	2.83
3 coats	1.000	S.F.	.081	.23	3.09	3.32
Painting, 1 coat	1.000	S.F.	.011	.13	.41	.54
2 coats	1.000	S.F.	.016	.21	.59	.80
Primer & 1 coat	1.000	S.F.	.013	.21	.48	.69
2 coats	1.000	S.F.	.018	.29	.67	.96
Lath, metal lath expanded 2.5 lb/S.Y., painted	1.000	S.F.	.010	.31	.39	.70
Galvanized	1.000	S.F.	.012	.34	.43	.77

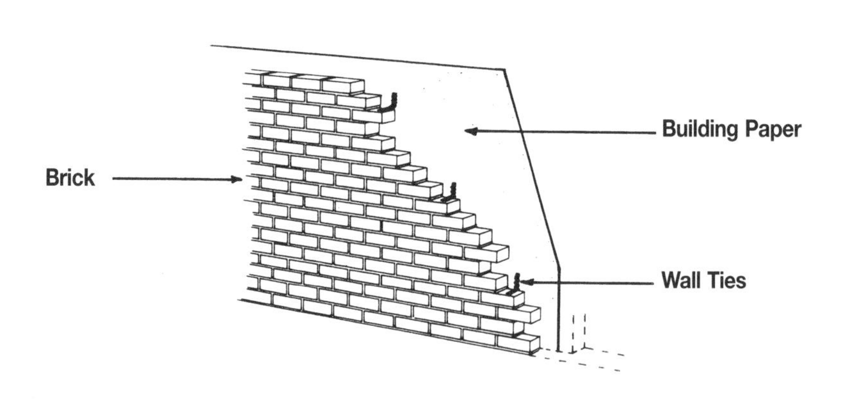

System Description	QUAN.	UNIT	LABOR HOURS	COST PER S.F.		
				MAT.	INST.	TOTAL
SELECT COMMON BRICK						
Brick, select common, running bond	1.000	S.F.	.174	3.61	6.70	10.31
Wall ties, 7/8" x 7", 22 gauge	1.000	Ea.	.008	.09	.33	.42
Building paper, spunbonded polypropylene	1.100	S.F.	.002	.13	.09	.22
Trim, pine, painted	.125	L.F.	.004	.10	.17	.27
TOTAL		S.F.	.188	3.93	7.29	11.22
RED FACED COMMON BRICK						
Brick, common, red faced, running bond	1.000	S.F.	.182	3.61	7	10.61
Wall ties, 7/8" x 7", 22 gauge	1.000	Ea.	.008	.09	.33	.42
Building paper, spundbonded polypropylene	1.100	S.F.	.002	.13	.09	.22
Trim, pine, painted	.125	L.F.	.004	.10	.17	.27
TOTAL		S.F.	.196	3.93	7.59	11.52
BUFF OR GREY FACE BRICK						
Brick, buff or grey	1.000	S.F.	.182	3.82	7	10.82
Wall ties, 7/8" x 7", 22 gauge	1.000	Ea.	.008	.09	.33	.42
Building paper, spundbonded polypropylene	1.100	S.F.	.002	.13	.09	.22
Trim, pine, painted	.125	L.F.	.004	.10	.17	.27
TOTAL		S.F.	.196	4.14	7.59	11.73
STONE WORK, ROUGH STONE, AVERAGE						
Field stone veneer	1.000	S.F.	.223	6.13	8.58	14.71
Wall ties, 7/8" x 7", 22 gauge	1.000	Ea.	.008	.09	.33	.42
Building paper, spundbonded polypropylene	1.000	S.F.	.002	.13	.09	.22
Trim, pine, painted	.125	L.F.	.004	.10	.17	.27
TOTAL		S.F.	.237	6.45	9.17	15.62

The costs in this system are based on a square foot of wall area. Do not subtract area for window & door openings.

Description	QUAN.	UNIT	LABOR HOURS	COST PER S.F.		
				MAT.	INST.	TOTAL

Important: See the Reference Section for critical supporting data - Reference Nos., Crews & Location Factors

Brick/Stone Veneer Price Sheet	QUAN.	UNIT	LABOR HOURS	COST PER S.F.		
				MAT.	INST.	TOTAL
Brick						
Select common, running bond	1.000	S.F.	.174	3.61	6.70	10.31
Red faced, running bond	1.000	S.F.	.182	3.61	7	10.61
Buff or grey faced, running bond	1.000	S.F.	.182	3.82	7	10.82
Header every 6th course	1.000	S.F.	.216	4.20	8.30	12.50
English bond	1.000	S.F.	.286	5.40	11	16.40
Flemish bond	1.000	S.F.	.195	3.81	7.50	11.31
Common bond	1.000	S.F.	.267	4.79	10.25	15.04
Stack bond	1.000	S.F.	.182	3.82	7	10.82
Jumbo, running bond	1.000	S.F.	.092	4.43	3.54	7.97
Norman, running bond	1.000	S.F.	.125	4.87	4.81	9.68
Norwegian, running bond	1.000	S.F.	.107	3.66	4.10	7.76
Economy, running bond	1.000	S.F.	.129	4.33	4.96	9.29
Engineer, running bond	1.000	S.F.	.154	3.67	5.90	9.57
Roman, running bond	1.000	S.F.	.160	5.60	6.15	11.75
Utility, running bond	1.000	S.F.	.089	4.12	3.42	7.54
Glazed, running bond	1.000	S.F.	.190	9.95	7.35	17.30
Stone work, rough stone, average	1.000	S.F.	.179	6.15	8.60	14.75
Maximum	1.000	S.F.	.267	9.15	12.80	21.95
Wall ties, galvanized, corrugated 7/8" x 7", 22 gauge	1.000	Ea.	.008	.09	.33	.42
16 gauge	1.000	Ea.	.008	.20	.33	.53
Cavity wall, every 3rd course 6" long Z type, 1/4" diameter	1.330	L.F.	.010	.47	.42	.89
3/16" diameter	1.330	L.F.	.010	.36	.42	.78
8" long, Z type, 1/4" diameter	1.330	L.F.	.010	1.12	.42	1.54
3/16" diameter	1.330	L.F.	.010	.41	.42	.83
Building paper, aluminum and kraft laminated foil, 1 side	1.000	S.F.	.002	.05	.09	.14
2 sides	1.000	S.F.	.002	.09	.09	.18
#15 asphalt paper	1.100	S.F.	.002	.04	.10	.14
Polyethylene, .002" thick	1.000	S.F.	.002	.01	.09	.10
.004" thick	1.000	S.F.	.002	.03	.09	.12
.006" thick	1.000	S.F.	.002	.04	.09	.13
.010" thick	1.000	S.F.	.002	.06	.09	.15
Trim, 1" x 4", cedar	.125	L.F.	.005	.23	.21	.44
Fir	.125	L.F.	.005	.11	.21	.32
Redwood	.125	L.F.	.005	.23	.21	.44
White pine	.125	L.F.	.005	.11	.21	.32

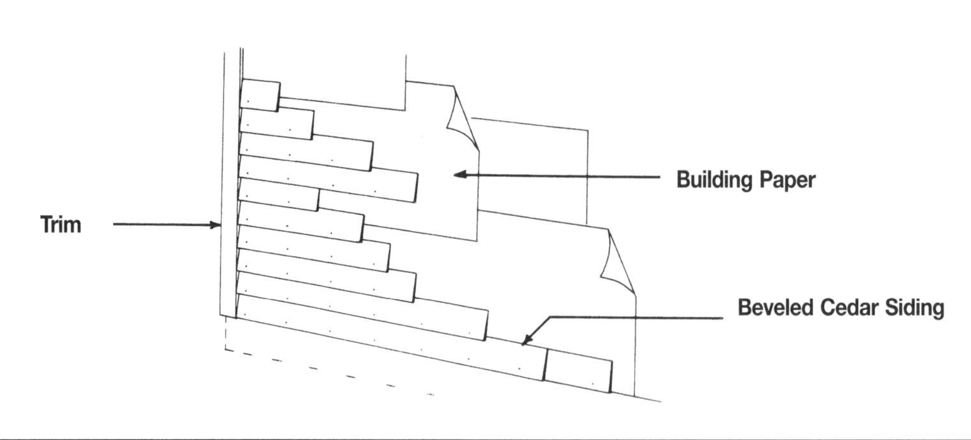

System Description	QUAN.	UNIT	LABOR HOURS	COST PER S.F.		
				MAT.	INST.	TOTAL
1/2" X 6" BEVELED CEDAR SIDING, "A" GRADE						
1/2" x 6" beveled cedar siding	1.000	S.F.	.032	3.52	1.35	4.87
Building wrap, spundbonded polypropylene	1.100	S.F.	.002	.13	.09	.22
Trim, cedar	.125	L.F.	.005	.23	.21	.44
Paint, primer & 2 coats	1.000	S.F.	.017	.20	.64	.84
TOTAL		S.F.	.056	4.08	2.29	6.37
1/2" X 8" BEVELED CEDAR SIDING, "A" GRADE						
1/2" x 8" beveled cedar siding	1.000	S.F.	.029	3.42	1.23	4.65
Building wrap, spundbonded polypropylene	1.100	S.F.	.002	.13	.09	.22
Trim, cedar	.125	L.F.	.005	.23	.21	.44
Paint, primer & 2 coats	1.000	S.F.	.017	.20	.64	.84
TOTAL		S.F.	.053	3.98	2.17	6.15
1" X 4" TONGUE & GROOVE, REDWOOD, VERTICAL GRAIN						
Redwood, clear, vertical grain, 1" x 10"	1.000	S.F.	.018	3.62	.78	4.40
Building wrap, spunbonded polypropylene	1.100	S.F.	.002	.13	.09	.22
Trim, redwood	.125	L.F.	.005	.23	.21	.44
Sealer, 1 coat, stain, 1 coat	1.000	S.F.	.013	.13	.49	.62
TOTAL		S.F.	.038	4.11	1.57	5.68
1" X 6" TONGUE & GROOVE, REDWOOD, VERTICAL GRAIN						
Redwood, clear, vertical grain, 1" x 10"	1.000	S.F.	.019	3.73	.80	4.53
Building wrap, spunbonded polypropylene	1.100	S.F.	.002	.13	.09	.22
Trim, redwood	.125	L.F.	.005	.23	.21	.44
Sealer, 1 coat, stain, 1 coat	1.000	S.F.	.013	.13	.49	.62
TOTAL		S.F.	.039	4.22	1.59	5.81

The costs in this system are based on a square foot of wall area.
Do not subtract area for door or window openings.

Description	QUAN.	UNIT	LABOR HOURS	COST PER S.F.		
				MAT.	INST.	TOTAL

Important: See the Reference Section for critical supporting data - Reference Nos., Crews & Location Factors

Wood Siding Price Sheet	QUAN.	UNIT	LABOR HOURS	COST PER S.F.		
				MAT.	INST.	TOTAL
Siding, beveled cedar, "A" grade, 1/2" x 6"	1.000	S.F.	.028	3.52	1.35	4.87
1/2" x 8"	1.000	S.F.	.023	3.42	1.23	4.65
"B" grade, 1/2" x 6"	1.000	S.F.	.032	3.91	1.50	5.41
1/2" x 8"	1.000	S.F.	.029	3.80	1.37	5.17
Clear grade, 1/2" x 6"	1.000	S.F.	.028	4.40	1.69	6.09
1/2" x 8"	1.000	S.F.	.023	4.28	1.54	5.82
Redwood, clear vertical grain, 1/2" x 6"	1.000	S.F.	.036	3.15	1.50	4.65
1/2" x 8"	1.000	S.F.	.032	2.54	1.35	3.89
Clear all heart vertical grain, 1/2" x 6"	1.000	S.F.	.028	3.50	1.67	5.17
1/2" x 8"	1.000	S.F.	.023	2.82	1.50	4.32
Siding board & batten, cedar, "B" grade, 1" x 10"	1.000	S.F.	.031	2.42	1.30	3.72
1" x 12"	1.000	S.F.	.031	2.42	1.30	3.72
Redwood, clear vertical grain, 1" x 6"	1.000	S.F.	.043	3.01	2.05	5.06
1" x 8"	1.000	S.F.	.018	2.75	1.80	4.55
White pine, #2 & better, 1" x 10"	1.000	S.F.	.029	.80	1.23	2.03
1" x 12"	1.000	S.F.	.029	.80	1.23	2.03
Siding vertical, tongue & groove, cedar "B" grade, 1" x 4"	1.000	S.F.	.033	2	.78	2.78
1" x 6"	1.000	S.F.	.024	2.06	.80	2.86
1" x 8"	1.000	S.F.	.024	2.12	.83	2.95
1" x 10"	1.000	S.F.	.021	2.18	.85	3.03
"A" grade, 1" x 4"	1.000	S.F.	.033	1.83	.71	2.54
1" x 6"	1.000	S.F.	.024	1.88	.73	2.61
1" x 8"	1.000	S.F.	.024	1.93	.75	2.68
1" x 10"	1.000	S.F.	.021	1.98	.77	2.75
Clear vertical grain, 1" x 4"	1.000	S.F.	.033	1.69	.66	2.35
1" x 6"	1.000	S.F.	.024	1.73	.67	2.40
1" x 8"	1.000	S.F.	.024	1.77	.69	2.46
1" x 10"	1.000	S.F.	.021	1.82	.71	2.53
Redwood, clear vertical grain, 1" x 4"	1.000	S.F.	.033	3.62	.78	4.40
1" x 6"	1.000	S.F.	.024	3.73	.80	4.53
1" x 8"	1.000	S.F.	.024	3.83	.83	4.66
1" x 10"	1.000	S.F.	.021	3.95	.85	4.80
Clear all heart vertical grain, 1" x 4"	1.000	S.F.	.033	3.32	.71	4.03
1" x 6"	1.000	S.F.	.024	3.41	.73	4.14
1" x 8"	1.000	S.F.	.024	3.50	.75	4.25
1" x 10"	1.000	S.F.	.021	3.59	.77	4.36
White pine, 1" x 10"	1.000	S.F.	.024	.77	.85	1.62
Siding plywood, texture 1-11 cedar, 3/8" thick	1.000	S.F.	.024	1.19	1	2.19
5/8" thick	1.000	S.F.	.024	2.62	1	3.62
Redwood, 3/8" thick	1.000	S.F.	.024	1.19	1	2.19
5/8" thick	1.000	S.F.	.024	1.98	1	2.98
Fir, 3/8" thick	1.000	S.F.	.024	.64	1	1.64
5/8" thick	1.000	S.F.	.024	1.11	1	2.11
Southern yellow pine, 3/8" thick	1.000	S.F.	.024	.64	1	1.64
5/8" thick	1.000	S.F.	.024	.89	1	1.89
Hard board, 7/16" thick primed, plain finish	1.000	S.F.	.025	1.24	1.04	2.28
Board finish	1.000	S.F.	.023	.92	.97	1.89
Polyvinyl coated, 3/8" thick	1.000	S.F.	.021	.97	.90	1.87
5/8" thick	1.000	S.F.	.024	.89	1	1.89
Paper, #15 asphalt felt	1.100	S.F.	.002	.04	.10	.14
Trim, cedar	.125	L.F.	.005	.23	.21	.44
Fir	.125	L.F.	.005	.11	.21	.32
Redwood	.125	L.F.	.005	.23	.21	.44
White pine	.125	L.F.	.005	.11	.21	.32
Painting, primer, & 1 coat	1.000	S.F.	.013	.13	.49	.62
2 coats	1.000	S.F.	.017	.20	.64	.84
Stain, sealer, & 1 coat	1.000	S.F.	.017	.10	.64	.74
2 coats	1.000	S.F.	.019	.15	.70	.85

12 Shingle Siding Systems

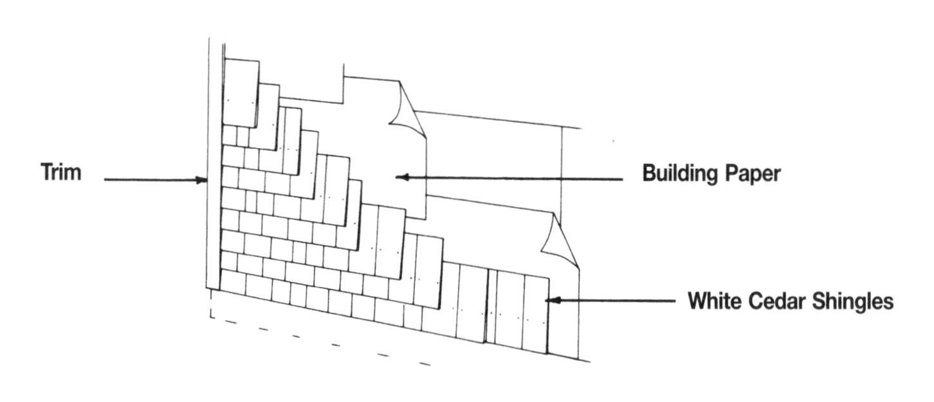

System Description	QUAN.	UNIT	LABOR HOURS	COST PER S.F.		
				MAT.	INST.	TOTAL
WHITE CEDAR SHINGLES, 5″ EXPOSURE						
White cedar shingles, 16″ long, grade "A", 5″ exposure	1.000	S.F.	.033	1.79	1.41	3.20
Building wrap, spunbonded polypropylene	1.100	S.F.	.002	.13	.09	.22
Trim, cedar	.125	S.F.	.005	.23	.21	.44
Paint, primer & 1 coat	1.000	S.F.	.017	.10	.64	.74
TOTAL		S.F.	.057	2.25	2.35	4.60
NO. 1 PERFECTIONS, 5-1/2″ EXPOSURE						
No. 1 perfections, red cedar, 5-1/2″ exposure	1.000	S.F.	.029	1.85	1.23	3.08
Building wrap, spunbonded polypropylene	1.100	S.F.	.002	.13	.09	.22
Trim, cedar	.125	S.F.	.005	.23	.21	.44
Stain, sealer & 1 coat	1.000	S.F.	.017	.10	.64	.74
TOTAL		S.F.	.053	2.31	2.17	4.48
RESQUARED & REBUTTED PERFECTIONS, 5-1/2″ EXPOSURE						
Resquared & rebutted perfections, 5-1/2″ exposure	1.000	S.F.	.027	2.31	1.13	3.44
Building wrap, spunbonded polypropylene	1.100	S.F.	.002	.13	.09	.22
Trim, cedar	.125	S.F.	.005	.23	.21	.44
Stain, sealer & 1 coat	1.000	S.F.	.017	.10	.64	.74
TOTAL		S.F.	.051	2.77	2.07	4.84
HAND-SPLIT SHAKES, 8-1/2″ EXPOSURE						
Hand-split red cedar shakes, 18″ long, 8-1/2″ exposure	1.000	S.F.	.040	.99	1.69	2.68
Building wrap, spunbonded polypropylene	1.100	S.F.	.002	.13	.09	.22
Trim, cedar	.125	S.F.	.005	.23	.21	.44
Stain, sealer & 1 coat	1.000	S.F.	.017	.10	.64	.74
TOTAL		S.F.	.064	1.45	2.63	4.08

The costs in this system are based on a square foot of wall area.
Do not subtract area for door or window openings.

Description	QUAN.	UNIT	LABOR HOURS	COST PER S.F.		
				MAT.	INST.	TOTAL

Important: See the Reference Section for critical supporting data - Reference Nos., Crews & Location Factors

Shingle Siding Price Sheet	QUAN.	UNIT	LABOR HOURS	COST PER S.F.		
				MAT.	INST.	TOTAL
Shingles wood, white cedar 16" long, "A" grade, 5" exposure	1.000	S.F.	.033	1.79	1.41	3.20
7" exposure	1.000	S.F.	.030	1.61	1.27	2.88
8-1/2" exposure	1.000	S.F.	.032	1.02	1.35	2.37
10" exposure	1.000	S.F.	.028	.90	1.18	2.08
"B" grade, 5" exposure	1.000	S.F.	.040	1.61	1.69	3.30
7" exposure	1.000	S.F.	.028	1.13	1.18	2.31
8-1/2" exposure	1.000	S.F.	.024	.97	1.01	1.98
10" exposure	1.000	S.F.	.020	.81	.85	1.66
Fire retardant, "A" grade, 5" exposure	1.000	S.F.	.033	2.18	1.41	3.59
7" exposure	1.000	S.F.	.028	1.40	1.18	2.58
8-1/2" exposure	1.000	S.F.	.032	1.22	1.35	2.57
10" exposure	1.000	S.F.	.025	.95	1.05	2
Fire retardant, 5" exposure	1.000	S.F.	.029	2.24	1.23	3.47
7" exposure	1.000	S.F.	.036	1.75	1.50	3.25
8-1/2" exposure	1.000	S.F.	.032	1.57	1.35	2.92
10" exposure	1.000	S.F.	.025	1.22	1.05	2.27
Resquared & rebutted, 5-1/2" exposure	1.000	S.F.	.027	2.31	1.13	3.44
7" exposure	1.000	S.F.	.024	2.08	1.02	3.10
8-1/2" exposure	1.000	S.F.	.021	1.85	.90	2.75
10" exposure	1.000	S.F.	.019	1.62	.79	2.41
Fire retardant, 5" exposure	1.000	S.F.	.027	2.70	1.13	3.83
7" exposure	1.000	S.F.	.024	2.43	1.02	3.45
8-1/2" exposure	1.000	S.F.	.021	2.16	.90	3.06
10" exposure	1.000	S.F.	.023	1.45	.97	2.42
Hand-split, red cedar, 24" long, 7" exposure	1.000	S.F.	.045	1.78	1.89	3.67
8-1/2" exposure	1.000	S.F.	.038	1.52	1.62	3.14
10" exposure	1.000	S.F.	.032	1.27	1.35	2.62
12" exposure	1.000	S.F.	.026	1.02	1.08	2.10
Fire retardant, 7" exposure	1.000	S.F.	.045	2.55	1.89	4.44
8-1/2" exposure	1.000	S.F.	.038	2.18	1.62	3.80
10" exposure	1.000	S.F.	.032	1.82	1.35	3.17
12" exposure	1.000	S.F.	.026	1.46	1.08	2.54
18" long, 5" exposure	1.000	S.F.	.068	1.67	2.87	4.54
7" exposure	1.000	S.F.	.048	1.18	2.03	3.21
8-1/2" exposure	1.000	S.F.	.040	.99	1.69	2.68
10" exposure	1.000	S.F.	.036	.89	1.52	2.41
Fire retardant, 5" exposure	1.000	S.F.	.068	2.51	2.87	5.38
7" exposure	1.000	S.F.	.048	1.77	2.03	3.80
8-1/2" exposure	1.000	S.F.	.040	1.49	1.69	3.18
10" exposure	1.000	S.F.	.036	1.34	1.52	2.86
Paper, #15 asphalt felt	1.100	S.F.	.002	.04	.09	.13
Trim, cedar	.125	S.F.	.005	.23	.21	.44
Fir	.125	S.F.	.005	.11	.21	.32
Redwood	.125	S.F.	.005	.23	.21	.44
White pine	.125	S.F.	.005	.11	.21	.32
Painting, primer, & 1 coat	1.000	S.F.	.013	.13	.49	.62
2 coats	1.000	S.F.	.017	.20	.64	.84
Staining, sealer, & 1 coat	1.000	S.F.	.017	.10	.64	.74
2 coats	1.000	S.F.	.019	.15	.70	.85

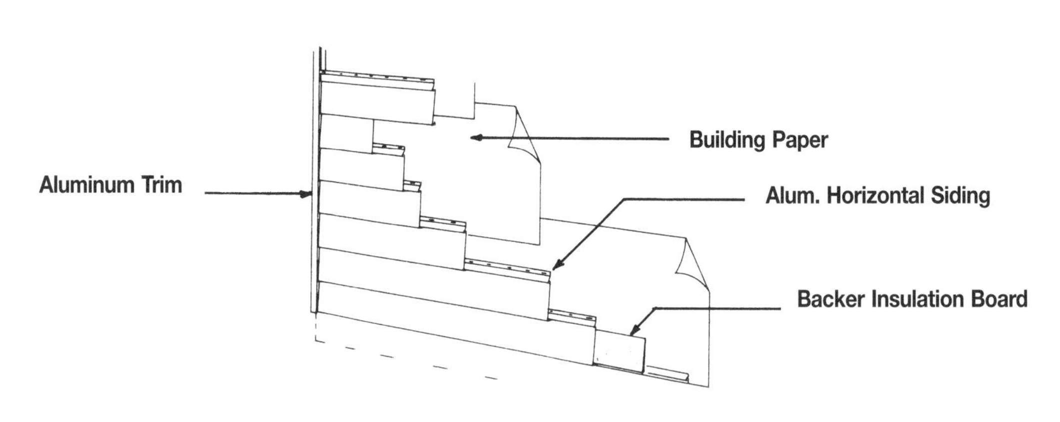

System Description	QUAN.	UNIT	LABOR HOURS	COST PER S.F.		
				MAT.	INST.	TOTAL
ALUMINUM CLAPBOARD SIDING, 8″ WIDE, WHITE						
Aluminum horizontal siding, 8″ clapboard	1.000	S.F.	.031	1.72	1.31	3.03
Backer, insulation board	1.000	S.F.	.008	.54	.34	.88
Trim, aluminum	.600	L.F.	.016	.81	.67	1.48
Building wrap, spunbonded polypropylene	1.100	S.F.	.002	.13	.09	.22
TOTAL		S.F.	.057	3.20	2.41	5.61
ALUMINUM VERTICAL BOARD & BATTEN, WHITE						
Aluminum vertical board & batten	1.000	S.F.	.027	1.88	1.15	3.03
Backer insulation board	1.000	S.F.	.008	.54	.34	.88
Trim, aluminum	.600	L.F.	.016	.81	.67	1.48
Building wrap, spunbonded polypropylene	1.100	S.F.	.002	.13	.09	.22
TOTAL		S.F.	.053	3.36	2.25	5.61
VINYL CLAPBOARD SIDING, 8″ WIDE, WHITE						
PVC vinyl horizontal siding, 8″ clapboard	1.000	S.F.	.032	.77	1.37	2.14
Backer, insulation board	1.000	S.F.	.008	.54	.34	.88
Trim, vinyl	.600	L.F.	.014	.46	.59	1.05
Building wrap, spunbonded polypropylene	1.100	S.F.	.002	.13	.09	.22
TOTAL		S.F.	.056	1.90	2.39	4.29
VINYL VERTICAL BOARD & BATTEN, WHITE						
PVC vinyl vertical board & batten	1.000	S.F.	.029	1.66	1.23	2.89
Backer, insulation board	1.000	S.F.	.008	.54	.34	.88
Trim, vinyl	.600	L.F.	.014	.46	.59	1.05
Building wrap, spunbonded polypropylene	1.100	S.F.	.002	.13	.09	.22
TOTAL		S.F.	.053	2.79	2.25	5.04

The costs in this system are on a square foot of wall basis.
subtract openings from wall area.

Description	QUAN.	UNIT	LABOR HOURS	COST PER S.F.		
				MAT.	INST.	TOTAL

Important: See the Reference Section for critical supporting data - Reference Nos., Crews & Location Factors

Metal & Plastic Siding Price Sheet

Metal & Plastic Siding Price Sheet	QUAN.	UNIT	LABOR HOURS	COST PER S.F. MAT.	INST.	TOTAL
Siding, aluminum, .024" thick, smooth, 8" wide, white	1.000	S.F.	.031	1.72	1.31	3.03
Color	1.000	S.F.	.031	1.83	1.31	3.14
Double 4" pattern, 8" wide, white	1.000	S.F.	.031	1.23	1.31	2.54
Color	1.000	S.F.	.031	1.34	1.31	2.65
Double 5" pattern, 10" wide, white	1.000	S.F.	.029	1.27	1.23	2.50
Color	1.000	S.F.	.029	1.38	1.23	2.61
Embossed, single, 8" wide, white	1.000	S.F.	.031	1.72	1.31	3.03
Color	1.000	S.F.	.031	1.83	1.31	3.14
Double 4" pattern, 8" wide, white	1.000	S.F.	.031	1.82	1.31	3.13
Color	1.000	S.F.	.031	1.93	1.31	3.24
Double 5" pattern, 10" wide, white	1.000	S.F.	.029	1.82	1.23	3.05
Color	1.000	S.F.	.029	1.93	1.23	3.16
Alum siding with insulation board, smooth, 8" wide, white	1.000	S.F.	.031	1.61	1.31	2.92
Color	1.000	S.F.	.031	1.72	1.31	3.03
Double 4" pattern, 8" wide, white	1.000	S.F.	.031	1.58	1.31	2.89
Color	1.000	S.F.	.031	1.69	1.31	3
Double 5" pattern, 10" wide, white	1.000	S.F.	.029	1.58	1.23	2.81
Color	1.000	S.F.	.029	1.69	1.23	2.92
Embossed, single, 8" wide, white	1.000	S.F.	.031	1.86	1.31	3.17
Color	1.000	S.F.	.031	1.97	1.31	3.28
Double 4" pattern, 8" wide, white	1.000	S.F.	.031	1.88	1.31	3.19
Color	1.000	S.F.	.031	1.99	1.31	3.30
Double 5" pattern, 10" wide, white	1.000	S.F.	.029	1.88	1.23	3.11
Color	1.000	S.F.	.029	1.99	1.23	3.22
Aluminum, shake finish, 10" wide, white	1.000	S.F.	.029	1.99	1.23	3.22
Color	1.000	S.F.	.029	2.10	1.23	3.33
Aluminum, vertical, 12" wide, white	1.000	S.F.	.027	1.88	1.15	3.03
Color	1.000	S.F.	.027	1.99	1.15	3.14
Vinyl siding, 8" wide, smooth, white	1.000	S.F.	.032	.77	1.37	2.14
Color	1.000	S.F.	.032	.87	1.37	2.24
10" wide, Dutch lap, smooth, white	1.000	S.F.	.029	.87	1.23	2.10
Color	1.000	S.F.	.029	.97	1.23	2.20
Double 4" pattern, 8" wide, white	1.000	S.F.	.032	.61	1.37	1.98
Color	1.000	S.F.	.032	.71	1.37	2.08
Double 5" pattern, 10" wide, white	1.000	S.F.	.029	.62	1.23	1.85
Color	1.000	S.F.	.029	.72	1.23	1.95
Embossed, single, 8" wide, white	1.000	S.F.	.032	.91	1.37	2.28
Color	1.000	S.F.	.032	1.01	1.37	2.38
10" wide, white	1.000	S.F.	.029	.74	1.23	1.97
Color	1.000	S.F.	.029	.84	1.23	2.07
Double 4" pattern, 8" wide, white	1.000	S.F.	.032	.72	1.37	2.09
Color	1.000	S.F.	.032	.82	1.37	2.19
Double 5" pattern, 10" wide, white	1.000	S.F.	.029	.72	1.23	1.95
Color	1.000	S.F.	.029	.82	1.23	2.05
Vinyl, shake finish, 10" wide, white	1.000	S.F.	.029	2.09	1.23	3.32
Color	1.000	S.F.	.029	2.19	1.23	3.42
Vinyl, vertical, double 5" pattern, 10" wide, white	1.000	S.F.	.029	1.66	1.23	2.89
Color	1.000	S.F.	.029	1.76	1.23	2.99
Backer board, installed in siding panels 8" or 10" wide	1.000	S.F.	.008	.54	.34	.88
4' x 8' sheets, polystyrene, 3/4" thick	1.000	S.F.	.010	.52	.42	.94
4' x 8' fiberboard, plain	1.000	S.F.	.008	.54	.34	.88
Trim, aluminum, white	.600	L.F.	.016	.81	.67	1.48
Color	.600	L.F.	.016	.87	.67	1.54
Vinyl, white	.600	L.F.	.014	.46	.59	1.05
Color	.600	L.F.	.014	.49	.59	1.08
Paper, #15 asphalt felt	1.100	S.F.	.002	.04	.10	.14
Kraft paper, plain	1.100	S.F.	.002	.06	.10	.16
Foil backed	1.100	S.F.	.002	.10	.10	.20

Description	QUAN.	UNIT	LABOR HOURS	COST PER S.F.		
				MAT.	INST.	TOTAL
Poured insulation, cellulose fiber, R3.8 per inch (1" thick)	1.000	S.F.	.003	.05	.14	.19
Fiberglass , R4.0 per inch (1" thick)	1.000	S.F.	.003	.04	.14	.18
Mineral wool, R3.0 per inch (1" thick)	1.000	S.F.	.003	.03	.14	.17
Polystyrene, R4.0 per inch (1" thick)	1.000	S.F.	.003	.26	.14	.40
Vermiculite, R2.7 per inch (1" thick)	1.000	S.F.	.003	.16	.14	.30
Perlite, R2.7 per inch (1" thick)	1.000	S.F.	.003	.16	.14	.30
Reflective insulation, aluminum foil reinforced with scrim	1.000	S.F.	.004	.16	.18	.34
Reinforced with woven polyolefin	1.000	S.F.	.004	.19	.18	.37
With single bubble air space, R8.8	1.000	S.F.	.005	.31	.23	.54
With double bubble air space, R9.8	1.000	S.F.	.005	.33	.23	.56
Rigid insulation, fiberglass, unfaced,						
1-1/2" thick, R6.2	1.000	S.F.	.008	.54	.34	.88
2" thick, R8.3	1.000	S.F.	.008	.65	.34	.99
2-1/2" thick, R10.3	1.000	S.F.	.010	.75	.42	1.17
3" thick, R12.4	1.000	S.F.	.010	.75	.42	1.17
Foil faced, 1" thick, R4.3	1.000	S.F.	.008	1.05	.34	1.39
1-1/2" thick, R6.2	1.000	S.F.	.008	1.42	.34	1.76
2" thick, R8.7	1.000	S.F.	.009	1.77	.38	2.15
2-1/2" thick, R10.9	1.000	S.F.	.010	2.09	.42	2.51
3" thick, R13.0	1.000	S.F.	.010	2.28	.42	2.70
Foam glass, 1-1/2" thick R2.64	1.000	S.F.	.010	1.38	.42	1.80
2" thick R5.26	1.000	S.F.	.011	3.31	.46	3.77
Perlite, 1" thick R2.77	1.000	S.F.	.010	.31	.42	.73
2" thick R5.55	1.000	S.F.	.011	.59	.46	1.05
Polystyrene, extruded, blue, 2.2#/C.F., 3/4" thick R4	1.000	S.F.	.010	.52	.42	.94
1-1/2" thick R8.1	1.000	S.F.	.011	1.02	.46	1.48
2" thick R10.8	1.000	S.F.	.011	1.43	.46	1.89
Molded bead board, white, 1" thick R3.85	1.000	S.F.	.010	.24	.42	.66
1-1/2" thick, R5.6	1.000	S.F.	.011	.61	.46	1.07
2" thick, R7.7	1.000	S.F.	.011	.75	.46	1.21
Non-rigid insulation, batts						
Fiberglass, kraft faced, 3-1/2" thick, R11, 11" wide	1.000	S.F.	.005	.32	.21	.53
15" wide	1.000	S.F.	.005	.32	.21	.53
23" wide	1.000	S.F.	.005	.32	.21	.53
6" thick, R19, 11" wide	1.000	S.F.	.006	.46	.25	.71
15" wide	1.000	S.F.	.006	.46	.25	.71
23" wide	1.000	S.F.	.006	.46	.25	.71
9" thick, R30, 15" wide	1.000	S.F.	.006	.78	.25	1.03
23" wide	1.000	S.F.	.006	.78	.25	1.03
12" thick, R38, 15" wide	1.000	S.F.	.006	.94	.25	1.19
23" wide	1.000	S.F.	.006	.94	.25	1.19
Fiberglass, foil faced, 3-1/2" thick, R11, 15" wide	1.000	S.F.	.005	.52	.21	.73
23" wide	1.000	S.F.	.005	.52	.21	.73
6" thick, R19, 15" thick	1.000	S.F.	.005	.53	.21	.74
23" wide	1.000	S.F.	.005	.53	.21	.74
9" thick, R30, 15" wide	1.000	S.F.	.006	.86	.25	1.11
23" wide	1.000	S.F.	.006	.86	.25	1.11

Important: See the Reference Section for critical supporting data - Reference Nos., Crews & Location Factors

Insulation Systems	QUAN.	UNIT	LABOR HOURS	COST PER S.F.		
				MAT.	INST.	TOTAL
Non-rigid insulation batts						
Fiberglass unfaced, 3-1/2" thick, R11, 15" wide	1.000	S.F.	.005	.36	.21	.57
23" wide	1.000	S.F.	.005	.36	.21	.57
6" thick, R19, 15" wide	1.000	S.F.	.006	.42	.25	.67
23" wide	1.000	S.F.	.006	.42	.25	.67
9" thick, R19, 15" wide	1.000	S.F.	.007	.73	.29	1.02
23" wide	1.000	S.F.	.007	.73	.29	1.02
12" thick, R38, 15" wide	1.000	S.F.	.007	.91	.29	1.20
23" wide	1.000	S.F.	.007	.91	.29	1.20
Mineral fiber batts, 3" thick, R11	1.000	S.F.	.005	.33	.21	.54
3-1/2" thick, R13	1.000	S.F.	.005	.33	.21	.54
6" thick, R19	1.000	S.F.	.005	.44	.21	.65
6-1/2" thick, R22	1.000	S.F.	.005	.44	.21	.65
10" thick, R30	1.000	S.F.	.006	.65	.25	.90

28 Double Hung Window Systems

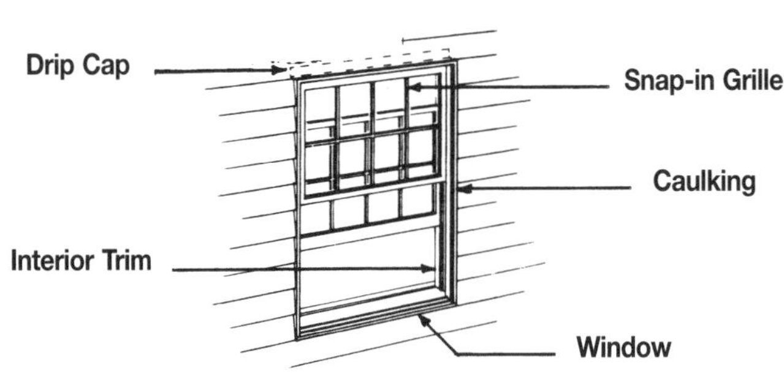

System Description	QUAN.	UNIT	LABOR HOURS	COST EACH		
				MAT.	INST.	TOTAL
BUILDER'S QUALITY WOOD WINDOW 2′ X 3′, DOUBLE HUNG						
Window, primed, builder's quality, 2′ x 3′, insulating glass	1.000	Ea.	.800	205	34	239
Trim, interior casing	11.000	L.F.	.367	15.51	15.51	31.02
Paint, interior & exterior, primer & 2 coats	2.000	Face	1.778	2.32	66	68.32
Caulking	10.000	L.F.	.323	1.70	13.70	15.40
Snap-in grille	1.000	Set	.333	48.50	14.10	62.60
Drip cap, metal	2.000	L.F.	.040	.66	1.70	2.36
TOTAL		Ea.	3.641	273.69	145.01	418.70
PLASTIC CLAD WOOD WINDOW 3′ X 4′, DOUBLE HUNG						
Window, plastic clad, premium, 3′ x 4′, insulating glass	1.000	Ea.	.889	385	37.50	422.50
Trim, interior casing	15.000	L.F.	.500	21.15	21.15	42.30
Paint, interior, primer & 2 coats	1.000	Face	.889	1.16	33	34.16
Caulking	14.000	L.F.	.452	2.38	19.18	21.56
Snap-in grille	1.000	Set	.333	48.50	14.10	62.60
TOTAL		Ea.	3.063	458.19	124.93	583.12
METAL CLAD WOOD WINDOW, 3′ X 5′, DOUBLE HUNG						
Window, metal clad, deluxe, 3′ x 5′, insulating glass	1.000	Ea.	1.000	350	42.50	392.50
Trim, interior casing	17.000	L.F.	.567	23.97	23.97	47.94
Paint, interior, primer & 2 coats	1.000	Face	.889	1.16	33	34.16
Caulking	16.000	L.F.	.516	2.72	21.92	24.64
Snap-in grille	1.000	Set	.235	136	9.95	145.95
Drip cap, metal	3.000	L.F.	.060	.99	2.55	3.54
TOTAL		Ea.	3.267	514.84	133.89	648.73

The cost of this system is on a cost per each window basis.

Description	QUAN.	UNIT	LABOR HOURS	COST EACH		
				MAT.	INST.	TOTAL

Double Hung Window Price Sheet	QUAN.	UNIT	LABOR HOURS	COST EACH		
				MAT.	INST.	TOTAL
Windows, double-hung, builder's quality, 2′ x 3′, single glass	1.000	Ea.	.800	225	34	259
Insulating glass	1.000	Ea.	.800	205	34	239
3′ x 4′, single glass	1.000	Ea.	.889	292	37.50	329.50
Insulating glass	1.000	Ea.	.889	274	37.50	311.50
4′ x 4′-6″, single glass	1.000	Ea.	1.000	340	42.50	382.50
Insulating glass	1.000	Ea.	1.000	365	42.50	407.50
Plastic clad premium insulating glass, 2′-6″ x 3′	1.000	Ea.	.800	240	34	274
3′ x 3′-6″	1.000	Ea.	.800	282	34	316
3′ x 4′	1.000	Ea.	.889	385	37.50	422.50
3′ x 4′-6″	1.000	Ea.	.889	325	37.50	362.50
3′ x 5′	1.000	Ea.	1.000	340	42.50	382.50
3′-6″ x 6′	1.000	Ea.	1.000	400	42.50	442.50
Metal clad deluxe insulating glass, 2′-6″ x 3′	1.000	Ea.	.800	240	34	274
3′ x 3′-6″	1.000	Ea.	.800	285	34	319
3′ x 4′	1.000	Ea.	.889	305	37.50	342.50
3′ x 4′-6″	1.000	Ea.	.889	325	37.50	362.50
3′ x 5′	1.000	Ea.	1.000	350	42.50	392.50
3′-6″ x 6′	1.000	Ea.	1.000	425	42.50	467.50
Trim, interior casing, window 2′ x 3′	11.000	L.F.	.367	15.50	15.50	31
2′-6″ x 3′	12.000	L.F.	.400	16.90	16.90	33.80
3′ x 3′-6″	14.000	L.F.	.467	19.75	19.75	39.50
3′ x 4′	15.000	L.F.	.500	21	21	42
3′ x 4′-6″	16.000	L.F.	.533	22.50	22.50	45
3′ x 5′	17.000	L.F.	.567	24	24	48
3′-6″ x 6′	20.000	L.F.	.667	28	28	56
4′ x 4′-6″	18.000	L.F.	.600	25.50	25.50	51
Paint or stain, interior or exterior, 2′ x 3′ window, 1 coat	1.000	Face	.444	.38	16.45	16.83
2 coats	1.000	Face	.727	.77	27	27.77
Primer & 1 coat	1.000	Face	.727	.77	27	27.77
Primer & 2 coats	1.000	Face	.889	1.16	33	34.16
3′ x 4′ window, 1 coat	1.000	Face	.667	.79	24.50	25.29
2 coats	1.000	Face	.667	.89	24.50	25.39
Primer & 1 coat	1.000	Face	.727	1.19	27	28.19
Primer & 2 coats	1.000	Face	.889	1.16	33	34.16
4′ x 4′-6″ window, 1 coat	1.000	Face	.667	.79	24.50	25.29
2 coats	1.000	Face	.667	.89	24.50	25.39
Primer & 1 coat	1.000	Face	.727	1.19	27	28.19
Primer & 2 coats	1.000	Face	.889	1.16	33	34.16
Caulking, window, 2′ x 3′	10.000	L.F.	.323	1.70	13.70	15.40
2′-6″ x 3′	11.000	L.F.	.355	1.87	15.05	16.92
3′ x 3′-6″	13.000	L.F.	.419	2.21	17.80	20.01
3′ x 4′	14.000	L.F.	.452	2.38	19.20	21.58
3′ x 4′-6″	15.000	L.F.	.484	2.55	20.50	23.05
3′ x 5′	16.000	L.F.	.516	2.72	22	24.72
3′-6″ x 6′	19.000	L.F.	.613	3.23	26	29.23
4′ x 4′-6″	17.000	L.F.	.548	2.89	23.50	26.39
Grilles, glass size to, 16″ x 24″ per sash	1.000	Set	.333	48.50	14.10	62.60
32″ x 32″ per sash	1.000	Set	.235	136	9.95	145.95
Drip cap, aluminum, 2′ long	2.000	L.F.	.040	.66	1.70	2.36
3′ long	3.000	L.F.	.060	.99	2.55	3.54
4′ long	4.000	L.F.	.080	1.32	3.40	4.72
Wood, 2′ long	2.000	L.F.	.067	2.82	2.82	5.64
3′ long	3.000	L.F.	.100	4.23	4.23	8.46
4′ long	4.000	L.F.	.133	5.65	5.65	11.30

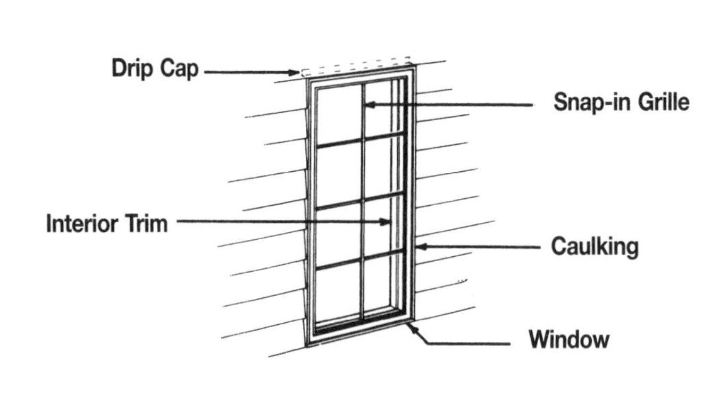

System Description	QUAN.	UNIT	LABOR HOURS	COST EACH		
				MAT.	INST.	TOTAL
BUILDER'S QUALITY WINDOW, WOOD, 2′ BY 3′, CASEMENT						
Window, primed, builder's quality, 2′ x 3′, insulating glass	1.000	Ea.	.800	248	34	282
Trim, interior casing	11.000	L.F.	.367	15.51	15.51	31.02
Paint, interior & exterior, primer & 2 coats	2.000	Face	1.778	2.32	66	68.32
Caulking	10.000	L.F.	.323	1.70	13.70	15.40
Snap-in grille	1.000	Ea.	.267	27.50	11.25	38.75
Drip cap, metal	2.000	L.F.	.040	.66	1.70	2.36
TOTAL		Ea.	3.575	295.69	142.16	437.85
PLASTIC CLAD WOOD WINDOW, 2′ X 4′, CASEMENT						
Window, plastic clad, premium, 2′ x 4′, insulating glass	1.000	Ea.	.889	340	37.50	377.50
Trim, interior casing	13.000	L.F.	.433	18.33	18.33	36.66
Paint, interior, primer & 2 coats	1.000	Ea.	.889	1.16	33	34.16
Caulking	12.000	L.F.	.387	2.04	16.44	18.48
Snap-in grille	1.000	Ea.	.267	27.50	11.25	38.75
TOTAL		Ea.	2.865	389.03	116.52	505.55
METAL CLAD WOOD WINDOW, 2′ X 5′, CASEMENT						
Window, metal clad, deluxe, 2′ x 5′, insulating glass	1.000	Ea.	1.000	310	42.50	352.50
Trim, interior casing	15.000	L.F.	.500	21.15	21.15	42.30
Paint, interior, primer & 2 coats	1.000	Ea.	.889	1.16	33	34.16
Caulking	14.000	L.F.	.452	2.38	19.18	21.56
Snap-in grille	1.000	Ea.	.250	39.50	10.55	50.05
Drip cap, metal	12.000	L.F.	.040	.66	1.70	2.36
TOTAL		Ea.	3.131	374.85	128.08	502.93

The cost of this system is on a cost per each window basis.

Description	QUAN.	UNIT	LABOR HOURS	COST EACH		
				MAT.	INST.	TOTAL

 Important: See the Reference Section for critical supporting data - Reference Nos., Crews & Location Factors

Casement Window Price Sheet

Casement Window Price Sheet	QUAN.	UNIT	LABOR HOURS	COST EACH		
				MAT.	INST.	TOTAL
Window, casement, builders quality, 2′ x 3′, single glass	1.000	Ea.	.800	205	34	239
Insulating glass	1.000	Ea.	.800	248	34	282
2′ x 4′-6″, single glass	1.000	Ea.	.727	855	30.50	885.50
Insulating glass	1.000	Ea.	.727	720	30.50	750.50
2′ x 6′, single glass	1.000	Ea.	.889	440	42.50	482.50
Insulating glass	1.000	Ea.	.889	1,100	37.50	1,137.50
Plastic clad premium insulating glass, 2′ x 3′	1.000	Ea.	.800	203	34	237
2′ x 4′	1.000	Ea.	.889	220	37.50	257.50
2′ x 5′	1.000	Ea.	1.000	300	42.50	342.50
2′ x 6′	1.000	Ea.	1.000	440	42.50	482.50
Metal clad deluxe insulating glass, 2′ x 3′	1.000	Ea.	.800	227	34	261
2′ x 4′	1.000	Ea.	.889	273	37.50	310.50
2′ x 5′	1.000	Ea.	1.000	310	42.50	352.50
2′ x 6′	1.000	Ea.	1.000	355	42.50	397.50
Trim, interior casing, window 2′ x 3′	11.000	L.F.	.367	15.50	15.50	31
2′ x 4′	13.000	L.F.	.433	18.35	18.35	36.70
2′ x 4′-6″	14.000	L.F.	.467	19.75	19.75	39.50
2′ x 5′	15.000	L.F.	.500	21	21	42
2′ x 6′	17.000	L.F.	.567	24	24	48
Paint or stain, interior or exterior, 2′ x 3′ window, 1 coat	1.000	Face	.444	.38	16.45	16.83
2 coats	1.000	Face	.727	.77	27	27.77
Primer & 1 coat	1.000	Face	.727	.77	27	27.77
Primer & 2 coats	1.000	Face	.889	1.16	33	34.16
2′ x 4′ window, 1 coat	1.000	Face	.444	.38	16.45	16.83
2 coats	1.000	Face	.727	.77	27	27.77
Primer & 1 coat	1.000	Face	.727	.77	27	27.77
Primer & 2 coats	1.000	Face	.889	1.16	33	34.16
2′ x 6′ window, 1 coat	1.000	Face	.667	.79	24.50	25.29
2 coats	1.000	Face	.667	.89	24.50	25.39
Primer & 1 coat	1.000	Face	.727	1.19	27	28.19
Primer & 2 coats	1.000	Face	.889	1.16	33	34.16
Caulking, window, 2′ x 3′	10.000	L.F.	.323	1.70	13.70	15.40
2′ x 4′	12.000	L.F.	.387	2.04	16.45	18.49
2′ x 4′-6″	13.000	L.F.	.419	2.21	17.80	20.01
2′ x 5′	14.000	L.F.	.452	2.38	19.20	21.58
2′ x 6′	16.000	L.F.	.516	2.72	22	24.72
Grilles, glass size, to 20″ x 36″	1.000	Ea.	.267	27.50	11.25	38.75
To 20″ x 56″	1.000	Ea.	.250	39.50	10.55	50.05
Drip cap, metal, 2′ long	2.000	L.F.	.040	.66	1.70	2.36
Wood, 2′ long	2.000	L.F.	.067	2.82	2.82	5.64

36 Awning Window Systems

System Description	QUAN.	UNIT	LABOR HOURS	COST EACH MAT.	COST EACH INST.	COST EACH TOTAL
BUILDER'S QUALITY WINDOW, WOOD, 34" X 22", AWNING						
Window, builder quality, 34" x 22", insulating glass	1.000	Ea.	.800	270	34	304
Trim, interior casing	10.500	L.F.	.350	14.81	14.81	29.62
Paint, interior & exterior, primer & 2 coats	2.000	Face	1.778	2.32	66	68.32
Caulking	9.500	L.F.	.306	1.62	13.02	14.64
Snap-in grille	1.000	Ea.	.267	22.50	11.25	33.75
Drip cap, metal	3.000	L.F.	.060	.99	2.55	3.54
TOTAL		Ea.	3.561	312.24	141.63	453.87
PLASTIC CLAD WOOD WINDOW, 40" X 28", AWNING						
Window, plastic clad, premium, 40" x 28", insulating glass	1.000	Ea.	.889	355	37.50	392.50
Trim interior casing	13.500	L.F.	.450	19.04	19.04	38.08
Paint, interior, primer & 2 coats	1.000	Face	.889	1.16	33	34.16
Caulking	12.500	L.F.	.403	2.13	17.13	19.26
Snap-in grille	1.000	Ea.	.267	22.50	11.25	33.75
TOTAL		Ea.	2.898	399.83	117.92	517.75
METAL CLAD WOOD WINDOW, 48" X 36", AWNING						
Window, metal clad, deluxe, 48" x 36", insulating glass	1.000	Ea.	1.000	375	42.50	417.50
Trim, interior casing	15.000	L.F.	.500	21.15	21.15	42.30
Paint, interior, primer & 2 coats	1.000	Face	.889	1.16	33	34.16
Caulking	14.000	L.F.	.452	2.38	19.18	21.56
Snap-in grille	1.000	Ea.	.250	32.50	10.55	43.05
Drip cap, metal	4.000	L.F.	.080	1.32	3.40	4.72
TOTAL		Ea.	3.171	433.51	129.78	563.29

The cost of this system is on a cost per each window basis.

Description	QUAN.	UNIT	LABOR HOURS	COST EACH MAT.	COST EACH INST.	COST EACH TOTAL

Awning Window Price Sheet	QUAN.	UNIT	LABOR HOURS	COST EACH		
				MAT.	INST.	TOTAL
Windows, awning, builder's quality, 34" x 22", insulated glass	1.000	Ea.	.800	256	34	290
Low E glass	1.000	Ea.	.800	270	34	304
40" x 28", insulated glass	1.000	Ea.	.889	325	37.50	362.50
Low E glass	1.000	Ea.	.889	345	37.50	382.50
48" x 36", insulated glass	1.000	Ea.	1.000	470	42.50	512.50
Low E glass	1.000	Ea.	1.000	495	42.50	537.50
Plastic clad premium insulating glass, 34" x 22"	1.000	Ea.	.800	272	34	306
40" x 22"	1.000	Ea.	.800	297	34	331
36" x 28"	1.000	Ea.	.889	315	37.50	352.50
36" x 36"	1.000	Ea.	.889	355	37.50	392.50
48" x 28"	1.000	Ea.	1.000	380	42.50	422.50
60" x 36"	1.000	Ea.	1.000	550	42.50	592.50
Metal clad deluxe insulating glass, 34" x 22"	1.000	Ea.	.800	254	34	288
40" x 22"	1.000	Ea.	.800	298	34	332
36" x 25"	1.000	Ea.	.889	276	37.50	313.50
40" x 30"	1.000	Ea.	.889	345	37.50	382.50
48" x 28"	1.000	Ea.	1.000	350	42.50	392.50
60" x 36"	1.000	Ea.	1.000	375	42.50	417.50
Trim, interior casing window, 34" x 22"	10.500	L.F.	.350	14.80	14.80	29.60
40" x 22"	11.500	L.F.	.383	16.20	16.20	32.40
36" x 28"	12.500	L.F.	.417	17.65	17.65	35.30
40" x 28"	13.500	L.F.	.450	19.05	19.05	38.10
48" x 28"	14.500	L.F.	.483	20.50	20.50	41
48" x 36"	15.000	L.F.	.500	21	21	42
Paint or stain, interior or exterior, 34" x 22", 1 coat	1.000	Face	.444	.38	16.45	16.83
2 coats	1.000	Face	.727	.77	27	27.77
Primer & 1 coat	1.000	Face	.727	.77	27	27.77
Primer & 2 coats	1.000	Face	.889	1.16	33	34.16
36" x 28", 1 coat	1.000	Face	.444	.38	16.45	16.83
2 coats	1.000	Face	.727	.77	27	27.77
Primer & 1 coat	1.000	Face	.727	.77	27	27.77
Primer & 2 coats	1.000	Face	.889	1.16	33	34.16
48" x 36", 1 coat	1.000	Face	.667	.79	24.50	25.29
2 coats	1.000	Face	.667	.89	24.50	25.39
Primer & 1 coat	1.000	Face	.727	1.19	27	28.19
Primer & 2 coats	1.000	Face	.889	1.16	33	34.16
Caulking, window, 34" x 22"	9.500	L.F.	.306	1.62	13	14.62
40" x 22"	10.500	L.F.	.339	1.79	14.40	16.19
36" x 28"	11.500	L.F.	.371	1.96	15.75	17.71
40" x 28"	12.500	L.F.	.403	2.13	17.15	19.28
48" x 28"	13.500	L.F.	.436	2.30	18.50	20.80
48" x 36"	14.000	L.F.	.452	2.38	19.20	21.58
Grilles, glass size, to 28" by 16"	1.000	Ea.	.267	22.50	11.25	33.75
To 44" by 24"	1.000	Ea.	.250	32.50	10.55	43.05
Drip cap, aluminum, 3' long	3.000	L.F.	.060	.99	2.55	3.54
3'-6" long	3.500	L.F.	.070	1.16	2.98	4.14
4' long	4.000	L.F.	.080	1.32	3.40	4.72
Wood, 3' long	3.000	L.F.	.100	4.23	4.23	8.46
3'-6" long	3.500	L.F.	.117	4.94	4.94	9.88
4' long	4.000	L.F.	.133	5.65	5.65	11.30

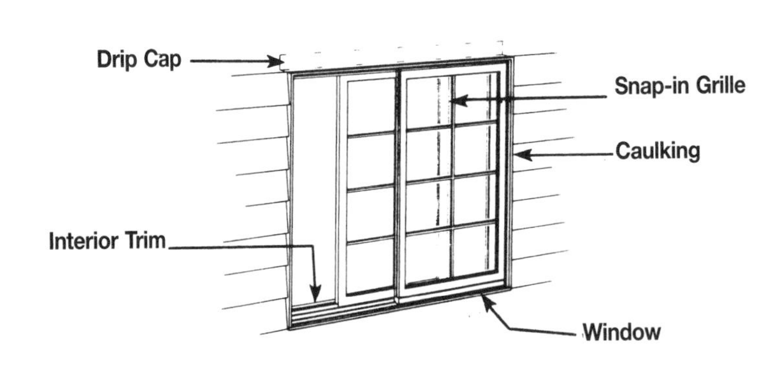

System Description	QUAN.	UNIT	LABOR HOURS	COST EACH		
				MAT.	INST.	TOTAL
BUILDER'S QUALITY WOOD WINDOW, 3′ X 2′, SLIDING						
Window, primed, builder's quality, 3′ x 2′, insul. glass	1.000	Ea.	.800	212	34	246
Trim, interior casing	11.000	L.F.	.367	15.51	15.51	31.02
Paint, interior & exterior, primer & 2 coats	2.000	Face	1.778	2.32	66	68.32
Caulking	10.000	L.F.	.323	1.70	13.70	15.40
Snap-in grille	1.000	Set	.333	27.50	14.10	41.60
Drip cap, metal	3.000	L.F.	.060	.99	2.55	3.54
TOTAL		Ea.	3.661	260.02	145.86	405.88
PLASTIC CLAD WOOD WINDOW, 4′ X 3′-6″, SLIDING						
Window, plastic clad, premium, 4′ x 3′-6″, insulating glass	1.000	Ea.	.889	690	37.50	727.50
Trim, interior casing	16.000	L.F.	.533	22.56	22.56	45.12
Paint, interior, primer & 2 coats	1.000	Face	.889	1.16	33	34.16
Caulking	17.000	L.F.	.548	2.89	23.29	26.18
Snap-in grille	1.000	Set	.333	27.50	14.10	41.60
TOTAL		Ea.	3.192	744.11	130.45	874.56
METAL CLAD WOOD WINDOW, 6′ X 5′, SLIDING						
Window, metal clad, deluxe, 6′ x 5′, insulating glass	1.000	Ea.	1.000	720	42.50	762.50
Trim, interior casing	23.000	L.F.	.767	32.43	32.43	64.86
Paint, interior, primer & 2 coats	1.000	Face	.889	1.16	33	34.16
Caulking	22.000	L.F.	.710	3.74	30.14	33.88
Snap-in grille	1.000	Set	.364	42	15.35	57.35
Drip cap, metal	6.000	L.F.	.120	1.98	5.10	7.08
TOTAL		Ea.	3.850	801.31	158.52	959.83

The cost of this system is on a cost per each window basis.

Description	QUAN.	UNIT	LABOR HOURS	COST EACH		
				MAT.	INST.	TOTAL

Important: See the Reference Section for critical supporting data - Reference Nos., Crews & Location Factors

Sliding Window Price Sheet	QUAN.	UNIT	LABOR HOURS	COST EACH		
				MAT.	INST.	TOTAL
Windows, sliding, builder's quality, 3′ x 3′, single glass	1.000	Ea.	.800	168	34	202
Insulating glass	1.000	Ea.	.800	212	34	246
4′ x 3′-6″, single glass	1.000	Ea.	.889	200	37.50	237.50
Insulating glass	1.000	Ea.	.889	250	37.50	287.50
6′ x 5′, single glass	1.000	Ea.	1.000	365	42.50	407.50
Insulating glass	1.000	Ea.	1.000	440	42.50	482.50
Plastic clad premium insulating glass, 3′ x 3′	1.000	Ea.	.800	560	34	594
4′ x 3′-6″	1.000	Ea.	.889	690	37.50	727.50
5′ x 4′	1.000	Ea.	.889	835	37.50	872.50
6′ x 5′	1.000	Ea.	1.000	1,050	42.50	1,092.50
Metal clad deluxe insulating glass, 3′ x 3′	1.000	Ea.	.800	320	34	354
4′ x 3′-6″	1.000	Ea.	.889	390	37.50	427.50
5′ x 4′	1.000	Ea.	.889	470	37.50	507.50
6′ x 5′	1.000	Ea.	1.000	720	42.50	762.50
Trim, interior casing, window 3′ x 2′	11.000	L.F.	.367	15.50	15.50	31
3′ x 3′	13.000	L.F.	.433	18.35	18.35	36.70
4′ x 3′-6″	16.000	L.F.	.533	22.50	22.50	45
5′ x 4′	19.000	L.F.	.633	27	27	54
6′ x 5′	23.000	L.F.	.767	32.50	32.50	65
Paint or stain, interior or exterior, 3′ x 2′ window, 1 coat	1.000	Face	.444	.38	16.45	16.83
2 coats	1.000	Face	.727	.77	27	27.77
Primer & 1 coat	1.000	Face	.727	.77	27	27.77
Primer & 2 coats	1.000	Face	.889	1.16	33	34.16
4′ x 3′-6″ window, 1 coat	1.000	Face	.667	.79	24.50	25.29
2 coats	1.000	Face	.667	.89	24.50	25.39
Primer & 1 coat	1.000	Face	.727	1.19	27	28.19
Primer & 2 coats	1.000	Face	.889	1.16	33	34.16
6′ x 5′ window, 1 coat	1.000	Face	.889	2.17	33	35.17
2 coats	1.000	Face	1.333	3.97	49.50	53.47
Primer & 1 coat	1.000	Face	1.333	4.09	49.50	53.59
Primer & 2 coats	1.000	Face	1.600	6.20	59.50	65.70
Caulking, window, 3′ x 2′	10.000	L.F.	.323	1.70	13.70	15.40
3′ x 3′	12.000	L.F.	.387	2.04	16.45	18.49
4′ x 3′-6″	15.000	L.F.	.484	2.55	20.50	23.05
5′ x 4′	18.000	L.F.	.581	3.06	24.50	27.56
6′ x 5′	22.000	L.F.	.710	3.74	30	33.74
Grilles, glass size, to 14″ x 36″	1.000	Set	.333	27.50	14.10	41.60
To 36″ x 36″	1.000	Set	.364	42	15.35	57.35
Drip cap, aluminum, 3′ long	3.000	L.F.	.060	.99	2.55	3.54
4′ long	4.000	L.F.	.080	1.32	3.40	4.72
5′ long	5.000	L.F.	.100	1.65	4.25	5.90
6′ long	6.000	L.F.	.120	1.98	5.10	7.08
Wood, 3′ long	3.000	L.F.	.100	4.23	4.23	8.46
4′ long	4.000	L.F.	.133	5.65	5.65	11.30
5′ long	5.000	L.F.	.167	7.05	7.05	14.10
6′ long	6.000	L.F.	.200	8.45	8.45	16.90

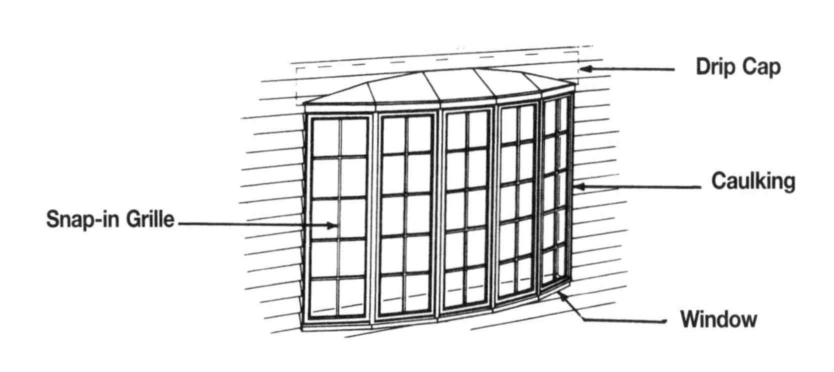

System Description	QUAN.	UNIT	LABOR HOURS	COST EACH		
				MAT.	INST.	TOTAL
AWNING TYPE BOW WINDOW, BUILDER'S QUALITY, 8′ X 5′						
Window, primed, builder's quality, 8′ x 5′, insulating glass	1.000	Ea.	1.600	1,325	67.50	1,392.50
Trim, interior casing	27.000	L.F.	.900	38.07	38.07	76.14
Paint, interior & exterior, primer & 1 coat	2.000	Face	3.200	12.40	119	131.40
Drip cap, vinyl	1.000	Ea.	.533	82.50	22.50	105
Caulking	26.000	L.F.	.839	4.42	35.62	40.04
Snap-in grilles	1.000	Set	1.067	110	45	155
TOTAL		Ea.	8.139	1,572.39	327.69	1,900.08
CASEMENT TYPE BOW WINDOW, PLASTIC CLAD, 10′ X 6′						
Window, plastic clad, premium, 10′ x 6′, insulating glass	1.000	Ea.	2.286	2,075	96.50	2,171.50
Trim, interior casing	33.000	L.F.	1.100	46.53	46.53	93.06
Paint, interior, primer & 1 coat	1.000	Face	1.778	2.32	66	68.32
Drip cap, vinyl	1.000	Ea.	.615	90	26	116
Caulking	32.000	L.F.	1.032	5.44	43.84	49.28
Snap-in grilles	1.000	Set	1.333	137.50	56.25	193.75
TOTAL		Ea.	8.144	2,356.79	335.12	2,691.91
DOUBLE HUNG TYPE, METAL CLAD, 9′ X 5′						
Window, metal clad, deluxe, 9′ x 5′, insulating glass	1.000	Ea.	2.667	1,400	113	1,513
Trim, interior casing	29.000	L.F.	.967	40.89	40.89	81.78
Paint, interior, primer & 1 coat	1.000	Face	1.778	2.32	66	68.32
Drip cap, vinyl	1.000	Set	.615	90	26	116
Caulking	28.000	L.F.	.903	4.76	38.36	43.12
Snap-in grilles	1.000	Set	1.067	110	45	155
TOTAL		Ea.	7.997	1,647.97	329.25	1,977.22

The cost of this system is on a cost per each window basis.

Description	QUAN.	UNIT	LABOR HOURS	COST EACH		
				MAT.	INST.	TOTAL

Important: See the Reference Section for critical supporting data - Reference Nos., Crews & Location Factors

Bow/Bay Window Price Sheet	QUAN.	UNIT	LABOR HOURS	COST EACH		
				MAT.	INST.	TOTAL
Windows, bow awning type, builder's quality, 8′ x 5′, insulating glass	1.000	Ea.	1.600	1,650	67.50	1,717.50
Low E glass	1.000	Ea.	1.600	1,325	67.50	1,392.50
12′ x 6′, insulating glass	1.000	Ea.	2.667	1,375	113	1,488
Low E glass	1.000	Ea.	2.667	1,450	113	1,563
Plastic clad premium insulating glass, 6′ x 4′	1.000	Ea.	1.600	1,500	67.50	1,567.50
9′ x 4′	1.000	Ea.	2.000	1,500	84.50	1,584.50
10′ x 5′	1.000	Ea.	2.286	2,425	96.50	2,521.50
12′ x 6′	1.000	Ea.	2.667	3,050	113	3,163
Metal clad deluxe insulating glass, 6′ x 4′	1.000	Ea.	1.600	940	67.50	1,007.50
9′ x 4′	1.000	Ea.	2.000	1,325	84.50	1,409.50
10′ x 5′	1.000	Ea.	2.286	1,825	96.50	1,921.50
12′ x 6′	1.000	Ea.	2.667	2,525	113	2,638
Bow casement type, builder's quality, 8′ x 5′, single glass	1.000	Ea.	1.600	1,825	67.50	1,892.50
Insulating glass	1.000	Ea.	1.600	2,225	67.50	2,292.50
12′ x 6′, single glass	1.000	Ea.	2.667	2,300	113	2,413
Insulating glass	1.000	Ea.	2.667	2,250	113	2,363
Plastic clad premium insulating glass, 8′ x 5′	1.000	Ea.	1.600	1,400	67.50	1,467.50
10′ x 5′	1.000	Ea.	2.000	2,000	84.50	2,084.50
10′ x 6′	1.000	Ea.	2.286	2,075	96.50	2,171.50
12′ x 6′	1.000	Ea.	2.667	2,475	113	2,588
Metal clad deluxe insulating glass, 8′ x 5′	1.000	Ea.	1.600	1,625	67.50	1,692.50
10′ x 5′	1.000	Ea.	2.000	1,750	84.50	1,834.50
10′ x 6′	1.000	Ea.	2.286	2,050	96.50	2,146.50
12′ x 6′	1.000	Ea.	2.667	2,850	113	2,963
Bow, double hung type, builder's quality, 8′ x 4′, single glass	1.000	Ea.	1.600	1,275	67.50	1,342.50
Insulating glass	1.000	Ea.	1.600	1,375	67.50	1,442.50
9′ x 5′, single glass	1.000	Ea.	2.667	1,400	113	1,513
Insulating glass	1.000	Ea.	2.667	1,475	113	1,588
Plastic clad premium insulating glass, 7′ x 4′	1.000	Ea.	1.600	1,325	67.50	1,392.50
8′ x 4′	1.000	Ea.	2.000	1,375	84.50	1,459.50
8′ x 5′	1.000	Ea.	2.286	1,425	96.50	1,521.50
9′ x 5′	1.000	Ea.	2.667	1,475	113	1,588
Metal clad deluxe insulating glass, 7′ x 4′	1.000	Ea.	1.600	1,225	67.50	1,292.50
8′ x 4′	1.000	Ea.	2.000	1,275	84.50	1,359.50
8′ x 5′	1.000	Ea.	2.286	1,325	96.50	1,421.50
9′ x 5′	1.000	Ea.	2.667	1,400	113	1,513
Trim, interior casing, window 7′ x 4′	1.000	Ea.	.767	32.50	32.50	65
8′ x 5′	1.000	Ea.	.900	38	38	76
10′ x 6′	1.000	Ea.	1.100	46.50	46.50	93
12′ x 6′	1.000	Ea.	1.233	52	52	104
Paint or stain, interior, or exterior, 7′ x 4′ window, 1 coat	1.000	Face	.889	2.17	33	35.17
Primer & 1 coat	1.000	Face	1.333	4.09	49.50	53.59
8′ x 5′ window, 1 coat	1.000	Face	.889	2.17	33	35.17
Primer & 1 coat	1.000	Face	1.333	4.09	49.50	53.59
10′ x 6′ window, 1 coat	1.000	Face	1.333	1.58	49	50.58
Primer & 1 coat	1.000	Face	1.778	2.32	66	68.32
12′ x 6′ window, 1 coat	1.000	Face	1.778	4.34	66	70.34
Primer & 1 coat	1.000	Face	2.667	8.20	99	107.20
Drip cap, vinyl moulded window, 7′ long	1.000	Ea.	.533	82.50	21.50	104
8′ long	1.000	Ea.	.533	82.50	22.50	105
10′ long	1.000	Ea.	.615	90	26	116
12′ long	1.000	Ea.	.615	90	26	116
Caulking, window, 7′ x 4′	1.000	Ea.	.710	3.74	30	33.74
8′ x 5′	1.000	Ea.	.839	4.42	35.50	39.92
10′ x 6′	1.000	Ea.	1.032	5.45	44	49.45
12′ x 6′	1.000	Ea.	1.161	6.10	49.50	55.60
Grilles, window, 7′ x 4′	1.000	Set	.800	82.50	34	116.50
8′ x 5′	1.000	Set	1.067	110	45	155
10′ x 6′	1.000	Set	1.333	138	56.50	194.50
12′ x 6′	1.000	Set	1.600	165	67.50	232.50

Fixed Window Systems

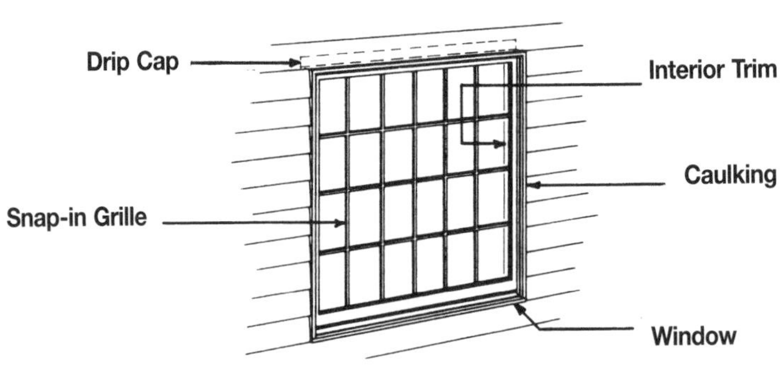

System Description	QUAN.	UNIT	LABOR HOURS	COST EACH MAT.	COST EACH INST.	COST EACH TOTAL
BUILDER'S QUALITY PICTURE WINDOW, 4′ X 4′						
Window, primed, builder's quality, 4′ x 4′, insulating glass	1.000	Ea.	1.333	340	56.50	396.50
Trim, interior casing	17.000	L.F.	.567	23.97	23.97	47.94
Paint, interior & exterior, primer & 2 coats	2.000	Face	1.778	2.32	66	68.32
Caulking	16.000	L.F.	.516	2.72	21.92	24.64
Snap-in grille	1.000	Ea.	.267	155	11.25	166.25
Drip cap, metal	4.000	L.F.	.080	1.32	3.40	4.72
TOTAL		Ea.	4.541	525.33	183.04	708.37
PLASTIC CLAD WOOD WINDOW, 4′-6″ X 6′-6″						
Window, plastic clad, prem., 4′-6″ x 6′-6″, insul. glass	1.000	Ea.	1.455	740	61.50	801.50
Trim, interior casing	23.000	L.F.	.767	32.43	32.43	64.86
Paint, interior, primer & 2 coats	1.000	Face	.889	1.16	33	34.16
Caulking	22.000	L.F.	.710	3.74	30.14	33.88
Snap-in grille	1.000	Ea.	.267	155	11.25	166.25
TOTAL		Ea.	4.088	932.33	168.32	1,100.65
METAL CLAD WOOD WINDOW, 6′-6″ X 6′-6″						
Window, metal clad, deluxe, 6′-6″ x 6′-6″, insulating glass	1.000	Ea.	1.600	665	67.50	732.50
Trim interior casing	27.000	L.F.	.900	38.07	38.07	76.14
Paint, interior, primer & 2 coats	1.000	Face	1.600	6.20	59.50	65.70
Caulking	26.000	L.F.	.839	4.42	35.62	40.04
Snap-in grille	1.000	Ea.	.267	155	11.25	166.25
Drip cap, metal	6.500	L.F.	.130	2.15	5.53	7.68
TOTAL		Ea.	5.336	870.84	217.47	1,088.31

The cost of this system is on a cost per each window basis.

Description	QUAN.	UNIT	LABOR HOURS	COST EACH MAT.	COST EACH INST.	COST EACH TOTAL

Important: See the Reference Section for critical supporting data - Reference Nos., Crews & Location Factors

Fixed Window Price Sheet	QUAN.	UNIT	LABOR HOURS	COST EACH		
				MAT.	INST.	TOTAL
Window-picture, builder's quality, 4′ x 4′, single glass	1.000	Ea.	1.333	310	56.50	366.50
Insulating glass	1.000	Ea.	1.333	340	56.50	396.50
4′ x 4′-6″, single glass	1.000	Ea.	1.455	345	61.50	406.50
Insulating glass	1.000	Ea.	1.455	380	61.50	441.50
5′ x 4′, single glass	1.000	Ea.	1.455	425	61.50	486.50
Insulating glass	1.000	Ea.	1.455	480	61.50	541.50
6′ x 4′-6″, single glass	1.000	Ea.	1.600	545	67.50	612.50
Insulating glass	1.000	Ea.	1.600	615	67.50	682.50
Plastic clad premium insulating glass, 4′ x 4′	1.000	Ea.	1.333	500	56.50	556.50
4′-6″ x 6′-6″	1.000	Ea.	1.455	740	61.50	801.50
5′-6″ x 6′-6″	1.000	Ea.	1.600	1,075	67.50	1,142.50
6′-6″ x 6′-6″	1.000	Ea.	1.600	1,075	67.50	1,142.50
Metal clad deluxe insulating glass, 4′ x 4′	1.000	Ea.	1.333	355	56.50	411.50
4′-6″ x 6′-6″	1.000	Ea.	1.455	525	61.50	586.50
5′-6″ x 6′-6″	1.000	Ea.	1.600	580	67.50	647.50
6′-6″ x 6′-6″	1.000	Ea.	1.600	665	67.50	732.50
Trim, interior casing, window 4′ x 4′	17.000	L.F.	.567	24	24	48
4′-6″ x 4′-6″	19.000	L.F.	.633	27	27	54
5′-0″ x 4′-0″	19.000	L.F.	.633	27	27	54
4′-6″ x 6′-6″	23.000	L.F.	.767	32.50	32.50	65
5′-6″ x 6′-6″	25.000	L.F.	.833	35.50	35.50	71
6′-6″ x 6′-6″	27.000	L.F.	.900	38	38	76
Paint or stain, interior or exterior, 4′ x 4′ window, 1 coat	1.000	Face	.667	.79	24.50	25.29
2 coats	1.000	Face	.667	.89	24.50	25.39
Primer & 1 coat	1.000	Face	.727	1.19	27	28.19
Primer & 2 coats	1.000	Face	.889	1.16	33	34.16
4′-6″ x 6′-6″ window, 1 coat	1.000	Face	.667	.79	24.50	25.29
2 coats	1.000	Face	.667	.89	24.50	25.39
Primer & 1 coat	1.000	Face	.727	1.19	27	28.19
Primer & 2 coats	1.000	Face	.889	1.16	33	34.16
6′-6″ x 6′-6″ window, 1 coat	1.000	Face	.889	2.17	33	35.17
2 coats	1.000	Face	1.333	3.97	49.50	53.47
Primer & 1 coat	1.000	Face	1.333	4.09	49.50	53.59
Primer & 2 coats	1.000	Face	1.600	6.20	59.50	65.70
Caulking, window, 4′ x 4′	1.000	Ea.	.516	2.72	22	24.72
4′-6″ x 4′-6″	1.000	Ea.	.581	3.06	24.50	27.56
5′-0″ x 4′-0″	1.000	Ea.	.581	3.06	24.50	27.56
4′-6″ x 6′-6″	1.000	Ea.	.710	3.74	30	33.74
5′-6″ x 6′-6″	1.000	Ea.	.774	4.08	33	37.08
6′-6″ x 6′-6″	1.000	Ea.	.839	4.42	35.50	39.92
Grilles, glass size, to 48″ x 48″	1.000	Ea.	.267	155	11.25	166.25
To 60″ x 68″	1.000	Ea.	.286	119	12.05	131.05
Drip cap, aluminum, 4′ long	4.000	L.F.	.080	1.32	3.40	4.72
4′-6″ long	4.500	L.F.	.090	1.49	3.83	5.32
5′ long	5.000	L.F.	.100	1.65	4.25	5.90
6′ long	6.000	L.F.	.120	1.98	5.10	7.08
Wood, 4′ long	4.000	L.F.	.133	5.65	5.65	11.30
4′-6″ long	4.500	L.F.	.150	6.35	6.35	12.70
5′ long	5.000	L.F.	.167	7.05	7.05	14.10
6′ long	6.000	L.F.	.200	8.45	8.45	16.90

52 Entrance Door Systems

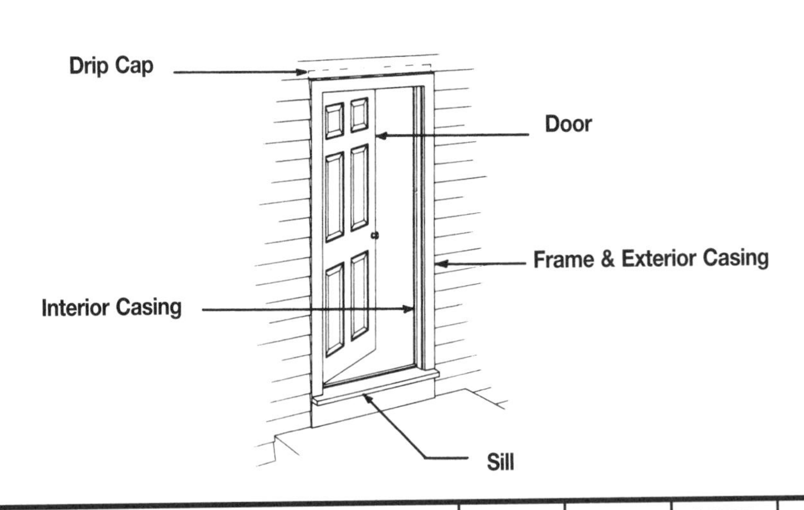

System Description	QUAN.	UNIT	LABOR HOURS	COST EACH		
				MAT.	INST.	TOTAL
COLONIAL, 6 PANEL, 3′ X 6′-8″, WOOD						
Door, 3′ x 6′-8″ x 1-3/4″ thick, pine, 6 panel colonial	1.000	Ea.	1.067	440	45	485
Frame, 5-13/16″ deep, incl. exterior casing & drip cap	17.000	L.F.	.725	148.75	30.60	179.35
Interior casing, 2-1/2″ wide	18.000	L.F.	.600	25.38	25.38	50.76
Sill, 8/4 x 8″ deep	3.000	L.F.	.480	51	20.25	71.25
Butt hinges, brass, 4-1/2″ x 4-1/2″	1.500	Pr.		21.53		21.53
Lockset	1.000	Ea.	.571	36.50	24	60.50
Weatherstripping, metal, spring type, bronze	1.000	Set	1.053	19.50	44.50	64
Paint, interior & exterior, primer & 2 coats	2.000	Face	1.778	12.30	66	78.30
TOTAL		Ea.	6.274	754.96	255.73	1,010.69
SOLID CORE BIRCH, FLUSH, 3′ X 6′-8″						
Door, 3′ x 6′-8″, 1-3/4″ thick, birch, flush solid core	1.000	Ea.	1.067	108	45	153
Frame, 5-13/16″ deep, incl. exterior casing & drip cap	17.000	L.F.	.725	148.75	30.60	179.35
Interior casing, 2-1/2″ wide	18.000	L.F.	.600	25.38	25.38	50.76
Sill, 8/4 x 8″ deep	3.000	L.F.	.480	51	20.25	71.25
Butt hinges, brass, 4-1/2″ x 4-1/2″	1.500	Pr.		21.53		21.53
Lockset	1.000	Ea.	.571	36.50	24	60.50
Weatherstripping, metal, spring type, bronze	1.000	Set	1.053	19.50	44.50	64
Paint, Interior & exterior, primer & 2 coats	2.000	Face	1.778	11.50	66	77.50
TOTAL		Ea.	6.274	422.16	255.73	677.89

These systems are on a cost per each door basis.

Description	QUAN.	UNIT	LABOR HOURS	COST EACH		
				MAT.	INST.	TOTAL

Important: See the Reference Section for critical supporting data - Reference Nos., Crews & Location Factors

Entrance Door Price Sheet	QUAN.	UNIT	LABOR HOURS	COST EACH		
				MAT.	INST.	TOTAL
Door exterior wood 1-3/4" thick, pine, dutch door, 2'-8" x 6'-8" minimum	1.000	Ea.	1.333	740	56.50	796.50
Maximum	1.000	Ea.	1.600	785	67.50	852.50
3'-0" x 6'-8", minimum	1.000	Ea.	1.333	755	56.50	811.50
Maximum	1.000	Ea.	1.600	825	67.50	892.50
Colonial, 6 panel, 2'-8" x 6'-8"	1.000	Ea.	1.000	395	42.50	437.50
3'-0" x 6'-8"	1.000	Ea.	1.067	440	45	485
8 panel, 2'-6" x 6'-8"	1.000	Ea.	1.000	655	42.50	697.50
3'-0" x 6'-8"	1.000	Ea.	1.067	590	45	635
Flush, birch, solid core, 2'-8" x 6'-8"	1.000	Ea.	1.000	104	42.50	146.50
3'-0" x 6'-8"	1.000	Ea.	1.067	108	45	153
Porch door, 2'-8" x 6'-8"	1.000	Ea.	1.000	365	42.50	407.50
3'-0" x 6'-8"	1.000	Ea.	1.067	290	45	335
Hand carved mahogany, 2'-8" x 6'-8"	1.000	Ea.	1.067	505	45	550
3'-0" x 6'-8"	1.000	Ea.	1.067	540	45	585
Rosewood, 2'-8" x 6'-8"	1.000	Ea.	1.067	785	45	830
3'-0" x 6-8"	1.000	Ea.	1.067	815	45	860
Door, metal clad wood 1-3/8" thick raised panel, 2'-8" x 6'-8"	1.000	Ea.	1.067	237	40	277
3'-0" x 6'-8"	1.000	Ea.	1.067	239	45	284
Deluxe metal door, 3'-0" x 6'-8"	1.000	Ea.	1.231	239	45	284
3'-0" x 6'-8"	1.000	Ea.	1.231	239	45	284
Frame, pine, including exterior trim & drip cap, 5/4, x 4-9/16" deep	17.000	L.F.	.725	95	30.50	125.50
5-13/16" deep	17.000	L.F.	.725	149	30.50	179.50
6-9/16" deep	17.000	L.F.	.725	147	30.50	177.50
Safety glass lites, add	1.000	Ea.		68.50		68.50
Interior casing, 2'-8" x 6'-8" door	18.000	L.F.	.600	25.50	25.50	51
3'-0" x 6'-8" door	19.000	L.F.	.633	27	27	54
Sill, oak, 8/4 x 8" deep	3.000	L.F.	.480	51	20.50	71.50
8/4 x 10" deep	3.000	L.F.	.533	69	22.50	91.50
Butt hinges, steel plated, 4-1/2" x 4-1/2", plain	1.500	Pr.		21.50		21.50
Ball bearing	1.500	Pr.		48		48
Bronze, 4-1/2" x 4-1/2", plain	1.500	Pr.		24.50		24.50
Ball bearing	1.500	Pr.		51		51
Lockset, minimum	1.000	Ea.	.571	36.50	24	60.50
Maximum	1.000	Ea.	1.000	153	42.50	195.50
Weatherstripping, metal, interlocking, zinc	1.000	Set	2.667	15.15	113	128.15
Bronze	1.000	Set	2.667	24	113	137
Spring type, bronze	1.000	Set	1.053	19.50	44.50	64
Rubber, minimum	1.000	Set	1.053	4.87	44.50	49.37
Maximum	1.000	Set	1.143	5.55	48.50	54.05
Felt minimum	1.000	Set	.571	2.26	24	26.26
Maximum	1.000	Set	.615	2.43	26	28.43
Paint or stain, flush door, interior or exterior, 1 coat	2.000	Face	.941	4.30	35	39.30
2 coats	2.000	Face	1.455	8.60	54	62.60
Primer & 1 coat	2.000	Face	1.455	7.40	54	61.40
Primer & 2 coats	2.000	Face	1.778	11.50	66	77.50
Paneled door, interior & exterior, 1 coat	2.000	Face	1.143	4.60	42	46.60
2 coats	2.000	Face	2.000	9.20	74	83.20
Primer & 1 coat	2.000	Face	1.455	7.90	54	61.90
Primer & 2 coats	2.000	Face	1.778	12.30	66	78.30

53 Sliding Door Systems

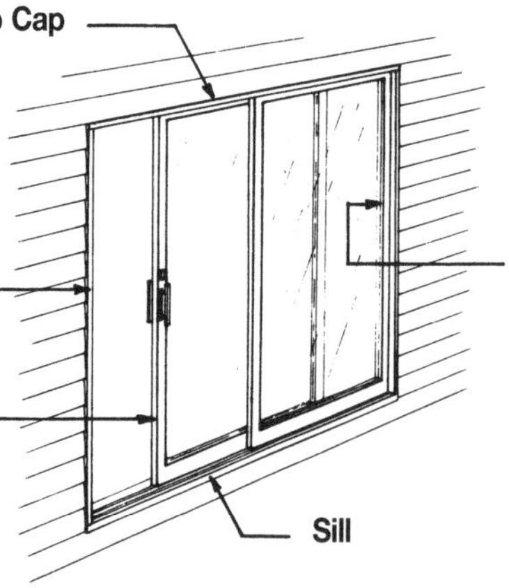

Drip Cap
Interior Casing
Frame & Exterior Casing
Door
Sill

System Description	QUAN.	UNIT	LABOR HOURS	COST EACH MAT.	COST EACH INST.	COST EACH TOTAL
WOOD SLIDING DOOR, 8′ WIDE, PREMIUM						
Wood, 5/8″ thick tempered insul. glass, 8′ wide, premium	1.000	Ea.	5.333	1,475	225	1,700
Interior casing	22.000	L.F.	.733	31.02	31.02	62.04
Exterior casing	22.000	L.F.	.733	31.02	31.02	62.04
Sill, oak, 8/4 x 8″ deep	8.000	L.F.	1.280	136	54	190
Drip cap	8.000	L.F.	.160	2.64	6.80	9.44
Paint, interior & exterior, primer & 2 coats	2.000	Face	2.816	17.60	104.72	122.32
TOTAL		Ea.	11.055	1,693.28	452.56	2,145.84
ALUMINUM SLIDING DOOR, 8′ WIDE, PREMIUM						
Aluminum, 5/8″ tempered insul. glass, 8′ wide, premium	1.000	Ea.	5.333	1,675	225	1,900
Interior casing	22.000	L.F.	.733	31.02	31.02	62.04
Exterior casing	22.000	L.F.	.733	31.02	31.02	62.04
Sill, oak, 8/4 x 8″ deep	8.000	L.F.	1.280	136	54	190
Drip cap	8.000	L.F.	.160	2.64	6.80	9.44
Paint, interior & exterior, primer & 2 coats	2.000	Face	2.816	17.60	104.72	122.32
TOTAL		Ea.	11.055	1,893.28	452.56	2,345.84

The cost of this system is on a cost per each door basis.

Description	QUAN.	UNIT	LABOR HOURS	COST EACH MAT.	COST EACH INST.	COST EACH TOTAL

Sliding Door Price Sheet	QUAN.	UNIT	LABOR HOURS	MAT.	INST.	TOTAL
Sliding door, wood, 5/8" thick, tempered insul. glass, 6' wide, premium	1.000	Ea.	4.000	1,275	169	1,444
Economy	1.000	Ea.	4.000	865	169	1,034
8'wide, wood premium	1.000	Ea.	5.333	1,475	225	1,700
Economy	1.000	Ea.	5.333	990	225	1,215
12' wide, wood premium	1.000	Ea.	6.400	3,175	270	3,445
Economy	1.000	Ea.	6.400	2,250	270	2,520
Aluminum, 5/8" thick, tempered insul. glass, 6'wide, premium	1.000	Ea.	4.000	1,475	169	1,644
Economy	1.000	Ea.	4.000	775	169	944
8'wide, premium	1.000	Ea.	5.333	1,675	225	1,900
Economy	1.000	Ea.	5.333	1,425	225	1,650
12' wide, premium	1.000	Ea.	6.400	2,700	270	2,970
Economy	1.000	Ea.	6.400	1,600	270	1,870
Interior casing, 6' wide door	20.000	L.F.	.667	28	28	56
8' wide door	22.000	L.F.	.733	31	31	62
12' wide door	26.000	L.F.	.867	36.50	36.50	73
Exterior casing, 6' wide door	20.000	L.F.	.667	28	28	56
8' wide door	22.000	L.F.	.733	31	31	62
12' wide door	26.000	L.F.	.867	36.50	36.50	73
Sill, oak, 8/4 x 8" deep, 6' wide door	6.000	L.F.	.960	102	40.50	142.50
8' wide door	8.000	L.F.	1.280	136	54	190
12' wide door	12.000	L.F.	1.920	204	81	285
8/4 x 10" deep, 6' wide door	6.000	L.F.	1.067	138	45	183
8' wide door	8.000	L.F.	1.422	184	60	244
12' wide door	12.000	L.F.	2.133	276	90	366
Drip cap, 6' wide door	6.000	L.F.	.120	1.98	5.10	7.08
8' wide door	8.000	L.F.	.160	2.64	6.80	9.44
12' wide door	12.000	L.F.	.240	3.96	10.20	14.16
Paint or stain, interior & exterior, 6' wide door, 1 coat	2.000	Face	1.600	5.60	59	64.60
2 coats	2.000	Face	1.600	5.60	59	64.60
Primer & 1 coat	2.000	Face	1.778	10.40	65.50	75.90
Primer & 2 coats	2.000	Face	2.560	16	95	111
8' wide door, 1 coat	2.000	Face	1.760	6.15	65	71.15
2 coats	2.000	Face	1.760	6.15	65	71.15
Primer & 1 coat	2.000	Face	1.955	11.45	72	83.45
Primer & 2 coats	2.000	Face	2.816	17.60	105	122.60
12' wide door, 1 coat	2.000	Face	2.080	7.30	77	84.30
2 coats	2.000	Face	2.080	7.30	77	84.30
Primer & 1 coat	2.000	Face	2.311	13.50	85.50	99
Primer & 2 coats	2.000	Face	3.328	21	124	145
Aluminum door, trim only, interior & exterior, 6' door, 1 coat	2.000	Face	.800	2.80	29.50	32.30
2 coats	2.000	Face	.800	2.80	29.50	32.30
Primer & 1 coat	2.000	Face	.889	5.20	33	38.20
Primer & 2 coats	2.000	Face	1.280	8	47.50	55.50
8' wide door, 1 coat	2.000	Face	.880	3.08	32.50	35.58
2 coats	2.000	Face	.880	3.08	32.50	35.58
Primer & 1 coat	2.000	Face	.978	5.70	36	41.70
Primer & 2 coats	2.000	Face	1.408	8.80	52.50	61.30
12' wide door, 1 coat	2.000	Face	1.040	3.64	38.50	42.14
2 coats	2.000	Face	1.040	3.64	38.50	42.14
Primer & 1 coat	2.000	Face	1.155	6.75	42.50	49.25
Primer & 2 coats	2.000	Face	1.664	10.40	62	72.40

56 Residential Garage Door Systems

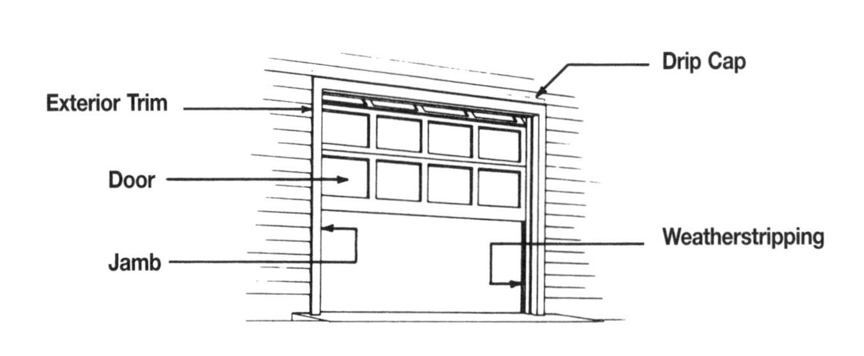

System Description	QUAN.	UNIT	LABOR HOURS	COST EACH MAT.	COST EACH INST.	COST EACH TOTAL
OVERHEAD, SECTIONAL GARAGE DOOR, 9′ X 7′						
Wood, overhead sectional door, std., incl. hardware, 9′ x 7′	1.000	Ea.	2.000	550	84.50	634.50
Jamb & header blocking, 2″ x 6″	25.000	L.F.	.901	16.25	38	54.25
Exterior trim	25.000	L.F.	.833	35.25	35.25	70.50
Paint, interior & exterior, primer & 2 coats	2.000	Face	3.556	24.60	132	156.60
Weatherstripping, molding type	1.000	Set	.767	32.43	32.43	64.86
Drip cap	9.000	L.F.	.180	2.97	7.65	10.62
TOTAL		Ea.	8.237	661.50	329.83	991.33
OVERHEAD, SECTIONAL GARAGE DOOR, 16′ X 7′						
Wood, overhead sectional, std., incl. hardware, 16′ x 7′	1.000	Ea.	2.667	1,100	113	1,213
Jamb & header blocking, 2″ x 6″	30.000	L.F.	1.081	19.50	45.60	65.10
Exterior trim	30.000	L.F.	1.000	42.30	42.30	84.60
Paint, interior & exterior, primer & 2 coats	2.000	Face	5.333	36.90	198	234.90
Weatherstripping, molding type	1.000	Set	1.000	42.30	42.30	84.60
Drip cap	16.000	L.F.	.320	5.28	13.60	18.88
TOTAL		Ea.	11.401	1,246.28	454.80	1,701.08
OVERHEAD, SWING-UP TYPE, GARAGE DOOR, 16′ X 7′						
Wood, overhead, swing-up, std., incl. hardware, 16′ x 7′	1.000	Ea.	2.667	720	113	833
Jamb & header blocking, 2″ x 6″	30.000	L.F.	1.081	19.50	45.60	65.10
Exterior trim	30.000	L.F.	1.000	42.30	42.30	84.60
Paint, interior & exterior, primer & 2 coats	2.000	Face	5.333	36.90	198	234.90
Weatherstripping, molding type	1.000	Set	1.000	42.30	42.30	84.60
Drip cap	16.000	L.F.	.320	5.28	13.60	18.88
TOTAL		Ea.	11.401	866.28	454.80	1,321.08

This system is on a cost per each door basis.

Description	QUAN.	UNIT	LABOR HOURS	COST EACH MAT.	COST EACH INST.	COST EACH TOTAL

Resi Garage Door Price Sheet	QUAN.	UNIT	LABOR HOURS	COST EACH		
				MAT.	INST.	TOTAL
Overhead, sectional, including hardware, fiberglass, 9′ x 7′, standard	1.000	Ea.	3.030	670	128	798
Deluxe	1.000	Ea.	3.030	850	128	978
16′ x 7′, standard	1.000	Ea.	2.667	1,200	113	1,313
Deluxe	1.000	Ea.	2.667	1,500	113	1,613
Hardboard, 9′ x 7′, standard	1.000	Ea.	2.000	445	84.50	529.50
Deluxe	1.000	Ea.	2.000	595	84.50	679.50
16′ x 7′, standard	1.000	Ea.	2.667	870	113	983
Deluxe	1.000	Ea.	2.667	1,025	113	1,138
Metal, 9′ x 7′, standard	1.000	Ea.	3.030	520	128	648
Deluxe	1.000	Ea.	2.000	700	84.50	784.50
16′ x 7′, standard	1.000	Ea.	5.333	665	225	890
Deluxe	1.000	Ea.	2.667	1,075	113	1,188
Wood, 9′ x 7′, standard	1.000	Ea.	2.000	550	84.50	634.50
Deluxe	1.000	Ea.	2.000	1,575	84.50	1,659.50
16′ x 7′, standard	1.000	Ea.	2.667	1,100	113	1,213
Deluxe	1.000	Ea.	2.667	2,300	113	2,413
Overhead swing-up type including hardware, fiberglass, 9′ x 7′, standard	1.000	Ea.	2.000	755	84.50	839.50
Deluxe	1.000	Ea.	2.000	795	84.50	879.50
16′ x 7′, standard	1.000	Ea.	2.667	955	113	1,068
Deluxe	1.000	Ea.	2.667	985	113	1,098
Hardboard, 9′ x 7′, standard	1.000	Ea.	2.000	345	84.50	429.50
Deluxe	1.000	Ea.	2.000	460	84.50	544.50
16′ x 7′, standard	1.000	Ea.	2.667	485	113	598
Deluxe	1.000	Ea.	2.667	720	113	833
Metal, 9′ x 7′, standard	1.000	Ea.	2.000	380	84.50	464.50
Deluxe	1.000	Ea.	2.000	670	84.50	754.50
16′ x 7′, standard	1.000	Ea.	2.667	595	113	708
Deluxe	1.000	Ea.	2.667	960	113	1,073
Wood, 9′ x 7′, standard	1.000	Ea.	2.000	415	84.50	499.50
Deluxe	1.000	Ea.	2.000	730	84.50	814.50
16′ x 7′, standard	1.000	Ea.	2.667	720	113	833
Deluxe	1.000	Ea.	2.667	1,025	113	1,138
Jamb & header blocking, 2″ x 6″, 9′ x 7′ door	25.000	L.F.	.901	16.25	38	54.25
16′ x 7′ door	30.000	L.F.	1.081	19.50	45.50	65
2″ x 8″, 9′ x 7′ door	25.000	L.F.	1.000	23.50	42.50	66
16′ x 7′ door	30.000	L.F.	1.200	28	50.50	78.50
Exterior trim, 9′ x 7′ door	25.000	L.F.	.833	35.50	35.50	71
16′ x 7′ door	30.000	L.F.	1.000	42.50	42.50	85
Paint or stain, interior & exterior, 9′ x 7′ door, 1 coat	1.000	Face	2.286	9.20	84	93.20
2 coats	1.000	Face	4.000	18.40	148	166.40
Primer & 1 coat	1.000	Face	2.909	15.85	108	123.85
Primer & 2 coats	1.000	Face	3.556	24.50	132	156.50
16′ x 7′ door, 1 coat	1.000	Face	3.429	13.80	126	139.80
2 coats	1.000	Face	6.000	27.50	222	249.50
Primer & 1 coat	1.000	Face	4.364	24	162	186
Primer & 2 coats	1.000	Face	5.333	37	198	235
Weatherstripping, molding type, 9′ x 7′ door	1.000	Set	.767	32.50	32.50	65
16′ x 7′ door	1.000	Set	1.000	42.50	42.50	85
Drip cap, 9′ door	9.000	L.F.	.180	2.97	7.65	10.62
16′ door	16.000	L.F.	.320	5.30	13.60	18.90
Garage door opener, economy	1.000	Ea.	1.000	315	42.50	357.50
Deluxe, including remote control	1.000	Ea.	1.000	460	42.50	502.50

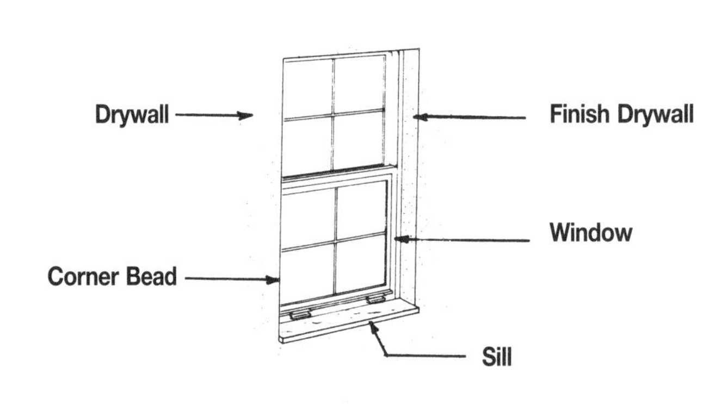

System Description	QUAN.	UNIT	LABOR HOURS	COST EACH		
				MAT.	INST.	TOTAL
SINGLE HUNG, 2′ X 3′ OPENING						
Window, 2′ x 3′ opening, enameled, insulating glass	1.000	Ea.	1.600	199	83	282
Blocking, 1″ x 3″ furring strip nailers	10.000	L.F.	.146	3.10	6.10	9.20
Drywall, 1/2″ thick, standard	5.000	S.F.	.040	1.50	1.70	3.20
Corner bead, 1″ x 1″, galvanized steel	8.000	L.F.	.160	1.20	6.80	8
Finish drywall, tape and finish corners inside and outside	16.000	L.F.	.269	1.28	11.36	12.64
Sill, slate	2.000	L.F.	.400	20.40	14.80	35.20
TOTAL		Ea.	2.615	226.48	123.76	350.24
SLIDING, 3′ X 2′ OPENING						
Window, 3′ x 2′ opening, enameled, insulating glass	1.000	Ea.	1.600	204	83	287
Blocking, 1″ x 3″ furring strip nailers	10.000	L.F.	.146	3.10	6.10	9.20
Drywall, 1/2″ thick, standard	5.000	S.F.	.040	1.50	1.70	3.20
Corner bead, 1″ x 1″, galvanized steel	7.000	L.F.	.140	1.05	5.95	7
Finish drywall, tape and finish corners inside and outside	14.000	L.F.	.236	1.12	9.94	11.06
Sill, slate	3.000	L.F.	.600	30.60	22.20	52.80
TOTAL		Ea.	2.762	241.37	128.89	370.26
AWNING, 3′-1″ X 3′-2″						
Window, 3′-1″ x 3′-2″ opening, enameled, insul. glass	1.000	Ea.	1.600	218	83	301
Blocking, 1″ x 3″ furring strip, nailers	12.500	L.F.	.182	3.88	7.63	11.51
Drywall, 1/2″ thick, standard	4.500	S.F.	.036	1.35	1.53	2.88
Corner bead, 1″ x 1″, galvanized steel	9.250	L.F.	.185	1.39	7.86	9.25
Finish drywall, tape and finish corners, inside and outside	18.500	L.F.	.312	1.48	13.14	14.62
Sill, slate	3.250	L.F.	.650	33.15	24.05	57.20
TOTAL		Ea.	2.965	259.25	137.21	396.46

Description	QUAN.	UNIT	LABOR HOURS	COST PER S.F.		
				MAT.	INST.	TOTAL

Important: See the Reference Section for critical supporting data - Reference Nos., Crews & Location Factors

Aluminum Window Price Sheet	QUAN.	UNIT	LABOR HOURS	COST EACH		
				MAT.	INST.	TOTAL
Window, aluminum, awning, 3′-1″ x 3′-2″, standard glass	1.000	Ea.	1.600	244	83	327
Insulating glass	1.000	Ea.	1.600	218	83	301
4′-5″ x 5′-3″, standard glass	1.000	Ea.	2.000	345	104	449
Insulating glass	1.000	Ea.	2.000	380	104	484
Casement, 3′-1″ x 3′-2″, standard glass	1.000	Ea.	1.600	340	83	423
Insulating glass	1.000	Ea.	1.600	315	83	398
Single hung, 2′ x 3′, standard glass	1.000	Ea.	1.600	164	83	247
Insulating glass	1.000	Ea.	1.600	199	83	282
2′-8″ x 6′-8″, standard glass	1.000	Ea.	2.000	350	104	454
Insulating glass	1.000	Ea.	2.000	450	104	554
3′-4″ x 5′-0″, standard glass	1.000	Ea.	1.778	224	92	316
Insulating glass	1.000	Ea.	1.778	315	92	407
Sliding, 3′ x 2′, standard glass	1.000	Ea.	1.600	185	83	268
Insulating glass	1.000	Ea.	1.600	204	83	287
5′ x 3′, standard glass	1.000	Ea.	1.778	234	92	326
Insulating glass	1.000	Ea.	1.778	330	92	422
8′ x 4′, standard glass	1.000	Ea.	2.667	335	138	473
Insulating glass	1.000	Ea.	2.667	540	138	678
Blocking, 1″ x 3″ furring, opening 3′ x 2′	10.000	L.F.	.146	3.10	6.10	9.20
3′ x 3′	12.500	L.F.	.182	3.88	7.65	11.53
3′ x 5′	16.000	L.F.	.233	4.96	9.75	14.71
4′ x 4′	16.000	L.F.	.233	4.96	9.75	14.71
4′ x 5′	18.000	L.F.	.262	5.60	11	16.60
4′ x 6′	20.000	L.F.	.291	6.20	12.20	18.40
4′ x 8′	24.000	L.F.	.349	7.45	14.65	22.10
6′-8″ x 2′-8″	19.000	L.F.	.276	5.90	11.60	17.50
Drywall, 1/2″ thick, standard, opening 3′ x 2′	5.000	S.F.	.040	1.50	1.70	3.20
3′ x 3′	6.000	S.F.	.048	1.80	2.04	3.84
3′ x 5′	8.000	S.F.	.064	2.40	2.72	5.12
4′ x 4′	8.000	S.F.	.064	2.40	2.72	5.12
4′ x 5′	9.000	S.F.	.072	2.70	3.06	5.76
4′ x 6′	10.000	S.F.	.080	3	3.40	6.40
4′ x 8′	12.000	S.F.	.096	3.60	4.08	7.68
6′-8″ x 2′	9.500	S.F.	.076	2.85	3.23	6.08
Corner bead, 1″ x 1″, galvanized steel, opening 3′ x 2′	7.000	L.F.	.140	1.05	5.95	7
3′ x 3′	9.000	L.F.	.180	1.35	7.65	9
3′ x 5′	11.000	L.F.	.220	1.65	9.35	11
4′ x 4′	12.000	L.F.	.240	1.80	10.20	12
4′ x 5′	13.000	L.F.	.260	1.95	11.05	13
4′ x 6′	14.000	L.F.	.280	2.10	11.90	14
4′ x 8′	16.000	L.F.	.320	2.40	13.60	16
6′-8″ x 2′	15.000	L.F.	.300	2.25	12.75	15
Tape and finish corners, inside and outside, opening 3′ x 2′	14.000	L.F.	.204	1.12	9.95	11.07
3′ x 3′	18.000	L.F.	.262	1.44	12.80	14.24
3′ x 5′	22.000	L.F.	.320	1.76	15.60	17.36
4′ x 4′	24.000	L.F.	.349	1.92	17.05	18.97
4′ x 5′	26.000	L.F.	.378	2.08	18.45	20.53
4′ x 6′	28.000	L.F.	.407	2.24	19.90	22.14
4′ x 8′	32.000	L.F.	.466	2.56	22.50	25.06
6′-8″ x 2′	30.000	L.F.	.437	2.40	21.50	23.90
Sill, slate, 2′ long	2.000	L.F.	.400	20.50	14.80	35.30
3′ long	3.000	L.F.	.600	30.50	22	52.50
4′ long	4.000	L.F.	.800	41	29.50	70.50
Wood, 1-5/8″ x 6-1/4″, 2′ long	2.000	L.F.	.128	14	5.40	19.40
3′ long	3.000	L.F.	.192	21	8.10	29.10
4′ long	4.000	L.F.	.256	28	10.80	38.80

System Description	QUAN.	UNIT	LABOR HOURS	COST EACH		
				MAT.	INST.	TOTAL
Storm door, aluminum, combination, storm & screen, anodized, 2'-6" x 6'-8"	1.000	Ea.	1.067	179	45	224
2'-8" x 6'-8"	1.000	Ea.	1.143	205	48.50	253.50
3'-0" x 6'-8"	1.000	Ea.	1.143	205	48.50	253.50
Mill finish, 2'-6" x 6'-8"	1.000	Ea.	1.067	237	45	282
2'-8" x 6'-8"	1.000	Ea.	1.143	237	48.50	285.50
3'-0" x 6'-8"	1.000	Ea.	1.143	257	48.50	305.50
Painted, 2'-6" x 6'-8"	1.000	Ea.	1.067	238	45	283
2'-8" x 6'-8"	1.000	Ea.	1.143	242	48.50	290.50
3'-0" x 6'-8"	1.000	Ea.	1.143	252	48.50	300.50
Wood, combination, storm & screen, crossbuck, 2'-6" x 6'-9"	1.000	Ea.	1.455	340	61.50	401.50
2'-8" x 6'-9"	1.000	Ea.	1.600	300	67.50	367.50
3'-0" x 6'-9"	1.000	Ea.	1.778	305	75	380
Full lite, 2'-6" x 6'-9"	1.000	Ea.	1.455	320	61.50	381.50
2'-8" x 6'-9"	1.000	Ea.	1.600	320	67.50	387.50
3'-0" x 6'-9"	1.000	Ea.	1.778	330	75	405
Windows, aluminum, combination storm & screen, basement, 1'-10" x 1'-0"	1.000	Ea.	.533	32.50	22.50	55
2'-9" x 1'-6"	1.000	Ea.	.533	35.50	22.50	58
3'-4" x 2'-0"	1.000	Ea.	.533	42.50	22.50	65
Double hung, anodized, 2'-0" x 3'-5"	1.000	Ea.	.533	84	22.50	106.50
2'-6" x 5'-0"	1.000	Ea.	.571	113	24	137
4'-0" x 6'-0"	1.000	Ea.	.640	238	27	265
Painted, 2'-0" x 3'-5"	1.000	Ea.	.533	100	22.50	122.50
2'-6" x 5'-0"	1.000	Ea.	.571	161	24	185
4'-0" x 6'-0"	1.000	Ea.	.640	288	27	315
Fixed window, anodized, 4'-6" x 4'-6"	1.000	Ea.	.640	128	27	155
5'-8" x 4'-6"	1.000	Ea.	.800	146	34	180
Painted, 4'-6" x 4'-6"	1.000	Ea.	.640	128	27	155
5'-8" x 4'-6"	1.000	Ea.	.800	146	34	180

Important: See the Reference Section for critical supporting data - Reference Nos., Crews & Location Factors

64 Shutters/Blinds Systems

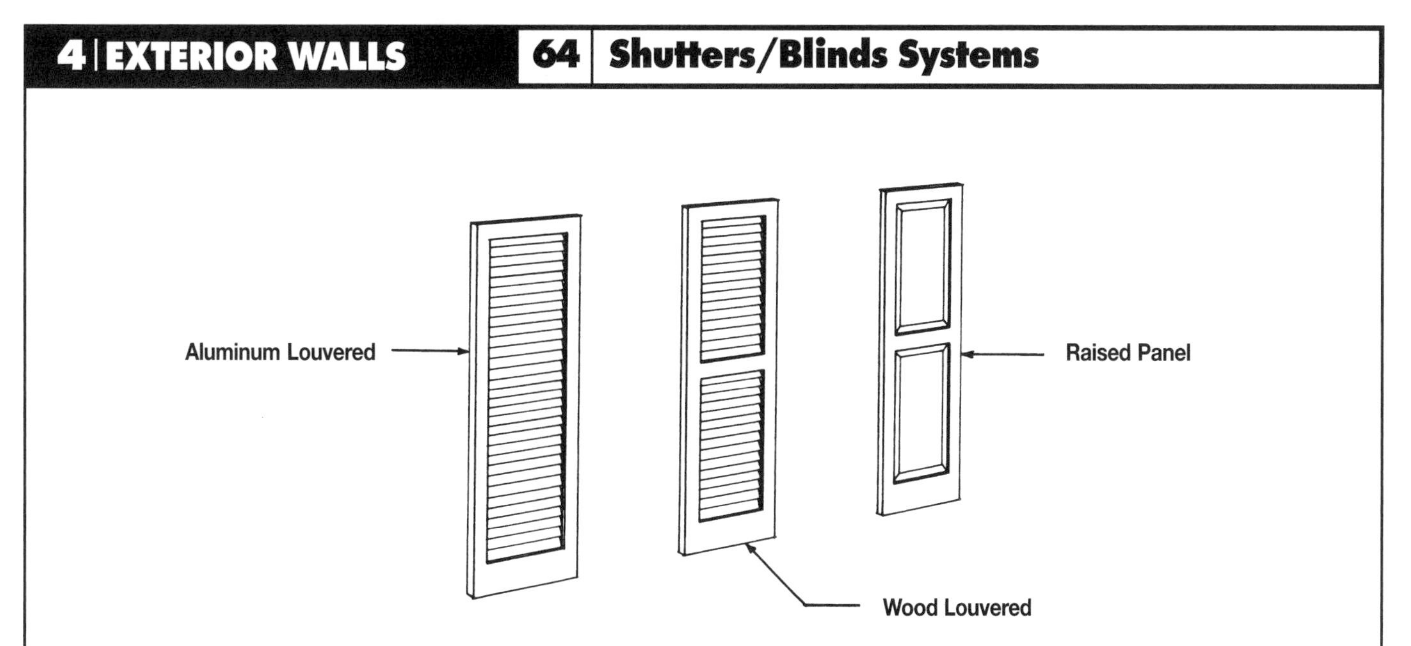

System Description	QUAN.	UNIT	LABOR HOURS	COST PER PAIR		
				MAT.	INST.	TOTAL
Shutters, exterior blinds, aluminum, louvered, 1′-4″ wide, 3″-0″ long	1.000	Set	.800	50	34	84
4′-0″ long	1.000	Set	.800	60	34	94
5′-4″ long	1.000	Set	.800	79	34	113
6′-8″ long	1.000	Set	.889	101	37.50	138.50
Wood, louvered, 1′-2″ wide, 3′-3″ long	1.000	Set	.800	95.50	34	129.50
4′-7″ long	1.000	Set	.800	129	34	163
5′-3″ long	1.000	Set	.800	146	34	180
1′-6″ wide, 3′-3″ long	1.000	Set	.800	101	34	135
4′-7″ long	1.000	Set	.800	142	34	176
Polystyrene, louvered, 1′-2″ wide, 3′-3″ long	1.000	Set	.800	28.50	34	62.50
4′-7″ long	1.000	Set	.800	35.50	34	69.50
5′-3″ long	1.000	Set	.800	40.50	34	74.50
6′-8″ long	1.000	Set	.889	48	37.50	85.50
Vinyl, louvered, 1′-2″ wide, 4′-7″ long	1.000	Set	.720	31.50	30.50	62
1′-4″ x 6′-8″ long	1.000	Set	.889	49.50	37.50	87

Division 5
Roofing

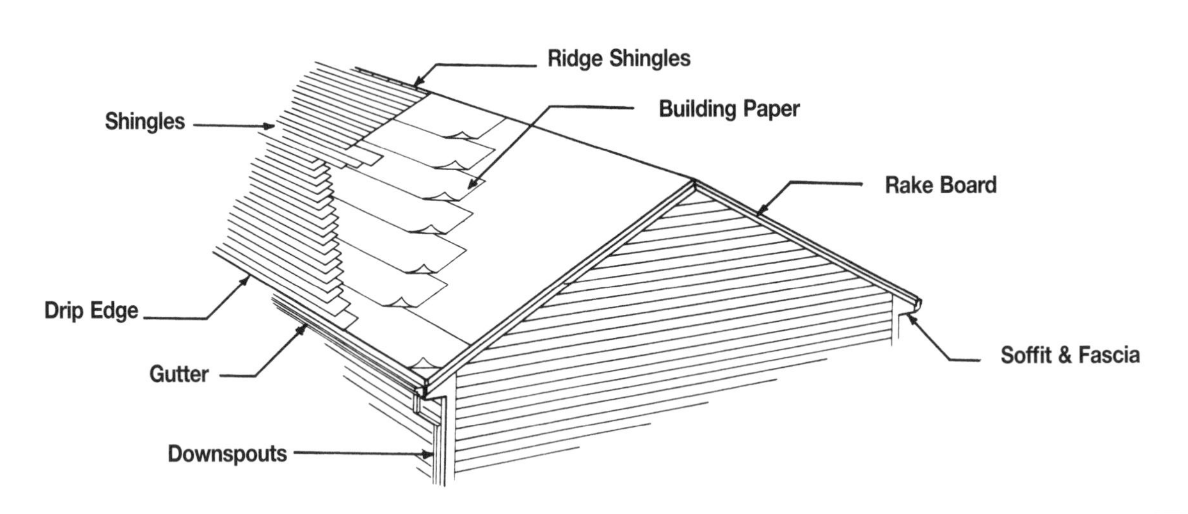

System Description	QUAN.	UNIT	LABOR HOURS	COST PER S.F.		
				MAT.	INST.	TOTAL
ASPHALT, ROOF SHINGLES, CLASS A						
Shingles, inorganic class A, 210-235 lb./sq., 4/12 pitch	1.160	S.F.	.017	.50	.68	1.18
Drip edge, metal, 5″ wide	.150	L.F.	.003	.07	.13	.20
Building paper, #15 felt	1.300	S.F.	.002	.05	.07	.12
Ridge shingles, asphalt	.042	L.F.	.001	.06	.04	.10
Soffit & fascia, white painted aluminum, 1′ overhang	.083	L.F.	.012	.25	.51	.76
Rake trim, 1″ x 6″	.040	L.F.	.002	.05	.07	.12
Rake trim, prime and paint	.040	L.F.	.002	.01	.07	.08
Gutter, seamless, aluminum painted	.083	L.F.	.006	.11	.26	.37
Downspouts, aluminum painted	.035	L.F.	.002	.07	.07	.14
TOTAL		S.F.	.047	1.17	1.90	3.07
WOOD, CEDAR SHINGLES NO. 1 PERFECTIONS, 18″ LONG						
Shingles, wood, cedar, No. 1 perfections, 4/12 pitch	1.160	S.F.	.035	2.22	1.48	3.70
Drip edge, metal, 5″ wide	.150	L.F.	.003	.07	.13	.20
Building paper, #15 felt	1.300	S.F.	.002	.05	.07	.12
Ridge shingles, cedar	.042	L.F.	.001	.12	.05	.17
Soffit & fascia, white painted aluminum, 1′ overhang	.083	L.F.	.012	.25	.51	.76
Rake trim, 1″ x 6″	.040	L.F.	.002	.05	.07	.12
Rake trim, prime and paint	.040	L.F.	.002	.01	.07	.08
Gutter, seamless, aluminum, painted	.083	L.F.	.006	.11	.26	.37
Downspouts, aluminum, painted	.035	L.F.	.002	.07	.07	.14
TOTAL		S.F.	.065	2.95	2.71	5.66

The prices in these systems are based on a square foot of plan area.
All quantities have been adjusted accordingly.

Description	QUAN.	UNIT	LABOR HOURS	COST PER S.F.		
				MAT.	INST.	TOTAL

 Important: See the Reference Section for critical supporting data - Reference Nos., Crews & Location Factors

Gable End Roofing Price Sheet	QUAN.	UNIT	LABOR HOURS	COST PER S.F.		
				MAT.	INST.	TOTAL
Shingles, asphalt, inorganic, class A, 210-235 lb./sq., 4/12 pitch	1.160	S.F.	.017	.50	.68	1.18
8/12 pitch	1.330	S.F.	.019	.55	.73	1.28
Laminated, multi-layered, 240-260 lb./sq., 4/12 pitch	1.160	S.F.	.021	.62	.83	1.45
8/12 pitch	1.330	S.F.	.023	.68	.90	1.58
Premium laminated, multi-layered, 260-300 lb./sq., 4/12 pitch	1.160	S.F.	.027	.83	1.06	1.89
8/12 pitch	1.330	S.F.	.030	.90	1.15	2.05
Clay tile, Spanish tile, red, 4/12 pitch	1.160	S.F.	.053	3.54	2.06	5.60
8/12 pitch	1.330	S.F.	.058	3.84	2.24	6.08
Mission tile, red, 4/12 pitch	1.160	S.F.	.083	8.60	3.24	11.84
8/12 pitch	1.330	S.F.	.090	9.30	3.51	12.81
French tile, red, 4/12 pitch	1.160	S.F.	.071	7.80	2.76	10.56
8/12 pitch	1.330	S.F.	.077	8.45	2.99	11.44
Slate, Buckingham, Virginia, black, 4/12 pitch	1.160	S.F.	.055	7.70	2.12	9.82
8/12 pitch	1.330	S.F.	.059	8.30	2.30	10.60
Vermont, black or grey, 4/12 pitch	1.160	S.F.	.055	4.74	2.12	6.86
8/12 pitch	1.330	S.F.	.059	5.15	2.30	7.45
Wood, No. 1 red cedar, 5X, 16″ long, 5″ exposure, 4/12 pitch	1.160	S.F.	.038	2.47	1.62	4.09
8/12 pitch	1.330	S.F.	.042	2.68	1.76	4.44
Fire retardant, 4/12 pitch	1.160	S.F.	.038	2.93	1.62	4.55
8/12 pitch	1.330	S.F.	.042	3.18	1.76	4.94
18″ long, No.1 perfections, 5″ exposure, 4/12 pitch	1.160	S.F.	.035	2.22	1.48	3.70
8/12 pitch	1.330	S.F.	.038	2.41	1.60	4.01
Fire retardant, 4/12 pitch	1.160	S.F.	.035	2.68	1.48	4.16
8/12 pitch	1.330	S.F.	.038	2.91	1.60	4.51
Resquared & rebutted, 18″ long, 6″ exposure, 4/12 pitch	1.160	S.F.	.032	2.77	1.36	4.13
8/12 pitch	1.330	S.F.	.035	3	1.47	4.47
Fire retardant, 4/12 pitch	1.160	S.F.	.032	3.23	1.36	4.59
8/12 pitch	1.330	S.F.	.035	3.50	1.47	4.97
Wood shakes hand split, 24″ long, 10″ exposure, 4/12 pitch	1.160	S.F.	.038	1.52	1.62	3.14
8/12 pitch	1.330	S.F.	.042	1.65	1.76	3.41
Fire retardant, 4/12 pitch	1.160	S.F.	.038	2.18	1.62	3.80
8/12 pitch	1.330	S.F.	.042	2.37	1.76	4.13
18″ long, 8″ exposure, 4/12 pitch	1.160	S.F.	.048	1.18	2.03	3.21
8/12 pitch	1.330	S.F.	.052	1.28	2.20	3.48
Fire retardant, 4/12 pitch	1.160	S.F.	.048	1.77	2.03	3.80
8/12 pitch	1.330	S.F.	.052	1.92	2.20	4.12
Drip edge, metal, 5″ wide	.150	L.F.	.003	.07	.13	.20
8″ wide	.150	L.F.	.003	.08	.13	.21
Building paper, #15 asphalt felt	1.300	S.F.	.002	.05	.07	.12
Ridge shingles, asphalt	.042	L.F.	.001	.06	.04	.10
Clay	.042	L.F.	.002	.41	.07	.48
Slate	.042	L.F.	.002	.42	.07	.49
Wood, shingles	.042	L.F.	.001	.12	.05	.17
Shakes	.042	L.F.	.001	.12	.05	.17
Soffit & fascia, aluminum, vented, 1′ overhang	.083	L.F.	.012	.25	.51	.76
2′ overhang	.083	L.F.	.013	.37	.56	.93
Vinyl, vented, 1′ overhang	.083	L.F.	.011	.14	.47	.61
2′ overhang	.083	L.F.	.012	.19	.51	.70
Wood, board fascia, plywood soffit, 1′ overhang	.083	L.F.	.004	.02	.14	.16
2′ overhang	.083	L.F.	.006	.03	.20	.23
Rake trim, painted, 1″ x 6″	.040	L.F.	.004	.06	.14	.20
1″ x 8″	.040	L.F.	.004	.09	.15	.24
Gutter, 5″ box, aluminum, seamless, painted	.083	L.F.	.006	.11	.26	.37
Vinyl	.083	L.F.	.006	.09	.25	.34
Downspout, 2″ x 3″, aluminum, one story house	.035	L.F.	.001	.05	.07	.12
Two story house	.060	L.F.	.003	.08	.12	.20
Vinyl, one story house	.035	L.F.	.002	.07	.07	.14
Two story house	.060	L.F.	.003	.08	.12	.20

08 Hip Roof Roofing Systems

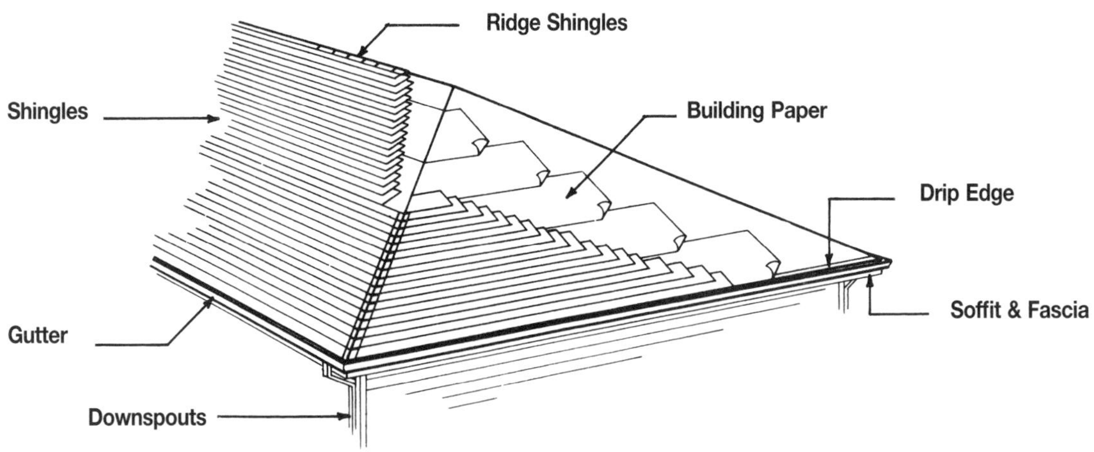

Ridge Shingles
Shingles
Building Paper
Drip Edge
Soffit & Fascia
Gutter
Downspouts

System Description	QUAN.	UNIT	LABOR HOURS	COST PER S.F.		
				MAT.	INST.	TOTAL
ASPHALT, ROOF SHINGLES, CLASS A						
Shingles, inorganic, class A, 210-235 lb./sq. 4/12 pitch	1.570	S.F.	.023	.67	.90	1.57
Drip edge, metal, 5" wide	.122	L.F.	.002	.05	.10	.15
Building paper, #15 asphalt felt	1.800	S.F.	.002	.07	.10	.17
Ridge shingles, asphalt	.075	L.F.	.002	.11	.07	.18
Soffit & fascia, white painted aluminum, 1′ overhang	.120	L.F.	.017	.36	.74	1.10
Gutter, seamless, aluminum, painted	.120	L.F.	.008	.17	.37	.54
Downspouts, aluminum, painted	.035	L.F.	.002	.07	.07	.14
TOTAL		S.F.	.056	1.50	2.35	3.85
WOOD, CEDAR SHINGLES, NO. 1 PERFECTIONS, 18" LONG						
Shingles, red cedar, No. 1 perfections, 5" exp., 4/12 pitch	1.570	S.F.	.047	2.96	1.97	4.93
Drip edge, metal, 5" wide	.122	L.F.	.002	.05	.10	.15
Building paper, #15 asphalt felt	1.800	S.F.	.002	.07	.10	.17
Ridge shingles, wood, cedar	.075	L.F.	.002	.22	.09	.31
Soffit & fascia, white painted aluminum, 1′ overhang	.120	L.F.	.017	.36	.74	1.10
Gutter, seamless, aluminum, painted	.120	L.F.	.008	.17	.37	.54
Downspouts, aluminum, painted	.035	L.F.	.002	.07	.07	.14
TOTAL		S.F.	.080	3.90	3.44	7.34

The prices in these systems are based on a square foot of plan area.
All quantities have been adjusted accordingly.

Description	QUAN.	UNIT	LABOR HOURS	COST PER S.F.		
				MAT.	INST.	TOTAL

 Important: See the Reference Section for critical supporting data - Reference Nos., Crews & Location Factors

Hip Roof - Roofing Price Sheet	QUAN.	UNIT	LABOR HOURS	COST PER S.F.		
				MAT.	INST.	TOTAL
Shingles, asphalt, inorganic, class A, 210-235 lb./sq., 4/12 pitch	1.570	S.F.	.023	.67	.90	1.57
8/12 pitch	1.850	S.F.	.028	.80	1.07	1.87
Laminated, multi-layered, 240-260 lb./sq., 4/12 pitch	1.570	S.F.	.028	.83	1.10	1.93
8/12 pitch	1.850	S.F.	.034	.99	1.31	2.30
Prem. laminated, multi-layered, 260-300 lb./sq., 4/12 pitch	1.570	S.F.	.037	1.10	1.42	2.52
8/12 pitch	1.850	S.F.	.043	1.31	1.68	2.99
Clay tile, Spanish tile, red, 4/12 pitch	1.570	S.F.	.071	4.72	2.75	7.47
8/12 pitch	1.850	S.F.	.084	5.60	3.27	8.87
Mission tile, red, 4/12 pitch	1.570	S.F.	.111	11.45	4.32	15.77
8/12 pitch	1.850	S.F.	.132	13.60	5.15	18.75
French tile, red, 4/12 pitch	1.570	S.F.	.095	10.40	3.68	14.08
8/12 pitch	1.850	S.F.	.113	12.35	4.37	16.72
Slate, Buckingham, Virginia, black, 4/12 pitch	1.570	S.F.	.073	10.25	2.83	13.08
8/12 pitch	1.850	S.F.	.087	12.15	3.36	15.51
Vermont, black or grey, 4/12 pitch	1.570	S.F.	.073	6.30	2.83	9.13
8/12 pitch	1.850	S.F.	.087	7.50	3.36	10.86
Wood, red cedar, No.1 5X, 16″ long, 5″ exposure, 4/12 pitch	1.570	S.F.	.051	3.30	2.16	5.46
8/12 pitch	1.850	S.F.	.061	3.91	2.57	6.48
Fire retardant, 4/12 pitch	1.570	S.F.	.051	3.92	2.16	6.08
8/12 pitch	1.850	S.F.	.061	4.64	2.57	7.21
18″ long, No.1 perfections, 5″ exposure, 4/12 pitch	1.570	S.F.	.047	2.96	1.97	4.93
8/12 pitch	1.850	S.F.	.055	3.52	2.34	5.86
Fire retardant, 4/12 pitch	1.570	S.F.	.047	3.58	1.97	5.55
8/12 pitch	1.850	S.F.	.055	4.25	2.34	6.59
Resquared & rebutted, 18″ long, 6″ exposure, 4/12 pitch	1.570	S.F.	.043	3.70	1.81	5.51
8/12 pitch	1.850	S.F.	.051	4.39	2.15	6.54
Fire retardant, 4/12 pitch	1.570	S.F.	.043	4.32	1.81	6.13
8/12 pitch	1.850	S.F.	.051	5.10	2.15	7.25
Wood shakes hand split, 24″ long, 10″ exposure, 4/12 pitch	1.570	S.F.	.051	2.03	2.16	4.19
8/12 pitch	1.850	S.F.	.061	2.41	2.57	4.98
Fire retardant, 4/12 pitch	1.570	S.F.	.051	2.91	2.16	5.07
8/12 pitch	1.850	S.F.	.061	3.46	2.57	6.03
18″ long, 8″ exposure, 4/12 pitch	1.570	S.F.	.064	1.58	2.70	4.28
8/12 pitch	1.850	S.F.	.076	1.87	3.21	5.08
Fire retardant, 4/12 pitch	1.570	S.F.	.064	2.37	2.70	5.07
8/12 pitch	1.850	S.F.	.076	2.81	3.21	6.02
Drip edge, metal, 5″ wide	.122	L.F.	.002	.05	.10	.15
8″ wide	.122	L.F.	.002	.07	.10	.17
Building paper, #15 asphalt felt	1.800	S.F.	.002	.07	.10	.17
Ridge shingles, asphalt	.075	L.F.	.002	.11	.07	.18
Clay	.075	L.F.	.003	.74	.12	.86
Slate	.075	L.F.	.003	.74	.12	.86
Wood, shingles	.075	L.F.	.002	.22	.09	.31
Shakes	.075	L.F.	.002	.22	.09	.31
Soffit & fascia, aluminum, vented, 1′ overhang	.120	L.F.	.017	.36	.74	1.10
2′ overhang	.120	L.F.	.019	.54	.81	1.35
Vinyl, vented, 1′ overhang	.120	L.F.	.016	.20	.68	.88
2′ overhang	.120	L.F.	.017	.28	.74	1.02
Wood, board fascia, plywood soffit, 1′ overhang	.120	L.F.	.004	.02	.14	.16
2′ overhang	.120	L.F.	.006	.03	.20	.23
Gutter, 5″ box, aluminum, seamless, painted	.120	L.F.	.008	.17	.37	.54
Vinyl	.120	L.F.	.009	.13	.37	.50
Downspout, 2″ x 3″, aluminum, one story house	.035	L.F.	.002	.07	.07	.14
Two story house	.060	L.F.	.003	.08	.12	.20
Vinyl, one story house	.035	L.F.	.001	.05	.07	.12
Two story house	.060	L.F.	.003	.08	.12	.20

12 Gambrel Roofing Systems

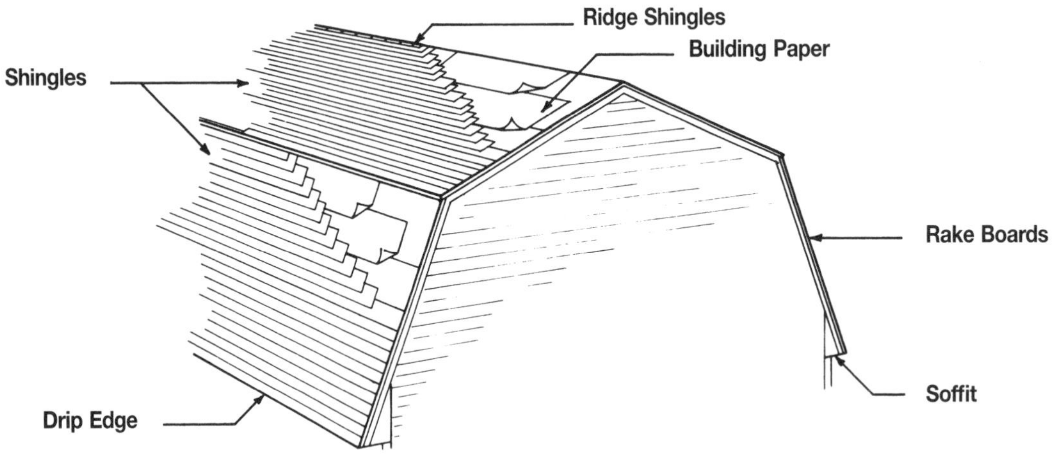

Ridge Shingles
Building Paper
Shingles
Rake Boards
Soffit
Drip Edge

System Description	QUAN.	UNIT	LABOR HOURS	COST PER S.F.		
				MAT.	INST.	TOTAL
ASPHALT, ROOF SHINGLES, CLASS A						
Shingles, asphalt, inorganic, class A, 210-235 lb./sq.	1.450	S.F.	.022	.63	.85	1.48
Drip edge, metal, 5″ wide	.146	L.F.	.003	.06	.12	.18
Building paper, #15 asphalt felt	1.500	S.F.	.002	.06	.08	.14
Ridge shingles, asphalt	.042	L.F.	.001	.06	.04	.10
Soffit & fascia, painted aluminum, 1′ overhang	.083	L.F.	.012	.25	.51	.76
Rake trim, 1″ x 6″	.063	L.F.	.003	.07	.11	.18
Rake trim, prime and paint	.063	L.F.	.003	.02	.10	.12
Gutter, seamless, alumunum, painted	.083	L.F.	.006	.11	.26	.37
Downspouts, aluminum, painted	.042	L.F.	.002	.08	.09	.17
TOTAL		S.F.	.054	1.34	2.16	3.50
WOOD, CEDAR SHINGLES, NO. 1 PERFECTIONS, 18″ LONG						
Shingles, wood, red cedar, No. 1 perfections, 5″ exposure	1.450	S.F.	.044	2.78	1.85	4.63
Drip edge, metal, 5″ wide	.146	L.F.	.003	.06	.12	.18
Building paper, #15 asphalt felt	1.500	S.F.	.002	.06	.08	.14
Ridge shingles, wood	.042	L.F.	.001	.12	.05	.17
Soffit & fascia, white painted aluminum, 1′ overhang	.083	L.F.	.012	.25	.51	.76
Rake trim, 1″ x 6″	.063	L.F.	.003	.07	.11	.18
Rake trim, prime and paint	.063	L.F.	.001	.01	.05	.06
Gutter, seamless, aluminum, painted	.083	L.F.	.006	.11	.26	.37
Downspouts, aluminum, painted	.042	L.F.	.002	.08	.09	.17
TOTAL		S.F.	.074	3.54	3.12	6.66

The prices in this system are based on a square foot of plan area.
All quantities have been adjusted accordingly.

Description	QUAN.	UNIT	LABOR HOURS	COST PER S.F.		
				MAT.	INST.	TOTAL

Gambrel Roofing Price Sheet	QUAN.	UNIT	LABOR HOURS	COST PER S.F.		
				MAT.	INST.	TOTAL
Shingles, asphalt, standard, inorganic, class A, 210-235 lb./sq.	1.450	S.F.	.022	.63	.85	1.48
Laminated, multi-layered, 240-260 lb./sq.	1.450	S.F.	.027	.78	1.04	1.82
Premium laminated, multi-layered, 260-300 lb./sq.	1.450	S.F.	.034	1.04	1.33	2.37
Slate, Buckingham, Virginia, black	1.450	S.F.	.069	9.60	2.66	12.26
Vermont, black or grey	1.450	S.F.	.069	5.95	2.66	8.61
Wood, red cedar, No.1 5X, 16" long, 5" exposure, plain	1.450	S.F.	.048	3.09	2.03	5.12
Fire retardant	1.450	S.F.	.048	3.67	2.03	5.70
18" long, No.1 perfections, 6" exposure, plain	1.450	S.F.	.044	2.78	1.85	4.63
Fire retardant	1.450	S.F.	.044	3.36	1.85	5.21
Resquared & rebutted, 18" long, 6" exposure, plain	1.450	S.F.	.040	3.47	1.70	5.17
Fire retardant	1.450	S.F.	.040	4.04	1.70	5.74
Shakes, hand split, 24" long, 10" exposure, plain	1.450	S.F.	.048	1.91	2.03	3.94
Fire retardant	1.450	S.F.	.048	2.74	2.03	4.77
18" long, 8" exposure, plain	1.450	S.F.	.060	1.48	2.54	4.02
Fire retardant	1.450	S.F.	.060	2.22	2.54	4.76
Drip edge, metal, 5" wide	.146	L.F.	.003	.06	.12	.18
8" wide	.146	L.F.	.003	.08	.12	.20
Building paper, #15 asphalt felt	1.500	S.F.	.002	.06	.08	.14
Ridge shingles, asphalt	.042	L.F.	.001	.06	.04	.10
Slate	.042	L.F.	.002	.42	.07	.49
Wood, shingles	.042	L.F.	.001	.12	.05	.17
Shakes	.042	L.F.	.001	.12	.05	.17
Soffit & fascia, aluminum, vented, 1' overhang	.083	L.F.	.012	.25	.51	.76
2' overhang	.083	L.F.	.013	.37	.56	.93
Vinyl vented, 1' overhang	.083	L.F.	.011	.14	.47	.61
2' overhang	.083	L.F.	.012	.19	.51	.70
Wood board fascia, plywood soffit, 1' overhang	.083	L.F.	.004	.02	.14	.16
2' overhang	.083	L.F.	.006	.03	.20	.23
Rake trim, painted, 1" x 6"	.063	L.F.	.006	.09	.21	.30
1" x 8"	.063	L.F.	.007	.12	.28	.40
Gutter, 5" box, aluminum, seamless, painted	.083	L.F.	.006	.11	.26	.37
Vinyl	.083	L.F.	.006	.09	.25	.34
Downspout 2" x 3", aluminum, one story house	.042	L.F.	.002	.06	.08	.14
Two story house	.070	L.F.	.003	.10	.14	.24
Vinyl, one story house	.042	L.F.	.002	.06	.08	.14
Two story house	.070	L.F.	.003	.10	.14	.24

16 Mansard Roofing Systems

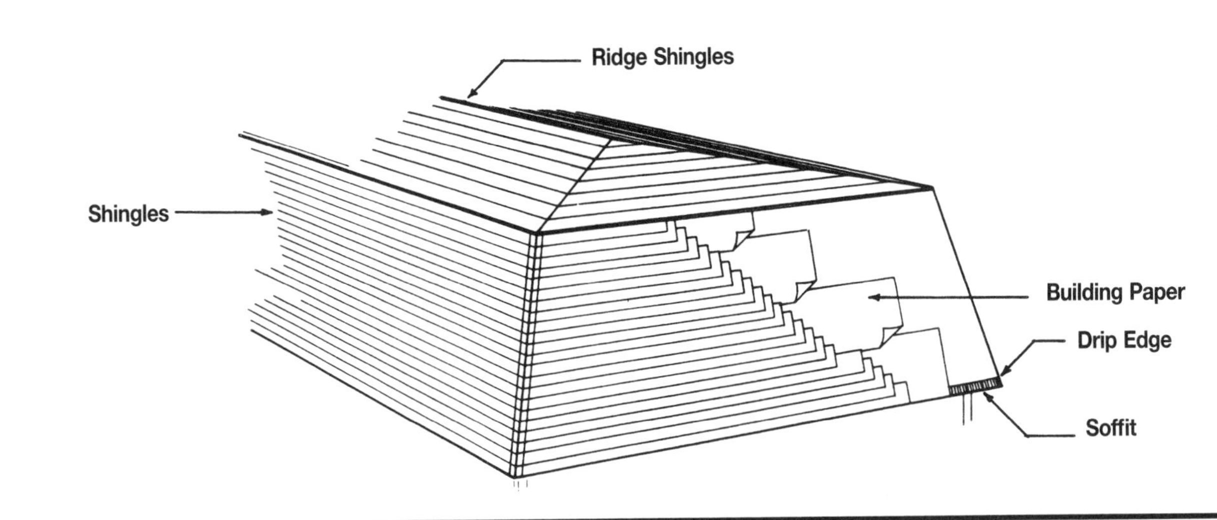

System Description	QUAN.	UNIT	LABOR HOURS	COST PER S.F. MAT.	COST PER S.F. INST.	COST PER S.F. TOTAL
ASPHALT, ROOF SHINGLES, CLASS A						
Shingles, standard inorganic class A 210-235 lb./sq.	2.210	S.F.	.032	.92	1.24	2.16
Drip edge, metal, 5″ wide	.122	L.F.	.002	.05	.10	.15
Building paper, #15 asphalt felt	2.300	S.F.	.003	.09	.12	.21
Ridge shingles, asphalt	.090	L.F.	.002	.13	.08	.21
Soffit & fascia, white painted aluminum, 1′ overhang	.122	L.F.	.018	.36	.75	1.11
Gutter, seamless, aluminum, painted	.122	L.F.	.008	.17	.38	.55
Downspouts, aluminum, painted	.042	L.F.	.002	.08	.09	.17
TOTAL		S.F.	.067	1.80	2.76	4.56
WOOD, CEDAR SHINGLES, NO. 1 PERFECTIONS, 18″ LONG						
Shingles, wood, red cedar, No. 1 perfections, 5″ exposure	2.210	S.F.	.064	4.07	2.71	6.78
Drip edge, metal, 5″ wide	.122	L.F.	.002	.05	.10	.15
Building paper, #15 asphalt felt	2.300	S.F.	.003	.09	.12	.21
Ridge shingles, wood	.090	L.F.	.003	.26	.11	.37
Soffit & fascia, white painted aluminum, 1′ overhang	.122	L.F.	.018	.36	.75	1.11
Gutter, seamless, aluminum, painted	.122	L.F.	.008	.17	.38	.55
Downspouts, aluminum, painted	.042	L.F.	.002	.08	.09	.17
TOTAL		S.F.	.100	5.08	4.26	9.34

The prices in these systems are based on a square foot of plan area.
All quantities have been adjusted accordingly.

Description	QUAN.	UNIT	LABOR HOURS	COST PER S.F. MAT.	COST PER S.F. INST.	COST PER S.F. TOTAL

Important: See the Reference Section for critical supporting data - Reference Nos., Crews & Location Factors

Mansard Roofing Price Sheet	QUAN.	UNIT	LABOR HOURS	COST PER S.F.		
				MAT.	INST.	TOTAL
Shingles, asphalt, standard, inorganic, class A, 210-235 lb./sq.	2.210	S.F.	.032	.92	1.24	2.16
Laminated, multi-layered, 240-260 lb./sq.	2.210	S.F.	.039	1.14	1.52	2.66
Premium laminated, multi-layered, 260-300 lb./sq.	2.210	S.F.	.050	1.52	1.95	3.47
Slate Buckingham, Virginia, black	2.210	S.F.	.101	14.10	3.89	17.99
Vermont, black or grey	2.210	S.F.	.101	8.70	3.89	12.59
Wood, red cedar, No.1 5X, 16″ long, 5″ exposure, plain	2.210	S.F.	.070	4.53	2.97	7.50
Fire retardant	2.210	S.F.	.070	5.40	2.97	8.37
18″ long, No.1 perfections 6″ exposure, plain	2.210	S.F.	.064	4.07	2.71	6.78
Fire retardant	2.210	S.F.	.064	4.92	2.71	7.63
Resquared & rebutted, 18″ long, 6″ exposure, plain	2.210	S.F.	.059	5.10	2.49	7.59
Fire retardant	2.210	S.F.	.059	5.95	2.49	8.44
Shakes, hand split, 24″ long 10″ exposure, plain	2.210	S.F.	.070	2.79	2.97	5.76
Fire retardant	2.210	S.F.	.070	4	2.97	6.97
18″ long, 8″ exposure, plain	2.210	S.F.	.088	2.17	3.72	5.89
Fire retardant	2.210	S.F.	.088	3.26	3.72	6.98
Drip edge, metal, 5″ wide	.122	S.F.	.002	.05	.10	.15
8″ wide	.122	S.F.	.002	.07	.10	.17
Building paper, #15 asphalt felt	2.300	S.F.	.003	.09	.12	.21
Ridge shingles, asphalt	.090	L.F.	.002	.13	.08	.21
Slate	.090	L.F.	.004	.89	.14	1.03
Wood, shingles	.090	L.F.	.003	.26	.11	.37
Shakes	.090	L.F.	.003	.26	.11	.37
Soffit & fascia, aluminum vented, 1′ overhang	.122	L.F.	.018	.36	.75	1.11
2′ overhang	.122	L.F.	.020	.55	.82	1.37
Vinyl vented, 1′ overhang	.122	L.F.	.016	.20	.69	.89
2′ overhang	.122	L.F.	.018	.29	.75	1.04
Wood board fascia, plywood soffit, 1′ overhang	.122	L.F.	.013	.37	.52	.89
2′ overhang	.122	L.F.	.019	.48	.77	1.25
Gutter, 5″ box, aluminum, seamless, painted	.122	L.F.	.008	.17	.38	.55
Vinyl	.122	L.F.	.009	.13	.37	.50
Downspout 2″ x 3″, aluminum, one story house	.042	L.F.	.002	.06	.08	.14
Two story house	.070	L.F.	.003	.10	.14	.24
Vinyl, one story house	.042	L.F.	.002	.06	.08	.14
Two story house	.070	L.F.	.003	.10	.14	.24

20 Shed Roofing Systems

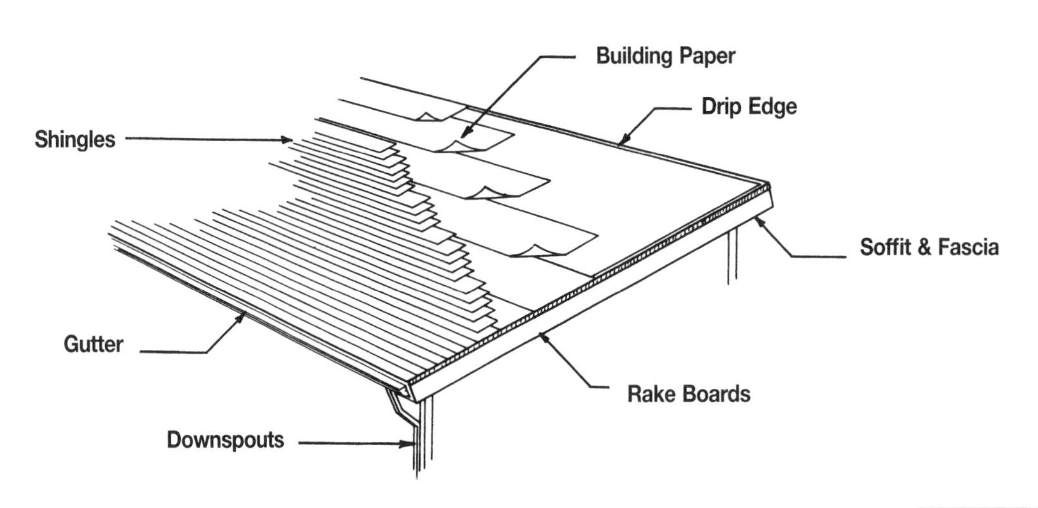

System Description	QUAN.	UNIT	LABOR HOURS	COST PER S.F.		
				MAT.	INST.	TOTAL
ASPHALT, ROOF SHINGLES, CLASS A						
Shingles, inorganic class A 210-235 lb./sq. 4/12 pitch	1.230	S.F.	.019	.55	.73	1.28
Drip edge, metal, 5″ wide	.100	L.F.	.002	.04	.09	.13
Building paper, #15 asphalt felt	1.300	S.F.	.002	.05	.07	.12
Soffit & fascia, white painted aluminum, 1′ overhang	.080	L.F.	.012	.24	.49	.73
Rake trim, 1″ x 6″	.043	L.F.	.002	.05	.07	.12
Rake trim, prime and paint	.043	L.F.	.002	.01	.07	.08
Gutter, seamless, aluminum, painted	.040	L.F.	.003	.06	.12	.18
Downspouts, painted aluminum	.020	L.F.	.001	.04	.04	.08
TOTAL		S.F.	.043	1.04	1.68	2.72
WOOD, CEDAR SHINGLES, NO. 1 PERFECTIONS, 18″ LONG						
Shingles, red cedar, No. 1 perfections, 5″ exp., 4/12 pitch	1.230	S.F.	.035	2.22	1.48	3.70
Drip edge, metal, 5″ wide	.100	L.F.	.002	.04	.09	.13
Building paper, #15 asphalt felt	1.300	S.F.	.002	.05	.07	.12
Soffit & fascia, white painted aluminum, 1′ overhang	.080	L.F.	.012	.24	.49	.73
Rake trim, 1″ x 6″	.043	L.F.	.002	.05	.07	.12
Rake trim, prime and paint	.043	L.F.	.001	.01	.03	.04
Gutter, seamless, aluminum, painted	.040	L.F.	.003	.06	.12	.18
Downspouts, painted aluminum	.020	L.F.	.001	.04	.04	.08
TOTAL		S.F.	.058	2.71	2.39	5.10

The prices in these systems are based on a square foot of plan area.
All quantities have been adjusted accordingly.

Description	QUAN.	UNIT	LABOR HOURS	COST PER S.F.		
				MAT.	INST.	TOTAL

Shed Roofing Price Sheet	QUAN.	UNIT	LABOR HOURS	COST PER S.F.		
				MAT.	INST.	TOTAL
Shingles, asphalt, inorganic, class A, 210-235 lb./sq., 4/12 pitch	1.230	S.F.	.017	.50	.68	1.18
8/12 pitch	1.330	S.F.	.019	.55	.73	1.28
Laminated, multi-layered, 240-260 lb./sq. 4/12 pitch	1.230	S.F.	.021	.62	.83	1.45
8/12 pitch	1.330	S.F.	.023	.68	.90	1.58
Premium laminated, multi-layered, 260-300 lb./sq. 4/12 pitch	1.230	S.F.	.027	.83	1.06	1.89
8/12 pitch	1.330	S.F.	.030	.90	1.15	2.05
Clay tile, Spanish tile, red, 4/12 pitch	1.230	S.F.	.053	3.54	2.06	5.60
8/12 pitch	1.330	S.F.	.058	3.84	2.24	6.08
Mission tile, red, 4/12 pitch	1.230	S.F.	.083	8.60	3.24	11.84
8/12 pitch	1.330	S.F.	.090	9.30	3.51	12.81
French tile, red, 4/12 pitch	1.230	S.F.	.071	7.80	2.76	10.56
8/12 pitch	1.330	S.F.	.077	8.45	2.99	11.44
Slate, Buckingham, Virginia, black, 4/12 pitch	1.230	S.F.	.055	7.70	2.12	9.82
8/12 pitch	1.330	S.F.	.059	8.30	2.30	10.60
Vermont, black or grey, 4/12 pitch	1.230	S.F.	.055	4.74	2.12	6.86
8/12 pitch	1.330	S.F.	.059	5.15	2.30	7.45
Wood, red cedar, No.1 5X, 16″ long, 5″ exposure, 4/12 pitch	1.230	S.F.	.038	2.47	1.62	4.09
8/12 pitch	1.330	S.F.	.042	2.68	1.76	4.44
Fire retardant, 4/12 pitch	1.230	S.F.	.038	2.93	1.62	4.55
8/12 pitch	1.330	S.F.	.042	3.18	1.76	4.94
18″ long, 6″ exposure, 4/12 pitch	1.230	S.F.	.035	2.22	1.48	3.70
8/12 pitch	1.330	S.F.	.038	2.41	1.60	4.01
Fire retardant, 4/12 pitch	1.230	S.F.	.035	2.68	1.48	4.16
8/12 pitch	1.330	S.F.	.038	2.91	1.60	4.51
Resquared & rebutted, 18″ long, 6″ exposure, 4/12 pitch	1.230	S.F.	.032	2.77	1.36	4.13
8/12 pitch	1.330	S.F.	.035	3	1.47	4.47
Fire retardant, 4/12 pitch	1.230	S.F.	.032	3.23	1.36	4.59
8/12 pitch	1.330	S.F.	.035	3.50	1.47	4.97
Wood shakes, hand split, 24″ long, 10″ exposure, 4/12 pitch	1.230	S.F.	.038	1.52	1.62	3.14
8/12 pitch	1.330	S.F.	.042	1.65	1.76	3.41
Fire retardant, 4/12 pitch	1.230	S.F.	.038	2.18	1.62	3.80
8/12 pitch	1.330	S.F.	.042	2.37	1.76	4.13
18″ long, 8″ exposure, 4/12 pitch	1.230	S.F.	.048	1.18	2.03	3.21
8/12 pitch	1.330	S.F.	.052	1.28	2.20	3.48
Fire retardant, 4/12 pitch	1.230	S.F.	.048	1.77	2.03	3.80
8/12 pitch	1.330	S.F.	.052	1.92	2.20	4.12
Drip edge, metal, 5″ wide	.100	L.F.	.002	.04	.09	.13
8″ wide	.100	L.F.	.002	.05	.09	.14
Building paper, #15 asphalt felt	1.300	S.F.	.002	.05	.07	.12
Soffit & fascia, aluminum vented, 1′ overhang	.080	L.F.	.012	.24	.49	.73
2′ overhang	.080	L.F.	.013	.36	.54	.90
Vinyl vented, 1′ overhang	.080	L.F.	.011	.13	.45	.58
2′ overhang	.080	L.F.	.012	.19	.49	.68
Wood board fascia, plywood soffit, 1′ overhang	.080	L.F.	.010	.26	.38	.64
2′ overhang	.080	L.F.	.014	.33	.58	.91
Rake, trim, painted, 1″ x 6″	.043	L.F.	.004	.06	.14	.20
1″ x 8″	.043	L.F.	.004	.06	.14	.20
Gutter, 5″ box, aluminum, seamless, painted	.040	L.F.	.003	.06	.12	.18
Vinyl	.040	L.F.	.003	.04	.12	.16
Downspout 2″ x 3″, aluminum, one story house	.020	L.F.	.001	.03	.04	.07
Two story house	.020	L.F.	.001	.05	.07	.12
Vinyl, one story house	.020	L.F.	.001	.03	.04	.07
Two story house	.020	L.F.	.001	.05	.07	.12

24 Gable Dormer Roofing Systems

Ridge Shingles
Shingles
Flashing
Building Paper
Rake Boards
Drip Edge
Soffit & Fascia

System Description	QUAN.	UNIT	LABOR HOURS	COST PER S.F.		
				MAT.	INST.	TOTAL
ASPHALT, ROOF SHINGLES, CLASS A						
Shingles, standard inorganic class A 210-235 lb./sq	1.400	S.F.	.020	.59	.79	1.38
Drip edge, metal, 5" wide	.220	L.F.	.004	.10	.19	.29
Building paper, #15 asphalt felt	1.500	S.F.	.002	.06	.08	.14
Ridge shingles, asphalt	.280	L.F.	.007	.41	.26	.67
Soffit & fascia, aluminum, vented	.220	L.F.	.032	.65	1.35	2
Flashing, aluminum, mill finish, .013" thick	1.500	S.F.	.083	.63	3.21	3.84
TOTAL		S.F.	.148	2.44	5.88	8.32
WOOD, CEDAR, NO. 1 PERFECTIONS						
Shingles, red cedar, No.1 perfections, 18" long, 5" exp.	1.400	S.F.	.041	2.59	1.72	4.31
Drip edge, metal, 5" wide	.220	L.F.	.004	.10	.19	.29
Building paper, #15 asphalt felt	1.500	S.F.	.002	.06	.08	.14
Ridge shingles, wood	.280	L.F.	.008	.82	.34	1.16
Soffit & fascia, aluminum, vented	.220	L.F.	.032	.65	1.35	2
Flashing, aluminum, mill finish, .013" thick	1.500	S.F.	.083	.63	3.21	3.84
TOTAL		S.F.	.170	4.85	6.89	11.74
SLATE, BUCKINGHAM, BLACK						
Shingles, Buckingham, Virginia, black	1.400	S.F.	.064	8.96	2.48	11.44
Drip edge, metal, 5" wide	.220	L.F.	.004	.10	.19	.29
Building paper, #15 asphalt felt	1.500	S.F.	.002	.06	.08	.14
Ridge shingles, slate	.280	L.F.	.011	2.77	.43	3.20
Soffit & fascia, aluminum, vented	.220	L.F.	.032	.65	1.35	2
Flashing, copper, 16 oz.	1.500	S.F.	.104	4.56	4.05	8.61
TOTAL		S.F.	.217	17.10	8.58	25.68

The prices in these systems are based on a square foot of plan area under the dormer roof.

Description	QUAN.	UNIT	LABOR HOURS	COST PER S.F.		
				MAT.	INST.	TOTAL

Gable Dormer Roofing Price Sheet	QUAN.	UNIT	LABOR HOURS	COST PER S.F.		
				MAT.	INST.	TOTAL
Shingles, asphalt, standard, inorganic, class A, 210-235 lb./sq.	1.400	S.F.	.020	.59	.79	1.38
Laminated, multi-layered, 240-260 lb./sq.	1.400	S.F.	.025	.73	.97	1.70
Premium laminated, multi-layered, 260-300 lb./sq.	1.400	S.F.	.032	.97	1.24	2.21
Clay tile, Spanish tile, red	1.400	S.F.	.062	4.13	2.41	6.54
Mission tile, red	1.400	S.F.	.097	10	3.78	13.78
French tile, red	1.400	S.F.	.083	9.10	3.22	12.32
Slate Buckingham, Virginia, black	1.400	S.F.	.064	8.95	2.48	11.43
Vermont, black or grey	1.400	S.F.	.064	5.55	2.48	8.03
Wood, red cedar, No.1 5X, 16″ long, 5″ exposure	1.400	S.F.	.045	2.88	1.89	4.77
Fire retardant	1.400	S.F.	.045	3.42	1.89	5.31
18″ long, No.1 perfections, 5″ exposure	1.400	S.F.	.041	2.59	1.72	4.31
Fire retardant	1.400	S.F.	.041	3.13	1.72	4.85
Resquared & rebutted, 18″ long, 5″ exposure	1.400	S.F.	.037	3.23	1.58	4.81
Fire retardant	1.400	S.F.	.037	3.77	1.58	5.35
Shakes hand split, 24″ long, 10″ exposure	1.400	S.F.	.045	1.78	1.89	3.67
Fire retardant	1.400	S.F.	.045	2.55	1.89	4.44
18″ long, 8″ exposure	1.400	S.F.	.056	1.38	2.37	3.75
Fire retardant	1.400	S.F.	.056	2.07	2.37	4.44
Drip edge, metal, 5″ wide	.220	L.F.	.004	.10	.19	.29
8″ wide	.220	L.F.	.004	.12	.19	.31
Building paper, #15 asphalt felt	1.500	S.F.	.002	.06	.08	.14
Ridge shingles, asphalt	.280	L.F.	.007	.41	.26	.67
Clay	.280	L.F.	.011	2.74	.43	3.17
Slate	.280	L.F.	.011	2.77	.43	3.20
Wood	.280	L.F.	.008	.82	.34	1.16
Soffit & fascia, aluminum, vented	.220	L.F.	.032	.65	1.35	2
Vinyl, vented	.220	L.F.	.029	.37	1.24	1.61
Wood, board fascia, plywood soffit	.220	L.F.	.026	.70	1.04	1.74
Flashing, aluminum, .013″ thick	1.500	S.F.	.083	.63	3.21	3.84
.032″ thick	1.500	S.F.	.083	1.94	3.21	5.15
.040″ thick	1.500	S.F.	.083	2.61	3.21	5.82
.050″ thick	1.500	S.F.	.083	3.32	3.21	6.53
Copper, 16 oz.	1.500	S.F.	.104	4.56	4.05	8.61
20 oz.	1.500	S.F.	.109	6.75	4.23	10.98
24 oz.	1.500	S.F.	.114	8.10	4.43	12.53
32 oz.	1.500	S.F.	.120	10.80	4.65	15.45

28 Shed Dormer Roofing Systems

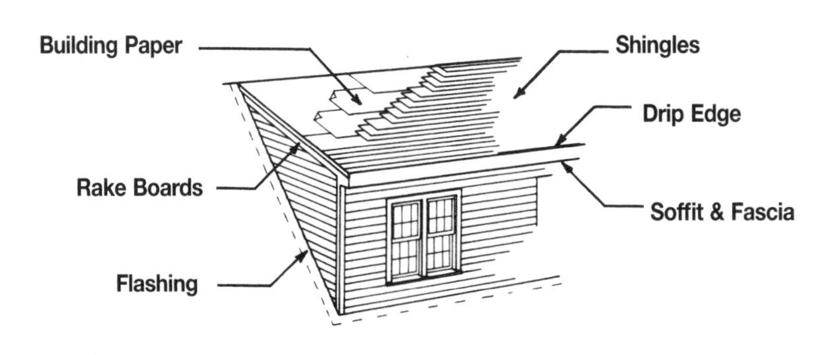

System Description	QUAN.	UNIT	LABOR HOURS	COST PER S.F.		
				MAT.	INST.	TOTAL
ASPHALT, ROOF SHINGLES, CLASS A						
Shingles, standard inorganic class A 210-235 lb./sq.	1.100	S.F.	.016	.46	.62	1.08
Drip edge, aluminum, 5″ wide	.250	L.F.	.005	.08	.21	.29
Building paper, #15 asphalt felt	1.200	S.F.	.002	.05	.06	.11
Soffit & fascia, aluminum, vented, 1′ overhang	.250	L.F.	.036	.74	1.54	2.28
Flashing, aluminum, mill finish, 0.013″ thick	.800	L.F.	.044	.34	1.71	2.05
TOTAL		S.F.	.103	1.67	4.14	5.81
WOOD, CEDAR, NO. 1 PERFECTIONS, 18″ LONG						
Shingles, wood, red cedar, #1 perfections, 5″ exposure	1.100	S.F.	.032	2.04	1.35	3.39
Drip edge, aluminum, 5″ wide	.250	L.F.	.005	.08	.21	.29
Building paper, #15 asphalt felt	1.200	S.F.	.002	.05	.06	.11
Soffit & fascia, aluminum, vented, 1′ overhang	.250	L.F.	.036	.74	1.54	2.28
Flashing, aluminum, mill finish, 0.013″ thick	.800	L.F.	.044	.34	1.71	2.05
TOTAL		S.F.	.119	3.25	4.87	8.12
SLATE, BUCKINGHAM, BLACK						
Shingles, slate, Buckingham, black	1.100	S.F.	.050	7.04	1.95	8.99
Drip edge, aluminum, 5″ wide	.250	L.F.	.005	.08	.21	.29
Building paper, #15 asphalt felt	1.200	S.F.	.002	.05	.06	.11
Soffit & fascia, aluminum, vented, 1′ overhang	.250	L.F.	.036	.74	1.54	2.28
Flashing, copper, 16 oz.	.800	L.F.	.056	2.43	2.16	4.59
TOTAL		S.F.	.149	10.34	5.92	16.26

The prices in this system are based on a square foot of plan area under the dormer roof.

Description	QUAN.	UNIT	LABOR HOURS	COST PER S.F.		
				MAT.	INST.	TOTAL

Important: See the Reference Section for critical supporting data - Reference Nos., Crews & Location Factors

Shed Dormer Roofing Price Sheet	QUAN.	UNIT	LABOR HOURS	COST PER S.F. MAT.	INST.	TOTAL
Shingles, asphalt, standard, inorganic, class A, 210-235 lb./sq.	1.100	S.F.	.016	.46	.62	1.08
Laminated, multi-layered, 240-260 lb./sq.	1.100	S.F.	.020	.57	.76	1.33
Premium laminated, multi-layered, 260-300 lb./sq.	1.100	S.F.	.025	.76	.97	1.73
Clay tile, Spanish tile, red	1.100	S.F.	.049	3.25	1.89	5.14
Mission tile, red	1.100	S.F.	.077	7.85	2.97	10.82
French tile, red	1.100	S.F.	.065	7.15	2.53	9.68
Slate Buckingham, Virginia, black	1.100	S.F.	.050	7.05	1.95	9
Vermont, black or grey	1.100	S.F.	.050	4.35	1.95	6.30
Wood, red cedar, No. 1 5X, 16" long, 5" exposure	1.100	S.F.	.035	2.27	1.49	3.76
Fire retardant	1.100	S.F.	.035	2.69	1.49	4.18
18" long, No.1 perfections, 5" exposure	1.100	S.F.	.032	2.04	1.35	3.39
Fire retardant	1.100	S.F.	.032	2.46	1.35	3.81
Resquared & rebutted, 18" long, 5" exposure	1.100	S.F.	.029	2.54	1.24	3.78
Fire retardant	1.100	S.F.	.029	2.96	1.24	4.20
Shakes hand split, 24" long, 10" exposure	1.100	S.F.	.035	1.40	1.49	2.89
Fire retardant	1.100	S.F.	.035	2.01	1.49	3.50
18" long, 8" exposure	1.100	S.F.	.044	1.08	1.86	2.94
Fire retardant	1.100	S.F.	.044	1.62	1.86	3.48
Drip edge, metal, 5" wide	.250	L.F.	.005	.08	.21	.29
8" wide	.250	L.F.	.005	.14	.21	.35
Building paper, #15 asphalt felt	1.200	S.F.	.002	.05	.06	.11
Soffit & fascia, aluminum, vented	.250	L.F.	.036	.74	1.54	2.28
Vinyl, vented	.250	L.F.	.033	.42	1.41	1.83
Wood, board fascia, plywood soffit	.250	L.F.	.030	.80	1.19	1.99
Flashing, aluminum, .013" thick	.800	L.F.	.044	.34	1.71	2.05
.032" thick	.800	L.F.	.044	1.03	1.71	2.74
.040" thick	.800	L.F.	.044	1.39	1.71	3.10
.050" thick	.800	L.F.	.044	1.77	1.71	3.48
Copper, 16 oz.	.800	L.F.	.056	2.43	2.16	4.59
20 oz.	.800	L.F.	.058	3.61	2.26	5.87
24 oz.	.800	L.F.	.061	4.32	2.36	6.68
32 oz.	.800	L.F.	.064	5.75	2.48	8.23

32 Skylight/Skywindow Systems

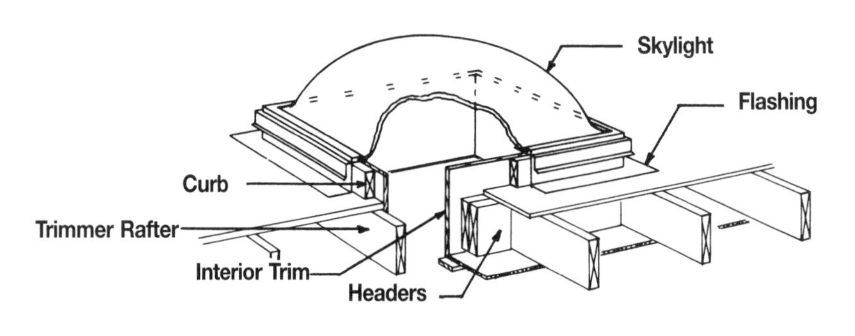

System Description	QUAN.	UNIT	LABOR HOURS	COST EACH		
				MAT.	INST.	TOTAL
SKYLIGHT, FIXED, 32" X 32"						
Skylight, fixed bubble, insulating, 32" x 32"	1.000	Ea.	1.422	109.15	54.75	163.90
Trimmer rafters, 2" x 6"	28.000	L.F.	.448	18.20	19.04	37.24
Headers, 2" x 6"	6.000	L.F.	.267	3.90	11.28	15.18
Curb, 2" x 4"	12.000	L.F.	.154	4.92	6.48	11.40
Flashing, aluminum, .013" thick	13.500	S.F.	.745	5.67	28.89	34.56
Trim, stock pine, 11/16" x 2-1/2"	12.000	L.F.	.400	16.92	16.92	33.84
Trim primer coat, oil base, brushwork	12.000	L.F.	.148	.36	5.52	5.88
Trim paint, 1 coat, brushwork	12.000	L.F.	.148	.36	5.52	5.88
TOTAL		Ea.	3.732	159.48	148.40	307.88
SKYLIGHT, FIXED, 48" X 48"						
Skylight, fixed bubble, insulating, 48" x 48"	1.000	Ea.	1.296	152.80	49.92	202.72
Trimmer rafters, 2" x 6"	28.000	L.F.	.448	18.20	19.04	37.24
Headers, 2" x 6"	8.000	L.F.	.356	5.20	15.04	20.24
Curb, 2" x 4"	16.000	L.F.	.205	6.56	8.64	15.20
Flashing, aluminum, .013" thick	16.000	S.F.	.883	6.72	34.24	40.96
Trim, stock pine, 11/16" x 2-1/2"	16.000	L.F.	.533	22.56	22.56	45.12
Trim primer coat, oil base, brushwork	16.000	L.F.	.197	.48	7.36	7.84
Trim paint, 1 coat, brushwork	16.000	L.F.	.197	.48	7.36	7.84
TOTAL		Ea.	4.115	213	164.16	377.16
SKYWINDOW, OPERATING, 24" X 48"						
Skywindow, operating, thermopane glass, 24" x 48"	1.000	Ea.	3.200	600	123	723
Trimmer rafters, 2" x 6"	28.000	L.F.	.448	18.20	19.04	37.24
Headers, 2" x 6"	8.000	L.F.	.267	3.90	11.28	15.18
Curb, 2" x 4"	14.000	L.F.	.179	5.74	7.56	13.30
Flashing, aluminum, .013" thick	14.000	S.F.	.772	5.88	29.96	35.84
Trim, stock pine, 11/16" x 2-1/2"	14.000	L.F.	.467	19.74	19.74	39.48
Trim primer coat, oil base, brushwork	14.000	L.F.	.172	.42	6.44	6.86
Trim paint, 1 coat, brushwork	14.000	L.F.	.172	.42	6.44	6.86
TOTAL		Ea.	5.677	654.30	223.46	877.76

The prices in these systems are on a cost each basis.

Description	QUAN.	UNIT	LABOR HOURS	COST EACH		
				MAT.	INST.	TOTAL

Skylight/Skywindow Price Sheet	QUAN.	UNIT	LABOR HOURS	COST EACH		
				MAT.	INST.	TOTAL
Skylight, fixed bubble insulating, 24″ x 24″	1.000	Ea.	.800	61.50	31	92.50
32″ x 32″	1.000	Ea.	1.422	109	55	164
32″ x 48″	1.000	Ea.	.864	102	33.50	135.50
48″ x 48″	1.000	Ea.	1.296	153	50	203
Ventilating bubble insulating, 36″ x 36″	1.000	Ea.	2.667	420	103	523
52″ x 52″	1.000	Ea.	2.667	625	103	728
28″ x 52″	1.000	Ea.	3.200	490	123	613
36″ x 52″	1.000	Ea.	3.200	530	123	653
Skywindow, operating, thermopane glass, 24″ x 48″	1.000	Ea.	3.200	600	123	723
32″ x 48″	1.000	Ea.	3.556	630	137	767
Trimmer rafters, 2″ x 6″	28.000	L.F.	.448	18.20	19.05	37.25
2″ x 8″	28.000	L.F.	.472	26.50	19.90	46.40
2″ x 10″	28.000	L.F.	.711	40	30	70
Headers, 24″ window, 2″ x 6″	4.000	L.F.	.178	2.60	7.50	10.10
2″ x 8″	4.000	L.F.	.188	3.76	7.95	11.71
2″ x 10″	4.000	L.F.	.200	5.70	8.45	14.15
32″ window, 2″ x 6″	6.000	L.F.	.267	3.90	11.30	15.20
2″ x 8″	6.000	L.F.	.282	5.65	11.95	17.60
2″ x 10″	6.000	L.F.	.300	8.50	12.65	21.15
48″ window, 2″ x 6″	8.000	L.F.	.356	5.20	15.05	20.25
2″ x 8″	8.000	L.F.	.376	7.50	15.90	23.40
2″ x 10″	8.000	L.F.	.400	11.35	16.90	28.25
Curb, 2″ x 4″, skylight, 24″ x 24″	8.000	L.F.	.102	3.28	4.32	7.60
32″ x 32″	12.000	L.F.	.154	4.92	6.50	11.42
32″ x 48″	14.000	L.F.	.179	5.75	7.55	13.30
48″ x 48″	16.000	L.F.	.205	6.55	8.65	15.20
Flashing, aluminum .013″ thick, skylight, 24″ x 24″	9.000	S.F.	.497	3.78	19.25	23.03
32″ x 32″	13.500	S.F.	.745	5.65	29	34.65
32″ x 48″	14.000	S.F.	.772	5.90	30	35.90
48″ x 48″	16.000	S.F.	.883	6.70	34	40.70
Copper 16 oz., skylight, 24″ x 24″	9.000	S.F.	.626	27.50	24.50	52
32″ x 32″	13.500	S.F.	.939	41	36.50	77.50
32″ x 48″	14.000	S.F.	.974	42.50	38	80.50
48″ x 48″	16.000	S.F.	1.113	48.50	43	91.50
Trim, interior casing painted, 24″ x 24″	8.000	L.F.	.347	12.10	14.25	26.35
32″ x 32″	12.000	L.F.	.520	18.10	21.50	39.60
32″ x 48″	14.000	L.F.	.607	21	25	46
48″ x 48″	16.000	L.F.	.693	24	28.50	52.50

Flashing
4″ x 4″ Cant
6″ x 2-1/4″ Wood Blocking
Gravel
Asphalt
Felt
Insulation Board

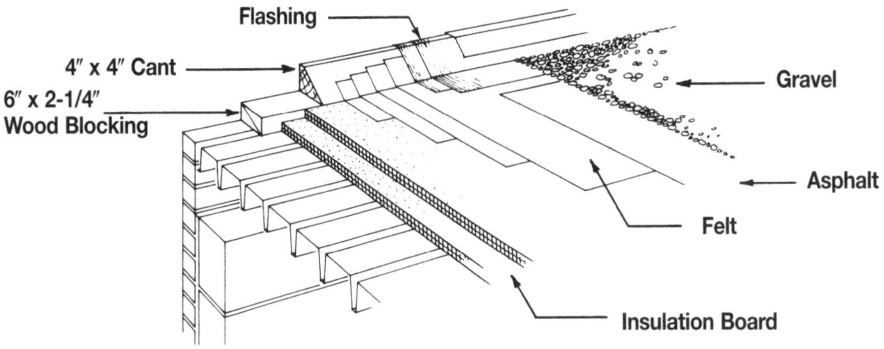

System Description	QUAN.	UNIT	LABOR HOURS	COST PER S.F.		
				MAT.	INST.	TOTAL
ASPHALT, ORGANIC, 4-PLY, INSULATED DECK						
Membrane, asphalt, 4-plies #15 felt, gravel surfacing	1.000	S.F.	.025	.73	1.10	1.83
Insulation board, 2-layers of 1-1/16″ glass fiber	2.000	S.F.	.016	1.84	.62	2.46
Wood blocking, 2″ x 6″	.040	L.F.	.004	.08	.18	.26
Treated 4″ x 4″ cant strip	.040	L.F.	.001	.06	.04	.10
Flashing, aluminum, 0.040″ thick	.050	S.F.	.003	.09	.11	.20
TOTAL		S.F.	.049	2.80	2.05	4.85
ASPHALT, INORGANIC, 3-PLY, INSULATED DECK						
Membrane, asphalt, 3-plies type IV glass felt, gravel surfacing	1.000	S.F.	.028	.70	1.21	1.91
Insulation board, 2-layers of 1-1/16″ glass fiber	2.000	S.F.	.016	1.84	.62	2.46
Wood blocking, 2″ x 6″	.040	L.F.	.004	.08	.18	.26
Treated 4″ x 4″ cant strip	.040	L.F.	.001	.06	.04	.10
Flashing, aluminum, 0.040″ thick	.050	S.F.	.003	.09	.11	.20
TOTAL		S.F.	.052	2.77	2.16	4.93
COAL TAR, ORGANIC, 4-PLY, INSULATED DECK						
Membrane, coal tar, 4-plies #15 felt, gravel surfacing	1.000	S.F.	.027	1.29	1.15	2.44
Insulation board, 2-layers of 1-1/16″ glass fiber	2.000	S.F.	.016	1.84	.62	2.46
Wood blocking, 2″ x 6″	.040	L.F.	.004	.08	.18	.26
Treated 4″ x 4″ cant strip	.040	L.F.	.001	.06	.04	.10
Flashing, aluminum, 0.040″ thick	.050	S.F.	.003	.09	.11	.20
TOTAL		S.F.	.051	3.36	2.10	5.46
COAL TAR, INORGANIC, 3-PLY, INSULATED DECK						
Membrane, coal tar, 3-plies type IV glass felt, gravel surfacing	1.000	S.F.	.029	1.06	1.27	2.33
Insulation board, 2-layers of 1-1/16″ glass fiber	2.000	S.F.	.016	1.84	.62	2.46
Wood blocking, 2″ x 6″	.040	L.F.	.004	.08	.18	.26
Treated 4″ x 4″ cant strip	.040	L.F.	.001	.06	.04	.10
Flashing, aluminum, 0.040″ thick	.050	S.F.	.003	.09	.11	.20
TOTAL		S.F.	.053	3.13	2.22	5.35

Built-Up Roofing Price Sheet	QUAN.	UNIT	LABOR HOURS	COST PER S.F.		
				MAT.	INST.	TOTAL
Membrane, asphalt, 4-plies #15 organic felt, gravel surfacing	1.000	S.F.	.025	.73	1.10	1.83
Asphalt base sheet & 3-plies #15 asphalt felt	1.000	S.F.	.025	.55	1.10	1.65
3-plies type IV glass fiber felt	1.000	S.F.	.028	.70	1.21	1.91
4-plies type IV glass fiber felt	1.000	S.F.	.028	.85	1.21	2.06
Coal tar, 4-plies #15 organic felt, gravel surfacing	1.000	S.F.	.027			
4-plies tarred felt	1.000	S.F.	.027	1.29	1.15	2.44
3-plies type IV glass fiber felt	1.000	S.F.	.029	1.06	1.27	2.33
4-plies type IV glass fiber felt	1.000	S.F.	.027	1.47	1.15	2.62
Roll, asphalt, 1-ply #15 organic felt, 2-plies mineral surfaced	1.000	S.F.	.021	.46	.90	1.36
3-plies type IV glass fiber, 1-ply mineral surfaced	1.000	S.F.	.022	.72	.97	1.69
Insulation boards, glass fiber, 1-1/16" thick	1.000	S.F.	.008	.92	.31	1.23
2-1/16" thick	1.000	S.F.	.010	1.35	.39	1.74
2-7/16" thick	1.000	S.F.	.010	1.54	.39	1.93
Expanded perlite, 1" thick	1.000	S.F.	.010	.35	.39	.74
1-1/2" thick	1.000	S.F.	.010	.44	.39	.83
2" thick	1.000	S.F.	.011	.70	.44	1.14
Fiberboard, 1" thick	1.000	S.F.	.010	.42	.39	.81
1-1/2" thick	1.000	S.F.	.010	.63	.39	1.02
2" thick	1.000	S.F.	.010	.84	.39	1.23
Extruded polystyrene, 15 PSI compressive strength, 2" thick R10	1.000	S.F.	.006	.73	.25	.98
3" thick R15	1.000	S.F.	.008	.95	.31	1.26
4" thick R20	1.000	S.F.	.008	1.47	.31	1.78
Tapered for drainage	1.000	S.F.	.005	.48	.21	.69
40 PSI compressive strength, 1" thick R5	1.000	S.F.	.005	.48	.21	.69
2" thick R10	1.000	S.F.	.006	.95	.25	1.20
3" thick R15	1.000	S.F.	.008	1.39	.31	1.70
4" thick R20	1.000	S.F.	.008	1.86	.31	2.17
Fiberboard high density, 1/2" thick R1.3	1.000	S.F.	.008	.22	.31	.53
1" thick R2.5	1.000	S.F.	.010	.44	.39	.83
1 1/2" thick R3.8	1.000	S.F.	.010	.66	.39	1.05
Polyisocyanurate, 1 1/2" thick R10.87	1.000	S.F.	.006	.76	.25	1.01
2" thick R14.29	1.000	S.F.	.007	1	.28	1.28
3 1/2" thick R25	1.000	S.F.	.008	1.63	.31	1.94
Tapered for drainage	1.000	S.F.	.006	.84	.22	1.06
Expanded polystyrene, 1" thick	1.000	S.F.	.005	.28	.21	.49
2" thick R10	1.000	S.F.	.006	.58	.25	.83
3" thick	1.000	S.F.	.006	.67	.25	.92
Wood blocking, treated, 6" x 2" & 4" x 4" cant	.040	L.F.	.002	.10	.10	.20
6" x 4-1/2" & 4" x 4" cant	.040	L.F.	.005	.16	.22	.38
6" x 5" & 4" x 4" cant	.040	L.F.	.007	.19	.28	.47
Flashing, aluminum, 0.019" thick	.050	S.F.	.003	.04	.11	.15
0.032" thick	.050	S.F.	.003	.06	.11	.17
0.040" thick	.050	S.F.	.003	.09	.11	.20
Copper sheets, 16 oz., under 500 lbs.	.050	S.F.	.003	.15	.14	.29
Over 500 lbs.	.050	S.F.	.003	.17	.10	.27
20 oz., under 500 lbs.	.050	S.F.	.004	.23	.14	.37
Over 500 lbs.	.050	S.F.	.003	.21	.11	.32
Stainless steel, 32 gauge	.050	S.F.	.003	.15	.10	.25
28 gauge	.050	S.F.	.003	.18	.10	.28
26 gauge	.050	S.F.	.003	.22	.10	.32
24 gauge	.050	S.F.	.003	.29	.10	.39

Division 6
Interiors

04 Drywall & Thincoat Wall Systems

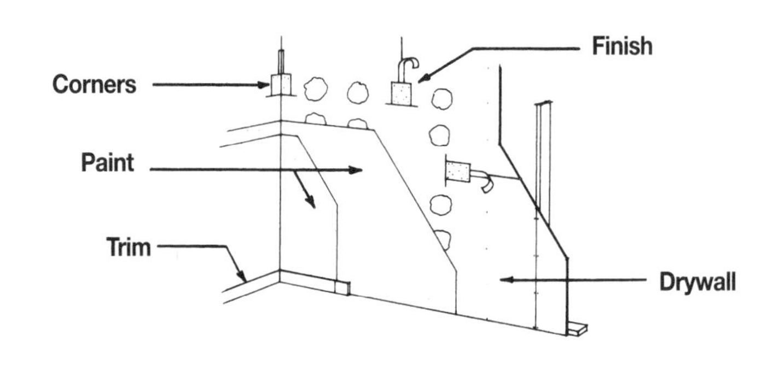

System Description	QUAN.	UNIT	LABOR HOURS	COST PER S.F. MAT.	COST PER S.F. INST.	COST PER S.F. TOTAL
1/2″ DRYWALL, TAPED & FINISHED						
Gypsum wallboard, 1/2″ thick, standard	1.000	S.F.	.008	.30	.34	.64
Finish, taped & finished joints	1.000	S.F.	.008	.04	.34	.38
Corners, taped & finished, 32 L.F. per 12′ x 12′ room	.083	L.F.	.002	.01	.06	.07
Painting, primer & 2 coats	1.000	S.F.	.011	.16	.39	.55
Paint trim, to 6″ wide, primer + 1 coat enamel	.125	L.F.	.001	.01	.05	.06
Trim, baseboard	.125	L.F.	.005	.33	.21	.54
TOTAL		S.F.	.035	.85	1.39	2.24
THINCOAT, SKIM-COAT, ON 1/2″ BACKER DRYWALL						
Gypsum wallboard, 1/2″ thick, thincoat backer	1.000	S.F.	.008	.30	.34	.64
Thincoat plaster	1.000	S.F.	.011	.08	.42	.50
Corners, taped & finished, 32 L.F. per 12′ x 12′ room	.083	L.F.	.002	.01	.06	.07
Painting, primer & 2 coats	1.000	S.F.	.011	.16	.39	.55
Paint trim, to 6″ wide, primer + 1 coat enamel	.125	L.F.	.001	.01	.05	.06
Trim, baseboard	.125	L.F.	.005	.33	.21	.54
TOTAL		S.F.	.038	.89	1.47	2.36
5/8″ DRYWALL, TAPED & FINISHED						
Gypsum wallboard, 5/8″ thick, standard	1.000	S.F.	.008	.33	.34	.67
Finish, taped & finished joints	1.000	S.F.	.008	.04	.34	.38
Corners, taped & finished, 32 L.F. per 12′ x 12′ room	.083	L.F.	.002	.01	.06	.07
Painting, primer & 2 coats	1.000	S.F.	.011	.16	.39	.55
Trim, baseboard	.125	L.F.	.005	.33	.21	.54
Paint trim, to 6″ wide, primer + 1 coat enamel	.125	L.F.	.001	.01	.05	.06
TOTAL		S.F.	.035	.88	1.39	2.27

The costs in this system are based on a square foot of wall.
Do not deduct for openings.

Description	QUAN.	UNIT	LABOR HOURS	COST PER S.F. MAT.	COST PER S.F. INST.	COST PER S.F. TOTAL

Important: See the Reference Section for critical supporting data - Reference Nos., Crews & Location Factors

Drywall & Thincoat Wall Price Sheet	QUAN.	UNIT	LABOR HOURS	COST PER S.F.		
				MAT.	INST.	TOTAL
Gypsum wallboard, 1/2" thick, standard	1.000	S.F.	.008	.30	.34	.64
Fire resistant	1.000	S.F.	.008	.33	.34	.67
Water resistant	1.000	S.F.	.008	.28	.34	.62
5/8" thick, standard	1.000	S.F.	.008	.33	.34	.67
Fire resistant	1.000	S.F.	.008	.33	.34	.67
Water resistant	1.000	S.F.	.008	.30	.34	.64
Gypsum wallboard backer for thincoat system, 1/2" thick	1.000	S.F.	.008	.30	.34	.64
5/8" thick	1.000	S.F.	.008	.33	.34	.67
Gypsum wallboard, taped & finished	1.000	S.F.	.008	.04	.34	.38
Texture spray	1.000	S.F.	.010	.06	.37	.43
Thincoat plaster, including tape	1.000	S.F.	.011	.08	.42	.50
Gypsum wallboard corners, taped & finished, 32 L.F. per 4' x 4' room	.250	L.F.	.004	.02	.18	.20
6' x 6' room	.110	L.F.	.002	.01	.08	.09
10' x 10' room	.100	L.F.	.001	.01	.07	.08
12' x 12' room	.083	L.F.	.001	.01	.06	.07
16' x 16' room	.063	L.F.	.001	.01	.04	.05
Thincoat system, 32 L.F. per 4' x 4' room	.250	L.F.	.003	.02	.11	.13
6' x 6' room	.110	L.F.	.001	.01	.04	.05
10' x 10' room	.100	L.F.	.001	.01	.04	.05
12' x 12' room	.083	L.F.	.001	.01	.03	.04
16' x 16' room	.063	L.F.	.001	.01	.02	.03
Painting, primer, & 1 coat	1.000	S.F.	.008	.11	.30	.41
& 2 coats	1.000	S.F.	.011	.16	.39	.55
Wallpaper, $7/double roll	1.000	S.F.	.013	.33	.47	.80
$17/double roll	1.000	S.F.	.015	.72	.56	1.28
$40/double roll	1.000	S.F.	.018	1.68	.68	2.36
Tile, ceramic adhesive thin set, 4 1/4" x 4 1/4" tiles	1.000	S.F.	.084	2.19	2.88	5.07
6" x 6" tiles	1.000	S.F.	.080	3	2.74	5.74
Pregrouted sheets	1.000	S.F.	.067	4.65	2.28	6.93
Trim, painted or stained, baseboard	.125	L.F.	.006	.34	.26	.60
Base shoe	.125	L.F.	.005	.17	.23	.40
Chair rail	.125	L.F.	.005	.21	.21	.42
Cornice molding	.125	L.F.	.004	.12	.18	.30
Cove base, vinyl	.125	L.F.	.003	.07	.12	.19
Paneling, not including furring or trim						
Plywood, prefinished, 1/4" thick, 4' x 8' sheets, vert. grooves						
Birch faced, minimum	1.000	S.F.	.032	.92	1.35	2.27
Average	1.000	S.F.	.038	1.41	1.61	3.02
Maximum	1.000	S.F.	.046	2.06	1.93	3.99
Mahogany, African	1.000	S.F.	.040	2.63	1.69	4.32
Philippine (lauan)	1.000	S.F.	.032	1.13	1.35	2.48
Oak or cherry, minimum	1.000	S.F.	.032	2.20	1.35	3.55
Maximum	1.000	S.F.	.040	3.38	1.69	5.07
Rosewood	1.000	S.F.	.050	4.80	2.11	6.91
Teak	1.000	S.F.	.040	3.38	1.69	5.07
Chestnut	1.000	S.F.	.043	4.99	1.80	6.79
Pecan	1.000	S.F.	.040	2.16	1.69	3.85
Walnut, minimum	1.000	S.F.	.032	2.88	1.35	4.23
Maximum	1.000	S.F.	.040	5.45	1.69	7.14

08 Drywall & Thincoat Ceiling Systems

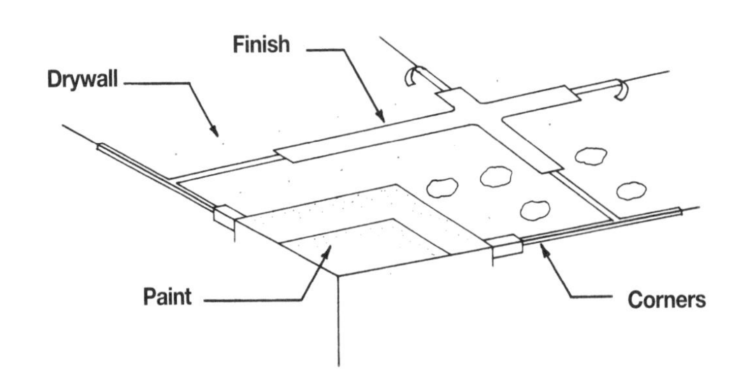

System Description	QUAN.	UNIT	LABOR HOURS	COST PER S.F. MAT.	COST PER S.F. INST.	COST PER S.F. TOTAL
1/2" SHEETROCK, TAPED & FINISHED						
Gypsum wallboard, 1/2" thick, standard	1.000	S.F.	.008	.30	.34	.64
Finish, taped & finished	1.000	S.F.	.008	.04	.34	.38
Corners, taped & finished, 12' x 12' room	.333	L.F.	.006	.03	.23	.26
Paint, primer & 2 coats	1.000	S.F.	.011	.16	.39	.55
TOTAL		S.F.	.033	.53	1.30	1.83
THINCOAT, SKIM COAT ON 1/2" GYPSUM WALLBOARD						
Gypsum wallboard, 1/2" thick, thincoat backer	1.000	S.F.	.008	.30	.34	.64
Thincoat plaster	1.000	S.F.	.011	.08	.42	.50
Corners, taped & finished, 12' x 12' room	.333	L.F.	.006	.03	.23	.26
Paint, primer & 2 coats	1.000	S.F.	.011	.16	.39	.55
TOTAL		S.F.	.036	.57	1.38	1.95
WATER-RESISTANT GYPSUM WALLBOARD, 1/2" THICK, TAPED & FINISHED						
Gypsum wallboard, 1/2" thick, water-resistant	1.000	S.F.	.008	.28	.34	.62
Finish, taped & finished	1.000	S.F.	.008	.04	.34	.38
Corners, taped & finished, 12' x 12' room	.333	L.F.	.006	.03	.23	.26
Paint, primer & 2 coats	1.000	S.F.	.011	.16	.39	.55
TOTAL		S.F.	.033	.51	1.30	1.81
5/8" GYPSUM WALLBOARD, TAPED & FINISHED						
Gypsum wallboard, 5/8" thick, standard	1.000	S.F.	.008	.33	.34	.67
Finish, taped & finished	1.000	S.F.	.008	.04	.34	.38
Corners, taped & finished, 12' x 12' room	.333	L.F.	.006	.03	.23	.26
Paint, primer & 2 coats	1.000	S.F.	.011	.16	.39	.55
TOTAL		S.F.	.033	.56	1.30	1.86

The costs in this system are based on a square foot of ceiling.

Description	QUAN.	UNIT	LABOR HOURS	COST PER S.F. MAT.	COST PER S.F. INST.	COST PER S.F. TOTAL

Drywall & Thincoat Ceilings	QUAN.	UNIT	LABOR HOURS	COST PER S.F.		
				MAT.	INST.	TOTAL
Gypsum wallboard ceilings, 1/2" thick, standard	1.000	S.F.	.008	.30	.34	.64
Fire resistant	1.000	S.F.	.008	.33	.34	.67
Water resistant	1.000	S.F.	.008	.28	.34	.62
5/8" thick, standard	1.000	S.F.	.008	.33	.34	.67
Fire resistant	1.000	S.F.	.008	.33	.34	.67
Water resistant	1.000	S.F.	.008	.30	.34	.64
Gypsum wallboard backer for thincoat ceiling system, 1/2" thick	1.000	S.F.	.016	.63	.68	1.31
5/8" thick	1.000	S.F.	.016	.66	.68	1.34
Gypsum wallboard ceilings, taped & finished	1.000	S.F.	.008	.04	.34	.38
Texture spray	1.000	S.F.	.010	.06	.37	.43
Thincoat plaster	1.000	S.F.	.011	.08	.42	.50
Corners taped & finished, 4' x 4' room	1.000	L.F.	.015	.08	.71	.79
6' x 6' room	.667	L.F.	.010	.05	.48	.53
10' x 10' room	.400	L.F.	.006	.03	.28	.31
12' x 12' room	.333	L.F.	.005	.03	.23	.26
16' x 16' room	.250	L.F.	.003	.01	.13	.14
Thincoat system, 4' x 4' room	1.000	L.F.	.011	.08	.42	.50
6' x 6' room	.667	L.F.	.007	.05	.28	.33
10' x 10' room	.400	L.F.	.004	.03	.17	.20
12' x 12' room	.333	L.F.	.004	.03	.14	.17
16' x 16' room	.250	L.F.	.002	.01	.08	.09
Painting, primer & 1 coat	1.000	S.F.	.008	.11	.30	.41
& 2 coats	1.000	S.F.	.011	.16	.39	.55
Wallpaper, double roll, solid pattern, avg. workmanship	1.000	S.F.	.013	.33	.47	.80
Basic pattern, avg. workmanship	1.000	S.F.	.015	.72	.56	1.28
Basic pattern, quality workmanship	1.000	S.F.	.018	1.68	.68	2.36
Tile, ceramic adhesive thin set, 4 1/4" x 4 1/4" tiles	1.000	S.F.	.084	2.19	2.88	5.07
6" x 6" tiles	1.000	S.F.	.080	3	2.74	5.74
Pregrouted sheets	1.000	S.F.	.067	4.65	2.28	6.93

12 Plaster & Stucco Wall Systems

Plaster
Paint
Trim
Lath

System Description	QUAN.	UNIT	LABOR HOURS	COST PER S.F. MAT.	INST.	TOTAL
PLASTER ON GYPSUM LATH						
Plaster, gypsum or perlite, 2 coats	1.000	S.F.	.053	.40	2.04	2.44
Lath, 3/8" gypsum	1.000	S.F.	.010	.66	.39	1.05
Corners, expanded metal, 32 L.F. per 12' x 12' room	.083	L.F.	.002	.01	.07	.08
Painting, primer & 2 coats	1.000	S.F.	.011	.16	.39	.55
Paint trim, to 6" wide, primer + 1 coat enamel	.125	L.F.	.001	.01	.05	.06
Trim, baseboard	.125	L.F.	.005	.33	.21	.54
TOTAL		S.F.	.082	1.57	3.15	4.72
PLASTER ON METAL LATH						
Plaster, gypsum or perlite, 2 coats	1.000	S.F.	.053	.40	2.04	2.44
Lath, 2.5 Lb. diamond, metal	1.000	S.F.	.010	.31	.39	.70
Corners, expanded metal, 32 L.F. per 12' x 12' room	.083	L.F.	.002	.01	.07	.08
Painting, primer & 2 coats	1.000	S.F.	.011	.16	.39	.55
Paint trim, to 6" wide, primer + 1 coat enamel	.125	L.F.	.001	.01	.05	.06
Trim, baseboard	.125	L.F.	.005	.33	.21	.54
TOTAL		S.F.	.082	1.22	3.15	4.37
STUCCO ON METAL LATH						
Stucco, 2 coats	1.000	S.F.	.041	.24	1.57	1.81
Lath, 2.5 Lb. diamond, metal	1.000	S.F.	.010	.31	.39	.70
Corners, expanded metal, 32 L.F. per 12' x 12' room	.083	L.F.	.002	.01	.07	.08
Painting, primer & 2 coats	1.000	S.F.	.011	.16	.39	.55
Paint trim, to 6" wide, primer + 1 coat enamel	.125	L.F.	.001	.01	.05	.06
Trim, baseboard	.125	L.F.	.005	.33	.21	.54
TOTAL		S.F.	.070	1.06	2.68	3.74

The costs in these systems are based on a per square foot of wall area.
Do not deduct for openings.

Description	QUAN.	UNIT	LABOR HOURS	COST PER S.F. MAT.	INST.	TOTAL

Plaster & Stucco Wall Price Sheet	QUAN.	UNIT	LABOR HOURS	COST PER S.F.		
				MAT.	INST.	TOTAL
Plaster, gypsum or perlite, 2 coats	1.000	S.F.	.053	.40	2.04	2.44
3 coats	1.000	S.F.	.065	.57	2.48	3.05
1/2" thick	1.000	S.F.	.013	.52	.47	.99
Fire resistant, 3/8" thick	1.000	S.F.	.013	.45	.47	.92
1/2" thick	1.000	S.F.	.014	.55	.51	1.06
Metal, diamond, 2.5 Lb.	1.000	S.F.	.010	.31	.39	.70
3.4 Lb.	1.000	S.F.	.012	.36	.44	.80
Rib, 2.75 Lb.	1.000	S.F.	.012	.33	.44	.77
3.4 Lb.	1.000	S.F.	.013	.48	.47	.95
Corners, expanded metal, 32 L.F. per 4' x 4' room	.250	L.F.	.005	.04	.21	.25
6' x 6' room	.110	L.F.	.002	.02	.09	.11
10' x 10' room	.100	L.F.	.002	.02	.09	.11
12' x 12' room	.083	L.F.	.002	.01	.07	.08
16' x 16' room	.063	L.F.	.001	.01	.05	.06
Painting, primer & 1 coats	1.000	S.F.	.008	.11	.30	.41
Primer & 2 coats	1.000	S.F.	.011	.16	.39	.55
Wallpaper, low price double roll	1.000	S.F.	.013	.33	.47	.80
Medium price double roll	1.000	S.F.	.015	.72	.56	1.28
High price double roll	1.000	S.F.	.018	1.68	.68	2.36
Tile, ceramic thin set, 4-1/4" x 4-1/4" tiles	1.000	S.F.	.084	2.19	2.88	5.07
6" x 6" tiles	1.000	S.F.	.080	3	2.74	5.74
Pregrouted sheets	1.000	S.F.	.067	4.65	2.28	6.93
Trim, painted or stained, baseboard	.125	L.F.	.006	.34	.26	.60
Base shoe	.125	L.F.	.005	.17	.23	.40
Chair rail	.125	L.F.	.005	.21	.21	.42
Cornice molding	.125	L.F.	.004	.12	.18	.30
Cove base, vinyl	.125	L.F.	.003	.07	.12	.19
Paneling not including furring or trim						
Plywood, prefinished, 1/4" thick, 4' x 8' sheets, vert. grooves						
Birch faced, minimum	1.000	S.F.	.032	.92	1.35	2.27
Average	1.000	S.F.	.038	1.41	1.61	3.02
Maximum	1.000	S.F.	.046	2.06	1.93	3.99
Mahogany, African	1.000	S.F.	.040	2.63	1.69	4.32
Philippine (lauan)	1.000	S.F.	.032	1.13	1.35	2.48
Oak or cherry, minimum	1.000	S.F.	.032	2.20	1.35	3.55
Maximum	1.000	S.F.	.040	3.38	1.69	5.07
Rosewood	1.000	S.F.	.050	4.80	2.11	6.91
Teak	1.000	S.F.	.040	3.38	1.69	5.07
Chestnut	1.000	S.F.	.043	4.99	1.80	6.79
Pecan	1.000	S.F.	.040	2.16	1.69	3.85
Walnut, minimum	1.000	S.F.	.032	2.88	1.35	4.23
Maximum	1.000	S.F.	.040	5.45	1.69	7.14

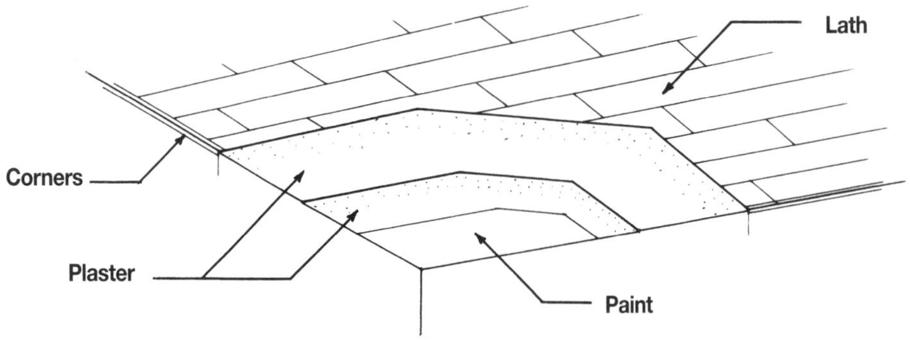

System Description	QUAN.	UNIT	LABOR HOURS	COST PER S.F.		
				MAT.	INST.	TOTAL
PLASTER ON GYPSUM LATH						
Plaster, gypsum or perlite, 2 coats	1.000	S.F.	.061	.40	2.33	2.73
Gypsum lath, plain or perforated, nailed, 3/8" thick	1.000	S.F.	.010	.66	.39	1.05
Gypsum lath, ceiling installation adder	1.000	S.F.	.004		.15	.15
Corners, expanded metal, 12' x 12' room	.330	L.F.	.007	.05	.28	.33
Painting, primer & 2 coats	1.000	S.F.	.011	.16	.39	.55
TOTAL		S.F.	.093	1.27	3.54	4.81
PLASTER ON METAL LATH						
Plaster, gypsum or perlite, 2 coats	1.000	S.F.	.061	.40	2.33	2.73
Lath, 2.5 Lb. diamond, metal	1.000	S.F.	.012	.31	.44	.75
Corners, expanded metal, 12' x 12' room	.330	L.F.	.007	.05	.28	.33
Painting, primer & 2 coats	1.000	S.F.	.011	.16	.39	.55
TOTAL		S.F.	.091	.92	3.44	4.36
STUCCO ON GYPSUM LATH						
Stucco, 2 coats	1.000	S.F.	.041	.24	1.57	1.81
Gypsum lath, plain or perforated, nailed, 3/8" thick	1.000	S.F.	.010	.66	.39	1.05
Gypsum lath, ceiling installation adder	1.000	S.F.	.004		.15	.15
Corners, expanded metal, 12' x 12' room	.330	L.F.	.007	.05	.28	.33
Painting, primer & 2 coats	1.000	S.F.	.011	.16	.39	.55
TOTAL		S.F.	.073	1.11	2.78	3.89
STUCCO ON METAL LATH						
Stucco, 2 coats	1.000	S.F.	.041	.24	1.57	1.81
Lath, 2.5 Lb. diamond, metal	1.000	S.F.	.012	.31	.44	.75
Corners, expanded metal, 12' x 12' room	.330	L.F.	.007	.05	.28	.33
Painting, primer & 2 coats	1.000	S.F.	.011	.16	.39	.55
TOTAL		S.F.	.071	.76	2.68	3.44

The costs in these systems are based on a square foot of ceiling area.

Description	QUAN.	UNIT	LABOR HOURS	COST PER S.F.		
				MAT.	INST.	TOTAL

 Important: See the Reference Section for critical supporting data - Reference Nos., Crews & Location Factors

Plaster & Stucco Ceiling Price Sheet	QUAN.	UNIT	LABOR HOURS	COST PER S.F.		
				MAT.	INST.	TOTAL
Plaster, gypsum or perlite, 2 coats	1.000	S.F.	.061	.40	2.33	2.73
3 coats	1.000	S.F.	.065	.57	2.48	3.05
Lath, gypsum, standard, 3/8″ thick	1.000	S.F.	.014	.66	.54	1.20
1/2″ thick	1.000	S.F.	.015	.52	.57	1.09
Fire resistant, 3/8″ thick	1.000	S.F.	.017	.45	.62	1.07
1/2″ thick	1.000	S.F.	.018	.55	.66	1.21
Metal, diamond, 2.5 Lb.	1.000	S.F.	.012	.31	.44	.75
3.4 Lb.	1.000	S.F.	.015	.36	.55	.91
Rib, 2.75 Lb.	1.000	S.F.	.012	.33	.44	.77
3.4 Lb.	1.000	S.F.	.013	.48	.47	.95
Corners expanded metal, 4′ x 4′ room	1.000	L.F.	.020	.15	.85	1
6′ x 6′ room	.667	L.F.	.013	.10	.57	.67
10′ x 10′ room	.400	L.F.	.008	.06	.34	.40
12′ x 12′ room	.333	L.F.	.007	.05	.28	.33
16′ x 16′ room	.250	L.F.	.004	.03	.16	.19
Painting, primer & 1 coat	1.000	S.F.	.008	.11	.30	.41
Primer & 2 coats	1.000	S.F.	.011	.16	.39	.55

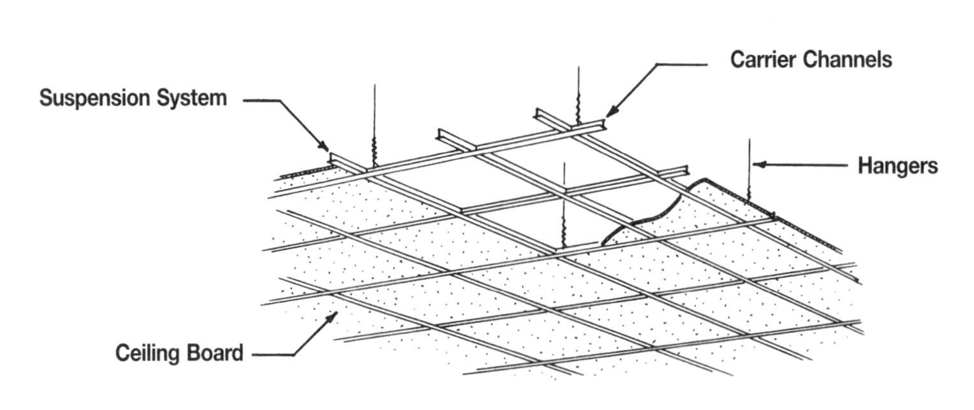

System Description	QUAN.	UNIT	LABOR HOURS	COST PER S.F. MAT.	COST PER S.F. INST.	COST PER S.F. TOTAL
2′ X 2′ GRID, FILM FACED FIBERGLASS, 5/8″ THICK						
Suspension system, 2′ x 2′ grid, T bar	1.000	S.F.	.012	.72	.52	1.24
Ceiling board, film faced fiberglass, 5/8″ thick	1.000	S.F.	.013	.57	.54	1.11
Carrier channels, 1-1/2″ x 3/4″	1.000	S.F.	.017	.11	.72	.83
Hangers, #12 wire	1.000	S.F.	.002	.06	.07	.13
TOTAL		S.F.	.044	1.46	1.85	3.31
2′ X 4′ GRID, FILM FACED FIBERGLASS, 5/8″ THICK						
Suspension system, 2′ x 4′ grid, T bar	1.000	S.F.	.010	.58	.42	1
Ceiling board, film faced fiberglass, 5/8″ thick	1.000	S.F.	.013	.57	.54	1.11
Carrier channels, 1-1/2″ x 3/4″	1.000	S.F.	.017	.11	.72	.83
Hangers, #12 wire	1.000	S.F.	.002	.06	.07	.13
TOTAL		S.F.	.042	1.32	1.75	3.07
2′ X 2′ GRID, MINERAL FIBER, REVEAL EDGE, 1″ THICK						
Suspension system, 2′ x 2′ grid, T bar	1.000	S.F.	.012	.72	.52	1.24
Ceiling board, mineral fiber, reveal edge, 1″ thick	1.000	S.F.	.013	1.38	.56	1.94
Carrier channels, 1-1/2″ x 3/4″	1.000	S.F.	.017	.11	.72	.83
Hangers, #12 wire	1.000	S.F.	.002	.06	.07	.13
TOTAL		S.F.	.044	2.27	1.87	4.14
2′ X 4′ GRID, MINERAL FIBER, REVEAL EDGE, 1″ THICK						
Suspension system, 2′ x 4′ grid, T bar	1.000	S.F.	.010	.58	.42	1
Ceiling board, mineral fiber, reveal edge, 1″ thick	1.000	S.F.	.013	1.38	.56	1.94
Carrier channels, 1-1/2″ x 3/4″	1.000	S.F.	.017	.11	.72	.83
Hangers, #12 wire	1.000	S.F.	.002	.06	.07	.13
TOTAL		S.F.	.042	2.13	1.77	3.90

Description	QUAN.	UNIT	LABOR HOURS	COST PER S.F. MAT.	COST PER S.F. INST.	COST PER S.F. TOTAL

Important: See the Reference Section for critical supporting data - Reference Nos., Crews & Location Factors

Suspended Ceiling Price Sheet

	QUAN.	UNIT	LABOR HOURS	COST PER S.F.		
				MAT.	INST.	TOTAL
Suspension systems, T bar, 2′ x 2′ grid	1.000	S.F.	.012	.72	.52	1.24
2′ x 4′ grid	1.000	S.F.	.010	.58	.42	1
Concealed Z bar, 12″ module	1.000	S.F.	.015	.52	.65	1.17
Ceiling boards, fiberglass, film faced, 2′ x 2′ or 2′ x 4′, 5/8″ thick	1.000	S.F.	.013	.57	.54	1.11
3/4″ thick	1.000	S.F.	.013	1.31	.56	1.87
3″ thick thermal R11	1.000	S.F.	.018	1.44	.75	2.19
Glass cloth faced, 3/4″ thick	1.000	S.F.	.016	1.89	.68	2.57
1″ thick	1.000	S.F.	.016	2.09	.70	2.79
1-1/2″ thick, nubby face	1.000	S.F.	.017	2.60	.71	3.31
Mineral fiber boards, 5/8″ thick, aluminum face 2′ x 2′	1.000	S.F.	.013	1.77	.56	2.33
2′ x 4′	1.000	S.F.	.012	1.21	.52	1.73
Standard faced, 2′ x 2′ or 2′ x 4′	1.000	S.F.	.012	.70	.50	1.20
Plastic coated face, 2′ x 2′ or 2′ x 4′	1.000	S.F.	.020	1.10	.85	1.95
Fire rated, 2 hour rating, 5/8″ thick	1.000	S.F.	.012	.96	.50	1.46
Tegular edge, 2′ x 2′ or 2′ x 4′, 5/8″ thick, fine textured	1.000	S.F.	.013	1.19	.72	1.91
Rough textured	1.000	S.F.	.015	1.54	.72	2.26
3/4″ thick, fine textured	1.000	S.F.	.016	1.68	.75	2.43
Rough textured	1.000	S.F.	.018	1.90	.75	2.65
Luminous panels, prismatic, acrylic	1.000	S.F.	.020	1.98	.85	2.83
Polystyrene	1.000	S.F.	.020	1.01	.85	1.86
Flat or ribbed, acrylic	1.000	S.F.	.020	3.44	.85	4.29
Polystyrene	1.000	S.F.	.020	2.35	.85	3.20
Drop pan, white, acrylic	1.000	S.F.	.020	5.05	.85	5.90
Polystyrene	1.000	S.F.	.020	4.22	.85	5.07
Carrier channels, 4′-0″ on center, 3/4″ x 1-1/2″	1.000	S.F.	.017	.11	.72	.83
1-1/2″ x 3-1/2″	1.000	S.F.	.017	.28	.72	1
Hangers, #12 wire	1.000	S.F.	.002	.06	.07	.13

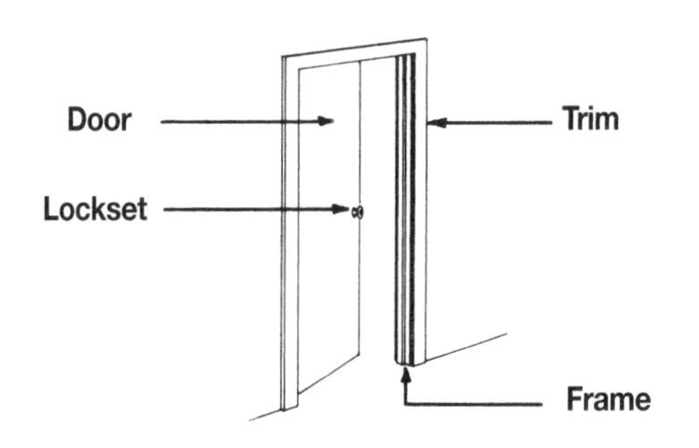

System Description	QUAN.	UNIT	LABOR HOURS	COST EACH		
				MAT.	INST.	TOTAL
LAUAN, FLUSH DOOR, HOLLOW CORE						
Door, flush, lauan, hollow core, 2′-8″ wide x 6′-8″ high	1.000	Ea.	.889	35.50	37.50	73
Frame, pine, 4-5/8″ jamb	17.000	L.F.	.725	108.80	30.60	139.40
Trim, stock pine, 11/16″ x 2-1/2″	34.000	L.F.	1.133	47.94	47.94	95.88
Paint trim, to 6″ wide, primer + 1 coat enamel	34.000	L.F.	.340	3.40	12.58	15.98
Butt hinges, chrome, 3-1/2″ x 3-1/2″	1.500	Pr.		39		39
Lockset, passage	1.000	Ea.	.500	16.35	21	37.35
Prime door & frame, oil, brushwork	2.000	Face	1.600	4.84	59	63.84
Paint door and frame, oil, 2 coats	2.000	Face	2.667	7.56	99	106.56
TOTAL		Ea.	7.854	263.39	307.62	571.01
BIRCH, FLUSH DOOR, HOLLOW CORE						
Door, flush, birch, hollow core, 2′-8″ wide x 6′-8″ high	1.000	Ea.	.889	50.50	37.50	88
Frame, pine, 4-5/8″ jamb	17.000	L.F.	.725	108.80	30.60	139.40
Trim, stock pine, 11/16″ x 2-1/2″	34.000	L.F.	1.133	47.94	47.94	95.88
Butt hinges, chrome, 3-1/2″ x 3-1/2″	1.500	Pr.		39		39
Lockset, passage	1.000	Ea.	.500	16.35	21	37.35
Prime door & frame, oil, brushwork	2.000	Face	1.600	4.84	59	63.84
Paint door and frame, oil, 2 coats	2.000	Face	2.667	7.56	99	106.56
TOTAL		Ea.	7.514	274.99	295.04	570.03
RAISED PANEL, SOLID, PINE DOOR						
Door, pine, raised panel, 2′-8″ wide x 6′-8″ high	1.000	Ea.	.889	167	37.50	204.50
Frame, pine, 4-5/8″ jamb	17.000	L.F.	.725	108.80	30.60	139.40
Trim, stock pine, 11/16″ x 2-1/2″	34.000	L.F.	1.133	47.94	47.94	95.88
Butt hinges, bronze, 3-1/2″ x 3-1/2″	1.500	Pr.		44.25		44.25
Lockset, passage	1.000	Ea.	.500	16.35	21	37.35
Prime door & frame, oil, brushwork	2.000		1.600	4.84	59	63.84
Paint door and frame, oil, 2 coats	2.000		2.667	7.56	99	106.56
TOTAL		Ea.	7.514	396.74	295.04	691.78

The costs in these systems are based on a cost per each door.

Description	QUAN.	UNIT	LABOR HOURS	COST EACH		
				MAT.	INST.	TOTAL

 Important: See the Reference Section for critical supporting data - Reference Nos., Crews & Location Factors

Interior Door Price Sheet	QUAN.	UNIT	LABOR HOURS	COST EACH		
				MAT.	INST.	TOTAL
Door, hollow core, lauan 1-3/8″ thick, 6′-8″ high x 1′-6″ wide	1.000	Ea.	.889	31	37.50	68.50
2′-0″ wide	1.000	Ea.	.889	30	37.50	67.50
2′-6″ wide	1.000	Ea.	.889	33.50	37.50	71
2′-8″ wide	1.000	Ea.	.889	35.50	37.50	73
3′-0″ wide	1.000	Ea.	.941	37	40	77
Birch 1-3/8″ thick, 6′-8″ high x 1′-6″ wide	1.000	Ea.	.889	39	37.50	76.50
2′-0″ wide	1.000	Ea.	.889	44	37.50	81.50
2′-6″ wide	1.000	Ea.	.889	49	37.50	86.50
2′-8″ wide	1.000	Ea.	.889	50.50	37.50	88
3′-0″ wide	1.000	Ea.	.941	55	40	95
Louvered pine 1-3/8″ thick, 6′-8″ high x 1′-6″ wide	1.000	Ea.	.842	110	35.50	145.50
2′-0″ wide	1.000	Ea.	.889	138	37.50	175.50
2′-6″ wide	1.000	Ea.	.889	150	37.50	187.50
2′-8″ wide	1.000	Ea.	.889	158	37.50	195.50
3′-0″ wide	1.000	Ea.	.941	170	40	210
Paneled pine 1-3/8″ thick, 6′-8″ high x 1′-6″ wide	1.000	Ea.	.842	119	35.50	154.50
2′-0″ wide	1.000	Ea.	.889	138	37.50	175.50
2′-6″ wide	1.000	Ea.	.889	154	37.50	191.50
2′-8″ wide	1.000	Ea.	.889	167	37.50	204.50
3′-0″ wide	1.000	Ea.	.941	174	40	214
Frame, pine, 1′-6″ thru 2′-0″ wide door, 3-5/8″ deep	16.000	L.F.	.683	76	29	105
4-5/8″ deep	16.000	L.F.	.683	102	29	131
5-5/8″ deep	16.000	L.F.	.683	69	29	98
2′-6″ thru 3′0″ wide door, 3-5/8″ deep	17.000	L.F.	.725	80.50	30.50	111
4-5/8″ deep	17.000	L.F.	.725	109	30.50	139.50
5-5/8″ deep	17.000	L.F.	.725	73.50	30.50	104
Trim, casing, painted, both sides, 1′-6″ thru 2′-6″ wide door	32.000	L.F.	1.855	47	74.50	121.50
2′-6″ thru 3′-0″ wide door	34.000	L.F.	1.971	50	79	129
Butt hinges 3-1/2″ x 3-1/2″, steel plated, chrome	1.500	Pr.		39		39
Bronze	1.500	Pr.		44.50		44.50
Locksets, passage, minimum	1.000	Ea.	.500	16.35	21	37.35
Maximum	1.000	Ea.	.575	18.80	24	42.80
Privacy, miniumum	1.000	Ea.	.625	20.50	26.50	47
Maximum	1.000	Ea.	.675	22	28.50	50.50
Paint 2 sides, primer & 2 cts., flush door, 1′-6″ to 2′-0″ wide	2.000	Face	5.547	14.25	206	220.25
2′-6″ thru 3′-0″ wide	2.000	Face	6.933	17.85	257	274.85
Louvered door, 1′-6″ thru 2′-0″ wide	2.000	Face	6.400	13.60	238	251.60
2′-6″ thru 3′-0″ wide	2.000	Face	8.000	17	297	314
Paneled door, 1′-6″ thru 2′-0″ wide	2.000	Face	6.400	13.60	238	251.60
2′-6″ thru 3′-0″ wide	2.000	Face	8.000	17	297	314

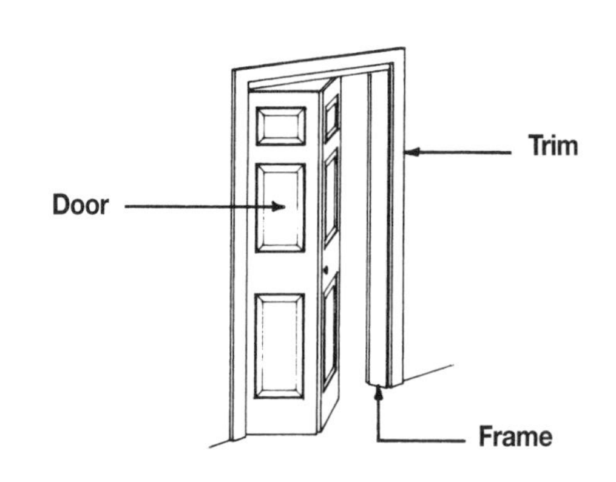

System Description	QUAN.	UNIT	LABOR HOURS	COST EACH		
				MAT.	INST.	TOTAL
BI-PASSING, FLUSH, LAUAN, HOLLOW CORE, 4′-0″ X 6′-8″						
Door, flush, lauan, hollow core, 4′-0″ x 6′-8″ opening	1.000	Ea.	1.333	181	56.50	237.50
Frame, pine, 4-5/8″ jamb	18.000	L.F.	.768	115.20	32.40	147.60
Trim, stock pine, 11/16″ x 2-1/2″	36.000	L.F.	1.200	50.76	50.76	101.52
Prime door & frame, oil, brushwork	2.000	Face	1.600	4.84	59	63.84
Paint door and frame, oil, 2 coats	2.000	Face	2.667	7.56	99	106.56
TOTAL		Ea.	7.568	359.36	297.66	657.02
BI-PASSING, FLUSH, BIRCH, HOLLOW CORE, 6′-0″ X 6′-8″						
Door, flush, birch, hollow core, 6′-0″ x 6′-8″ opening	1.000	Ea.	1.600	265	67.50	332.50
Frame, pine, 4-5/8″ jamb	19.000	L.F.	.811	121.60	34.20	155.80
Trim, stock pine, 11/16″ x 2-1/2″	38.000	L.F.	1.267	53.58	53.58	107.16
Prime door & frame, oil, brushwork	2.000	Face	2.000	6.05	73.75	79.80
Paint door and frame, oil, 2 coats	2.000	Face	3.333	9.45	123.75	133.20
TOTAL		Ea.	9.011	455.68	352.78	808.46
BI-FOLD, PINE, PANELED, 3′-0″ X 6′-8″						
Door, pine, paneled, 3′-0″ x 6′-8″ opening	1.000	Ea.	1.231	173	52	225
Frame, pine, 4-5/8″ jamb	17.000	L.F.	.725	108.80	30.60	139.40
Trim, stock pine, 11/16″ x 2-1/2″	34.000	L.F.	1.133	47.94	47.94	95.88
Prime door & frame, oil, brushwork	2.000	Face	1.600	4.84	59	63.84
Paint door and frame, oil, 2 coats	2.000	Face	2.667	7.56	99	106.56
TOTAL		Ea.	7.356	342.14	288.54	630.68
BI-FOLD, PINE, LOUVERED, 6′-0″ X 6′-8″						
Door, pine, louvered, 6′-0″ x 6′-8″ opening	1.000	Ea.	1.600	268	67.50	335.50
Frame, pine, 4-5/8″ jamb	19.000	L.F.	.811	121.60	34.20	155.80
Trim, stock pine, 11/16″ x 2-1/2″	38.000	L.F.	1.267	53.58	53.58	107.16
Prime door & frame, oil, brushwork	2.500	Face	2.000	6.05	73.75	79.80
Paint door and frame, oil, 2 coats	2.500	Face	3.333	9.45	123.75	133.20
TOTAL		Ea.	9.011	458.68	352.78	811.46

The costs in this system are based on a cost per each door.

Description	QUAN.	UNIT	LABOR HOURS	COST EACH		
				MAT.	INST.	TOTAL

Closet Door Price Sheet	QUAN.	UNIT	LABOR HOURS	COST EACH		
				MAT.	INST.	TOTAL
Doors, bi-passing, pine, louvered, 4'-0" x 6'-8" opening	1.000	Ea.	1.333	440	56.50	496.50
6'-0" x 6'-8" opening	1.000	Ea.	1.600	540	67.50	607.50
Paneled, 4'-0" x 6'-8" opening	1.000	Ea.	1.333	420	56.50	476.50
6'-0" x 6'-8" opening	1.000	Ea.	1.600	520	67.50	587.50
Flush, birch, hollow core, 4'-0" x 6'-8" opening	1.000	Ea.	1.333	221	56.50	277.50
6'-0" x 6'-8" opening	1.000	Ea.	1.600	265	67.50	332.50
Flush, lauan, hollow core, 4'-0" x 6'-8" opening	1.000	Ea.	1.333	181	56.50	237.50
6'-0" x 6'-8" opening	1.000	Ea.	1.600	212	67.50	279.50
Bi-fold, pine, louvered, 3'-0" x 6'-8" opening	1.000	Ea.	1.231	173	52	225
6'-0" x 6'-8" opening	1.000	Ea.	1.600	268	67.50	335.50
Paneled, 3'-0" x 6'-8" opening	1.000	Ea.	1.231	173	52	225
6'-0" x 6'-8" opening	1.000	Ea.	1.600	268	67.50	335.50
Flush, birch, hollow core, 3'-0" x 6'-8" opening	1.000	Ea.	1.231	58	52	110
6'-0" x 6'-8" opening	1.000	Ea.	1.600	114	67.50	181.50
Flush, lauan, hollow core, 3'-0" x 6'8" opening	1.000	Ea.	1.231	215	52	267
6'-0" x 6'-8" opening	1.000	Ea.	1.600	435	67.50	502.50
Frame pine, 3'-0" door, 3-5/8" deep	17.000	L.F.	.725	80.50	30.50	111
4-5/8" deep	17.000	L.F.	.725	109	30.50	139.50
5-5/8" deep	17.000	L.F.	.725	73.50	30.50	104
4'-0" door, 3-5/8" deep	18.000	L.F.	.768	85.50	32.50	118
4-5/8" deep	18.000	L.F.	.768	115	32.50	147.50
5-5/8" deep	18.000	L.F.	.768	77.50	32.50	110
6'-0" door, 3-5/8" deep	19.000	L.F.	.811	90	34	124
4-5/8" deep	19.000	L.F.	.811	122	34	156
5-5/8" deep	19.000	L.F.	.811	82	34	116
Trim both sides, painted 3'-0" x 6'-8" door	34.000	L.F.	1.971	50	79	129
4'-0" x 6'-8" door	36.000	L.F.	2.086	53	84	137
6'-0" x 6'-8" door	38.000	L.F.	2.203	56	88.50	144.50
Paint 2 sides, primer & 2 cts., flush door & frame, 3' x 6'-8" opng	2.000	Face	2.914	9.30	119	128.30
4'-0" x 6'-8" opening	2.000	Face	3.886	12.40	158	170.40
6'-0" x 6'-8" opening	2.000	Face	4.857	15.50	198	213.50
Paneled door & frame, 3'-0" x 6'-8" opening	2.000	Face	6.000	12.75	223	235.75
4'-0" x 6'-8" opening	2.000	Face	8.000	17	297	314
6'-0" x 6'-8" opening	2.000	Face	10.000	21.50	370	391.50
Louvered door & frame, 3'-0" x 6'-8" opening	2.000	Face	6.000	12.75	223	235.75
4'-0" x 6'-8" opening	2.000	Face	8.000	17	297	314
6'-0" x 6'-8" opening	2.000	Face	10.000	21.50	370	391.50

System Description	QUAN.	UNIT	LABOR HOURS	COST PER S.F.		
				MAT.	INST.	TOTAL
Carpet, direct glue-down, nylon, level loop, 26 oz.	1.000	S.F.	.018	2.15	.69	2.84
32 oz.	1.000	S.F.	.018	3.04	.69	3.73
40 oz.	1.000	S.F.	.018	4.52	.69	5.21
Nylon, plush, 20 oz.	1.000	S.F.	.018	1.43	.69	2.12
24 oz.	1.000	S.F.	.018	1.51	.69	2.20
30 oz.	1.000	S.F.	.018	2.25	.69	2.94
42 oz.	1.000	S.F.	.022	2.71	.83	3.54
48 oz.	1.000	S.F.	.022	4.05	.83	4.88
54 oz.	1.000	S.F.	.022	4.58	.83	5.41
Olefin, 15 oz.	1.000	S.F.	.018	.79	.69	1.48
22 oz.	1.000	S.F.	.018	.95	.69	1.64
Tile, foam backed, needle punch	1.000	S.F.	.014	3.47	.54	4.01
Tufted loop or shag	1.000	S.F.	.014	1.46	.54	2
Wool, 36 oz., level loop	1.000	S.F.	.018	10.35	.69	11.04
32 oz., patterned	1.000	S.F.	.020	10.20	.77	10.97
48 oz., patterned	1.000	S.F.	.020	10.40	.77	11.17
Padding, sponge rubber cushion, minimum	1.000	S.F.	.006	.44	.23	.67
Maximum	1.000	S.F.	.006	1.01	.23	1.24
Felt, 32 oz. to 56 oz., minimum	1.000	S.F.	.006	.44	.23	.67
Maximum	1.000	S.F.	.006	.82	.23	1.05
Bonded urethane, 3/8" thick, minimum	1.000	S.F.	.006	.48	.23	.71
Maximum	1.000	S.F.	.006	.83	.23	1.06
Prime urethane, 1/4" thick, minimum	1.000	S.F.	.006	.28	.23	.51
Maximum	1.000	S.F.	.006	.51	.23	.74
Stairs, for stairs, add to above carpet prices	1.000	Riser	.267		10.30	10.30
Underlayment plywood, 3/8" thick	1.000	S.F.	.011	.86	.45	1.31
1/2" thick	1.000	S.F.	.011	.96	.47	1.43
5/8" thick	1.000	S.F.	.011	1.34	.48	1.82
3/4" thick	1.000	S.F.	.012	1.45	.52	1.97
Particle board, 3/8" thick	1.000	S.F.	.011	.36	.45	.81
1/2" thick	1.000	S.F.	.011	.41	.47	.88
5/8" thick	1.000	S.F.	.011	.45	.48	.93
3/4" thick	1.000	S.F.	.012	.62	.52	1.14
Hardboard, 4' x 4', 0.215" thick	1.000	S.F.	.011	.43	.45	.88

System Description	QUAN.	UNIT	LABOR HOURS	COST PER S.F. MAT.	INST.	TOTAL
Resilient flooring, asphalt tile on concrete, 1/8" thick						
Color group B	1.000	S.F.	.020	1.17	.77	1.94
Color group C & D	1.000	S.F.	.020	1.28	.77	2.05
Asphalt tile on wood subfloor, 1/8" thick						
Color group B	1.000	S.F.	.020	1.38	.77	2.15
Color group C & D	1.000	S.F.	.020	1.49	.77	2.26
Vinyl composition tile, 12" x 12", 1/16" thick	1.000	S.F.	.016	.95	.62	1.57
Embossed	1.000	S.F.	.016	1.49	.62	2.11
Marbleized	1.000	S.F.	.016	1.49	.62	2.11
Plain	1.000	S.F.	.016	1.65	.62	2.27
.080" thick, embossed	1.000	S.F.	.016	1.13	.62	1.75
Marbleized	1.000	S.F.	.016	1.65	.62	2.27
Plain	1.000	S.F.	.016	2.16	.62	2.78
1/8" thick, marbleized	1.000	S.F.	.016	1.29	.62	1.91
Plain	1.000	S.F.	.016	2.26	.62	2.88
Vinyl tile, 12" x 12", .050" thick, minimum	1.000	S.F.	.016	2.31	.62	2.93
Maximum	1.000	S.F.	.016	4.35	.62	4.97
1/8" thick, minimum	1.000	S.F.	.016	3.03	.62	3.65
Maximum	1.000	S.F.	.016	5.55	.62	6.17
1/8" thick, solid colors	1.000	S.F.	.016	4.35	.62	4.97
Florentine pattern	1.000	S.F.	.016	4.86	.62	5.48
Marbleized or travertine pattern	1.000	S.F.	.016	10	.62	10.62
Vinyl sheet goods, backed, .070" thick, minimum	1.000	S.F.	.032	2.43	1.24	3.67
Maximum	1.000	S.F.	.040	2.98	1.54	4.52
.093" thick, minimum	1.000	S.F.	.035	2.63	1.34	3.97
Maximum	1.000	S.F.	.040	3.74	1.54	5.28
.125" thick, minimum	1.000	S.F.	.035	2.96	1.34	4.30
Maximum	1.000	S.F.	.040	4.68	1.54	6.22
Wood, oak, finished in place, 25/32" x 2-1/2" clear	1.000	S.F.	.074	4.08	2.81	6.89
Select	1.000	S.F.	.074	3.84	2.81	6.65
No. 1 common	1.000	S.F.	.074	4.87	2.81	7.68
Prefinished, oak, 2-1/2" wide	1.000	S.F.	.047	6.80	1.99	8.79
3-1/4" wide	1.000	S.F.	.043	8.75	1.83	10.58
Ranch plank, oak, random width	1.000	S.F.	.055	8.45	2.33	10.78
Parquet, 5/16" thick, finished in place, oak, minimum	1.000	S.F.	.077	4.55	2.93	7.48
Maximum	1.000	S.F.	.107	6.65	4.20	10.85
Teak, minimum	1.000	S.F.	.077	5.75	2.93	8.68
Maximum	1.000	S.F.	.107	9.50	4.20	13.70
Sleepers, treated, 16" O.C., 1" x 2"	1.000	S.F.	.007	.11	.29	.40
1" x 3"	1.000	S.F.	.008	.21	.34	.55
2" x 4"	1.000	S.F.	.011	.53	.45	.98
2" x 6"	1.000	S.F.	.012	.82	.52	1.34
Subfloor, plywood, 1/2" thick	1.000	S.F.	.011	.68	.45	1.13
5/8" thick	1.000	S.F.	.012	.73	.50	1.23
3/4" thick	1.000	S.F.	.013	.90	.54	1.44
Ceramic tile, color group 2, 1" x 1"	1.000	S.F.	.087	4.70	2.99	7.69
2" x 2" or 2" x 1"	1.000	S.F.	.084	4.93	2.88	7.81
Color group 1, 8" x 8"		S.F.	.064	3.45	2.19	5.64
12" x 12"		S.F.	.049	4.33	1.69	6.02
16" x 16"		S.F.	.029	5.95	1	6.95

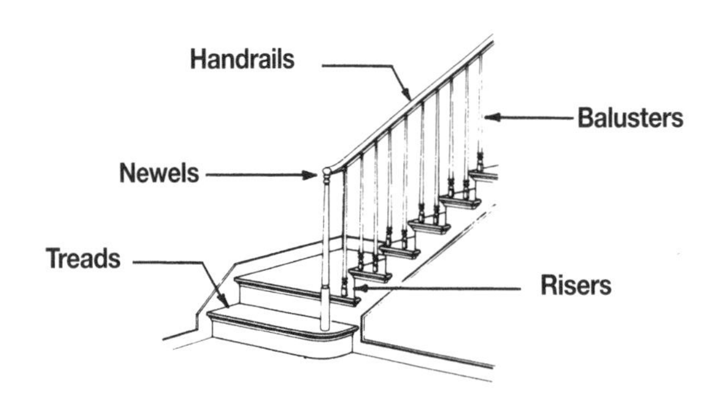

System Description	QUAN.	UNIT	LABOR HOURS	COST EACH		
				MAT.	INST.	TOTAL
7 RISERS, OAK TREADS, BOX STAIRS						
Treads, oak, 1-1/4″ x 10″ wide, 3′ long	6.000	Ea.	2.667	192	112.80	304.80
Risers, 3/4″ thick, beech	7.000	Ea.	2.625	133.35	111.30	244.65
Balusters, birch, 30″ high	12.000	Ea.	3.429	83.40	144.60	228
Newels, 3-1/4″ wide	2.000	Ea.	2.286	84	97	181
Handrails, oak laminated	7.000	L.F.	.933	245	39.55	284.55
Stringers, 2″ x 10″, 3 each	21.000	L.F.	.306	8.61	12.81	21.42
TOTAL		Ea.	12.246	746.36	518.06	1,264.42
14 RISERS, OAK TREADS, BOX STAIRS						
Treads, oak, 1-1/4″ x 10″ wide, 3′ long	13.000	Ea.	5.778	416	244.40	660.40
Risers, 3/4″ thick, beech	14.000	Ea.	5.250	266.70	222.60	489.30
Balusters, birch, 30″ high	26.000	Ea.	7.428	180.70	313.30	494
Newels, 3-1/4″ wide	2.000	Ea.	2.286	84	97	181
Handrails, oak, laminated	14.000	L.F.	1.867	490	79.10	569.10
Stringers, 2″ x 10″, 3 each	42.000	L.F.	5.169	59.64	218.40	278.04
TOTAL		Ea.	27.778	1,497.04	1,174.80	2,671.84
14 RISERS, PINE TREADS, BOX STAIRS						
Treads, pine, 9-1/2″ x 3/4″ thick	13.000	Ea.	5.778	273	244.40	517.40
Risers, 3/4″ thick, pine	14.000	Ea.	5.091	141.12	214.20	355.32
Balusters, pine, 30″ high	26.000	Ea.	7.428	105.82	313.30	419.12
Newels, 3-1/4″ wide	2.000	Ea.	2.286	84	97	181
Handrails, oak, laminated	14.000	L.F.	1.867	490	79.10	569.10
Stringers, 2″ x 10″, 3 each	42.000	L.F.	5.169	59.64	218.40	278.04
TOTAL		Ea.	27.619	1,153.58	1,166.40	2,319.98

Description	QUAN.	UNIT	LABOR HOURS	COST EACH		
				MAT.	INST.	TOTAL

 Important: See the Reference Section for critical supporting data - Reference Nos., Crews & Location Factors

Stairway Price Sheet	QUAN.	UNIT	LABOR HOURS	COST EACH		
				MAT.	INST.	TOTAL
Treads, oak, 1-1/16" x 9-1/2", 3' long, 7 riser stair	6.000	Ea.	2.667	192	113	305
14 riser stair	13.000	Ea.	5.778	415	244	659
1-1/16" x 11-1/2", 3' long, 7 riser stair	6.000	Ea.	2.667	204	113	317
14 riser stair	13.000	Ea.	5.778	440	244	684
Pine, 3/4" x 9-1/2", 3' long, 7 riser stair	6.000	Ea.	2.667	126	113	239
14 riser stair	13.000	Ea.	5.778	273	244	517
3/4" x 11-1/4", 3' long, 7 riser stair	6.000	Ea.	2.667	153	113	266
14 riser stair	13.000	Ea.	5.778	330	244	574
Risers, oak, 3/4" x 7-1/2" high, 7 riser stair	7.000	Ea.	2.625	141	111	252
14 riser stair	14.000	Ea.	5.250	281	223	504
Beech, 3/4" x 7-1/2" high, 7 riser stair	7.000	Ea.	2.625	133	111	244
14 riser stair	14.000	Ea.	5.250	267	223	490
Baluster, turned, 30" high, pine, 7 riser stair	12.000	Ea.	3.429	49	145	194
14 riser stair	26.000	Ea.	7.428	106	315	421
30" birch, 7 riser stair	12.000	Ea.	3.429	83.50	145	228.50
14 riser stair	26.000	Ea.	7.428	181	315	496
42" pine, 7 riser stair	12.000	Ea.	3.556	65	150	215
14 riser stair	26.000	Ea.	7.704	140	325	465
42" birch, 7 riser stair	12.000	Ea.	3.556	141	150	291
14 riser stair	26.000	Ea.	7.704	305	325	630
Newels, 3-1/4" wide, starting, 7 riser stair	2.000	Ea.	2.286	84	97	181
14 riser stair	2.000	Ea.	2.286	84	97	181
Landing, 7 riser stair	2.000	Ea.	3.200	228	135	363
14 riser stair	2.000	Ea.	3.200	228	135	363
Handrails, oak, laminated, 7 riser stair	7.000	L.F.	.933	245	39.50	284.50
14 riser stair	14.000	L.F.	1.867	490	79	569
Stringers, fir, 2" x 10" 7 riser stair	21.000	L.F.	2.585	30	109	139
14 riser stair	42.000	L.F.	5.169	59.50	218	277.50
2" x 12", 7 riser stair	21.000	L.F.	2.585	37	109	146
14 riser stair	42.000	L.F.	5.169	74	218	292

Special Stairways	QUAN.	UNIT	LABOR HOURS	COST EACH		
				MAT.	INST.	TOTAL
Basement stairs, open risers	1.000	Flight	4.000	655	169	824
Spiral stairs, oak, 4'-6" diameter, prefabricated, 9' high	1.000	Flight	10.667	4,850	450	5,300
Aluminum, 5'-0" diameter stock unit	1.000	Flight	9.956	6,100	555	6,655
Custom unit	1.000	Flight	9.956	11,500	555	12,055
Cast iron, 4'-0" diameter, minimum	1.000	Flight	9.956	5,450	555	6,005
Maximum	1.000	Flight	17.920	7,425	1,000	8,425
Steel, industrial, pre-erected, 3'-6" wide, bar rail	1.000	Flight	7.724	5,525	645	6,170
Picket rail	1.000	Flight	7.724	6,225	645	6,870

Division 7
Specialties

Soffit Drywall
Soffit Framing
Top Cabinets
Counter Top
Bottom Cabinets

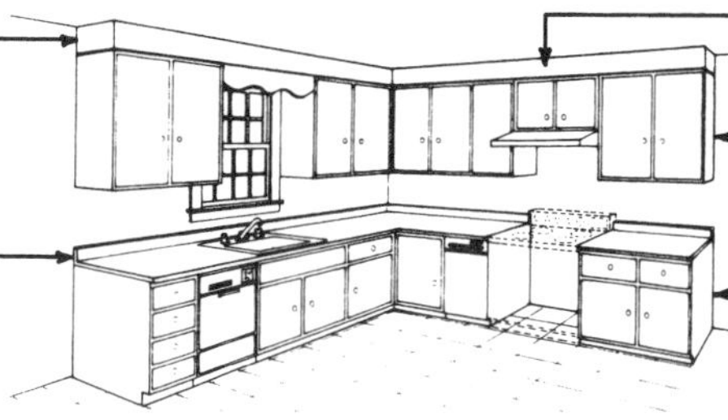

System Description	QUAN.	UNIT	LABOR HOURS	COST PER L.F.		
				MAT.	INST.	TOTAL
KITCHEN, ECONOMY GRADE						
Top cabinets, economy grade	1.000	L.F.	.171	33.60	7.20	40.80
Bottom cabinets, economy grade	1.000	L.F.	.256	50.40	10.80	61.20
Square edge, plastic face countertop	1.000		.267	24.50	11.25	35.75
Blocking, wood, 2″ x 4″	1.000	L.F.	.032	.41	1.35	1.76
Soffit, framing, wood, 2″ x 4″	4.000	L.F.	.071	1.64	3	4.64
Soffit drywall	2.000	S.F.	.047	.76	2	2.76
Drywall painting	2.000	S.F.	.013	.10	.56	.66
TOTAL		L.F.	.857	111.41	36.16	147.57
AVERAGE GRADE						
Top cabinets, average grade	1.000	L.F.	.213	42	9	51
Bottom cabinets, average grade	1.000	L.F.	.320	63	13.50	76.50
Solid surface countertop, solid color	1.000	L.F.	.800	65.50	34	99.50
Blocking, wood, 2″ x 4″	1.000	L.F.	.032	.41	1.35	1.76
Soffit framing, wood, 2″ x 4″	4.000	L.F.	.071	1.64	3	4.64
Soffit drywall	2.000	S.F.	.047	.76	2	2.76
Drywall painting	2.000	S.F.	.013	.10	.56	.66
TOTAL		L.F.	1.496	173.41	63.41	236.82
CUSTOM GRADE						
Top cabinets, custom grade	1.000	L.F.	.256	110	10.80	120.80
Bottom cabinets, custom grade	1.000	L.F.	.384	165	16.20	181.20
Solid surface countertop, premium patterned color	1.000	L.F.	1.067	114	45	159
Blocking, wood, 2″ x 4″	1.000	L.F.	.032	.41	1.35	1.76
Soffit framing, wood, 2″ x 4″	4.000	L.F.	.071	1.64	3	4.64
Soffit drywall	2.000	S.F.	.047	.76	2	2.76
Drywall painting	2.000	S.F.	.013	.10	.56	.66
TOTAL		L.F.	1.870	391.91	78.91	470.82

Description	QUAN.	UNIT	LABOR HOURS	COST PER L.F.		
				MAT.	INST.	TOTAL

Important: See the Reference Section for critical supporting data - Reference Nos., Crews & Location Factors

Kitchen Price Sheet	QUAN.	UNIT	LABOR HOURS	COST PER L.F.		
				MAT.	INST.	TOTAL
Top cabinets, economy grade	1.000	L.F.	.171	33.50	7.20	40.70
Average grade	1.000	L.F.	.213	42	9	51
Custom grade	1.000	L.F.	.256	110	10.80	120.80
Bottom cabinets, economy grade	1.000	L.F.	.256	50.50	10.80	61.30
Average grade	1.000	L.F.	.320	63	13.50	76.50
Custom grade	1.000	L.F.	.384	165	16.20	181.20
Counter top, laminated plastic, 7/8″ thick, no splash	1.000	L.F.	.267	19.45	11.25	30.70
With backsplash	1.000	L.F.	.267	25.50	11.25	36.75
1-1/4″ thick, no splash	1.000	L.F.	.286	22	12.05	34.05
With backsplash	1.000	L.F.	.286	28	12.05	40.05
Post formed, laminated plastic	1.000	L.F.	.267	10	11.25	21.25
Marble, with backsplash, minimum	1.000	L.F.	.471	35.50	20	55.50
Maximum	1.000	L.F.	.615	89	26	115
Maple, solid laminated, no backsplash	1.000	L.F.	.286	58	12.05	70.05
With backsplash	1.000	L.F.	.286	69	12.05	81.05
Blocking, wood, 2″ x 4″	1.000	L.F.	.032	.41	1.35	1.76
2″ x 6″	1.000	L.F.	.036	.65	1.52	2.17
2″ x 8″	1.000	L.F.	.040	.94	1.69	2.63
Soffit framing, wood, 2″ x 3″	4.000	L.F.	.064	1.68	2.72	4.40
2″ x 4″	4.000	L.F.	.071	1.64	3	4.64
Soffit, drywall, painted	2.000	S.F.	.060	.86	2.56	3.42
Paneling, standard	2.000	S.F.	.064	1.84	2.70	4.54
Deluxe	2.000	S.F.	.091	4.12	3.86	7.98
Sinks, porcelain on cast iron, single bowl, 21″ x 24″	1.000	Ea.	10.334	400	430	830
21″ x 30″	1.000	Ea.	10.334	380	430	810
Double bowl, 20″ x 32″	1.000	Ea.	10.810	490	450	940
Stainless steel, single bowl, 16″ x 20″	1.000	Ea.	10.334	560	430	990
22″ x 25″	1.000	Ea.	10.334	610	430	1,040
Double bowl, 20″ x 32″	1.000	Ea.	10.810	310	450	760

Kitchen Price Sheet	QUAN.	UNIT	LABOR HOURS	COST PER L.F.		
				MAT.	INST.	TOTAL
Range, free standing, minimum	1.000	Ea.	3.600	355	141	496
Maximum	1.000	Ea.	6.000	1,775	213	1,988
Built-in, minimum	1.000	Ea.	3.333	590	153	743
Maximum	1.000	Ea.	10.000	1,575	430	2,005
Counter top range, 4-burner, minimum	1.000	Ea.	3.333	294	153	447
Maximum	1.000	Ea.	4.667	675	214	889
Compactor, built-in, minimum	1.000	Ea.	2.215	490	96	586
Maximum	1.000	Ea.	3.282	550	142	692
Dishwasher, built-in, minimum	1.000	Ea.	6.735	385	315	700
Maximum	1.000	Ea.	9.235	430	430	860
Garbage disposer, minimum	1.000	Ea.	2.810	89.50	131	220.50
Maximum	1.000	Ea.	2.810	207	131	338
Microwave oven, minimum	1.000	Ea.	2.615	112	121	233
Maximum	1.000	Ea.	4.615	465	213	678
Range hood, ducted, minimum	1.000	Ea.	4.658	81.50	204	285.50
Maximum	1.000	Ea.	5.991	720	261	981
Ductless, minimum	1.000	Ea.	2.615	77.50	115	192.50
Maximum	1.000	Ea.	3.948	715	172	887
Refrigerator, 16 cu.ft., minimum	1.000	Ea.	2.000	625	60.50	685.50
Maximum	1.000	Ea.	3.200	1,000	97	1,097
16 cu.ft. with icemaker, minimum	1.000	Ea.	4.210	790	157	947
Maximum	1.000	Ea.	5.410	1,175	193	1,368
19 cu.ft., minimum	1.000	Ea.	2.667	775	80.50	855.50
Maximum	1.000	Ea.	4.667	1,350	141	1,491
19 cu.ft. with icemaker, minimum	1.000	Ea.	5.143	1,000	185	1,185
Maximum	1.000	Ea.	7.143	1,600	245	1,845
Sinks, porcelain on cast iron single bowl, 21″ x 24″	1.000	Ea.	10.334	400	430	830
21″ x 30″	1.000	Ea.	10.334	380	430	810
Double bowl, 20″ x 32″	1.000	Ea.	10.810	490	450	940
Stainless steel, single bowl 16″ x 20″	1.000	Ea.	10.334	560	430	990
22″ x 25″	1.000	Ea.	10.334	610	430	1,040
Double bowl, 20″ x 32″	1.000	Ea.	10.810	310	450	760
Water heater, electric, 30 gallon	1.000	Ea.	3.636	320	169	489
40 gallon	1.000	Ea.	4.000	345	186	531
Gas, 30 gallon	1.000	Ea.	4.000	690	186	876
75 gallon	1.000	Ea.	5.333	900	247	1,147
Wall, packaged terminal heater/air conditioner cabinet, wall sleeve, louver, electric heat, thermostat, manual changeover, 208V		Ea.				
6000 BTUH cooling, 8800 BTU heating	1.000	Ea.	2.667	1,050	112	1,162
9000 BTUH cooling, 13,900 BTU heating	1.000	Ea.	3.200	1,100	135	1,235
12,000 BTUH cooling, 13,900 BTU heating	1.000	Ea.	4.000	1,225	168	1,393
15,000 BTUH cooling, 13,900 BTU heating	1.000	Ea.	5.333	1,450	224	1,674

Important: See the Reference Section for critical supporting data - Reference Nos., Crews & Location Factors

System Description	QUAN.	UNIT	LABOR HOURS	COST EACH		
				MAT.	INST.	TOTAL
Curtain rods, stainless, 1" diameter, 3' long	1.000	Ea.	.615	33.50	26	59.50
5' long	1.000	Ea.	.615	33.50	26	59.50
Grab bar, 1" diameter, 12" long	1.000	Ea.	.283	18.70	12	30.70
36" long	1.000	Ea.	.340	19.55	14.35	33.90
1-1/4" diameter, 12" long	1.000	Ea.	.333	22	14.10	36.10
36" long	1.000	Ea.	.400	23	16.90	39.90
1-1/2" diameter, 12" long	1.000	Ea.	.383	25.50	16.20	41.70
36" long	1.000	Ea.	.460	26.50	19.45	45.95
Mirror, 18" x 24"	1.000	Ea.	.400	72.50	16.90	89.40
72" x 24"	1.000	Ea.	1.333	216	56.50	272.50
Medicine chest with mirror, 18" x 24"	1.000	Ea.	.400	172	16.90	188.90
36" x 24"	1.000	Ea.	.600	258	25.50	283.50
Toilet tissue dispenser, surface mounted, minimum	1.000	Ea.	.267	12.85	11.25	24.10
Maximum	1.000	Ea.	.400	19.30	16.90	36.20
Flush mounted, minimum	1.000	Ea.	.293	14.15	12.40	26.55
Maximum	1.000	Ea.	.427	20.50	18	38.50
Towel bar, 18" long, minimum	1.000	Ea.	.278	31	11.75	42.75
Maximum	1.000	Ea.	.348	38.50	14.70	53.20
24" long, minimum	1.000	Ea.	.313	34.50	13.25	47.75
Maximum	1.000	Ea.	.383	42.50	16.15	58.65
36" long, minimum	1.000	Ea.	.381	78.50	16.10	94.60
Maximum	1.000	Ea.	.419	86.50	17.70	104.20

24 | Masonry Fireplace Systems

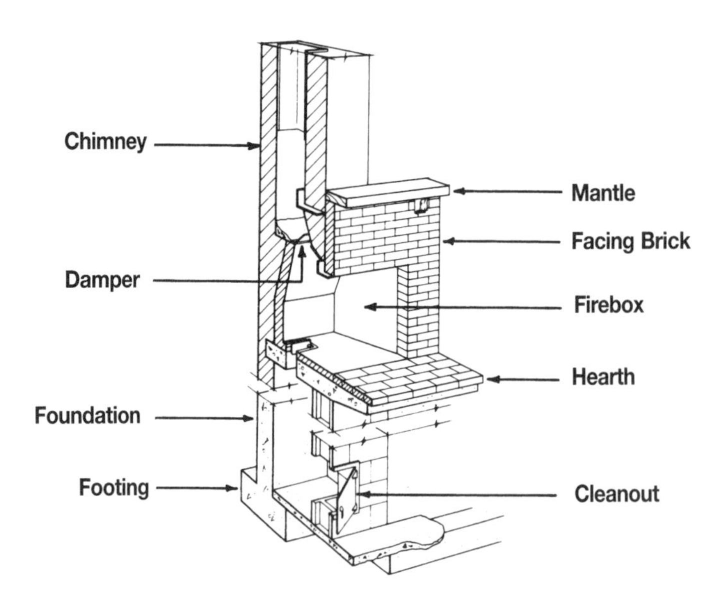

System Description	QUAN.	UNIT	LABOR HOURS	COST EACH		
				MAT.	INST.	TOTAL
MASONRY FIREPLACE						
Footing, 8" thick, concrete, 4' x 7'	.700	C.Y.	2.110	95.90	76.23	172.13
Foundation, concrete block, 32" x 60" x 4' deep	1.000	Ea.	5.275	141	202.80	343.80
Fireplace, brick firebox, 30" x 29" opening	1.000	Ea.	40.000	455	1,500	1,955
Damper, cast iron, 30" opening	1.000	Ea.	1.333	78	56.50	134.50
Facing brick, standard size brick, 6' x 5'	30.000	S.F.	5.217	108.30	201	309.30
Hearth, standard size brick, 3' x 6'	1.000	Ea.	8.000	168	300	468
Chimney, standard size brick, 8" x 12" flue, one story house	12.000	V.L.F.	12.000	354	450	804
Mantle, 4" x 8", wood	6.000	L.F.	1.333	33.90	56.40	90.30
Cleanout, cast iron, 8" x 8"	1.000	Ea.	.667	33	28.50	61.50
TOTAL		Ea.	75.935	1,467.10	2,871.43	4,338.53

The costs in this system are on a cost each basis.

Description	QUAN.	UNIT	LABOR HOURS	COST EACH		
				MAT.	INST.	TOTAL

Masonry Fireplace Price Sheet	QUAN.	UNIT	LABOR HOURS	COST EACH		
				MAT.	INST.	TOTAL
Footing 8″ thick, 3′ x 6′	.440	C.Y.	1.326	60.50	48	108.50
4′ x 7′	.700	C.Y.	2.110	96	76	172
5′ x 8′	1.000	C.Y.	3.014	137	109	246
1′ thick, 3′ x 6′	.670	C.Y.	2.020	92	73	165
4′ x 7′	1.030	C.Y.	3.105	141	112	253
5′ x 8′	1.480	C.Y.	4.461	203	161	364
Foundation-concrete block, 24″ x 48″, 4′ deep	1.000	Ea.	4.267	113	162	275
8′ deep	1.000	Ea.	8.533	226	325	551
24″ x 60″, 4′ deep	1.000	Ea.	4.978	132	189	321
8′ deep	1.000	Ea.	9.956	263	380	643
32″ x 48″, 4′ deep	1.000	Ea.	4.711	125	179	304
8′ deep	1.000	Ea.	9.422	249	360	609
32″ x 60″, 4′ deep	1.000	Ea.	5.333	141	203	344
8′ deep	1.000	Ea.	10.845	287	410	697
32″ x 72″, 4′ deep	1.000	Ea.	6.133	162	233	395
8′ deep	1.000	Ea.	12.267	325	465	790
Fireplace, brick firebox 30″ x 29″ opening	1.000	Ea.	40.000	455	1,500	1,955
48″ x 30″ opening	1.000	Ea.	60.000	685	2,250	2,935
Steel fire box with registers, 25″ opening	1.000	Ea.	26.667	835	1,025	1,860
48″ opening	1.000	Ea.	44.000	1,400	1,675	3,075
Damper, cast iron, 30″ opening	1.000	Ea.	1.333	78	56.50	134.50
36″ opening	1.000	Ea.	1.556	91	66	157
Steel, 30″ opening	1.000	Ea.	1.333	70	56.50	126.50
36″ opening	1.000	Ea.	1.556	81.50	66	147.50
Facing for fireplace, standard size brick, 6′ x 5′	30.000	S.F.	5.217	108	201	309
7′ x 5′	35.000	S.F.	6.087	126	235	361
8′ x 6′	48.000	S.F.	8.348	173	320	493
Fieldstone, 6′ x 5′	30.000	S.F.	5.217	470	201	671
7′ x 5′	35.000	S.F.	6.087	545	235	780
8′ x 6′	48.000	S.F.	8.348	750	320	1,070
Sheetrock on metal, studs, 6′ x 5′	30.000	S.F.	.980	21.50	41.50	63
7′ x 5′	35.000	S.F.	1.143	25	48.50	73.50
8′ x 6′	48.000	S.F.	1.568	34	66	100
Hearth, standard size brick, 3′ x 6′	1.000	Ea.	8.000	168	300	468
3′ x 7′	1.000	Ea.	9.280	195	350	545
3′ x 8′	1.000	Ea.	10.640	223	400	623
Stone, 3′ x 6′	1.000	Ea.	8.000	181	300	481
3′ x 7′	1.000	Ea.	9.280	210	350	560
3′ x 8′	1.000	Ea.	10.640	241	400	641
Chimney, standard size brick , 8″ x 12″ flue, one story house	12.000	V.L.F.	12.000	355	450	805
Two story house	20.000	V.L.F.	20.000	590	750	1,340
Mantle wood, beams, 4″ x 8″	6.000	L.F.	1.333	34	56.50	90.50
4″ x 10″	6.000	L.F.	1.371	41	58	99
Ornate, prefabricated, 6′ x 3′-6″ opening, minimum	1.000	Ea.	1.600	153	67.50	220.50
Maximum	1.000	Ea.	1.600	190	67.50	257.50
Cleanout, door and frame, cast iron, 8″ x 8″	1.000	Ea.	.667	33	28.50	61.50
12″ x 12″	1.000	Ea.	.800	40	34	74

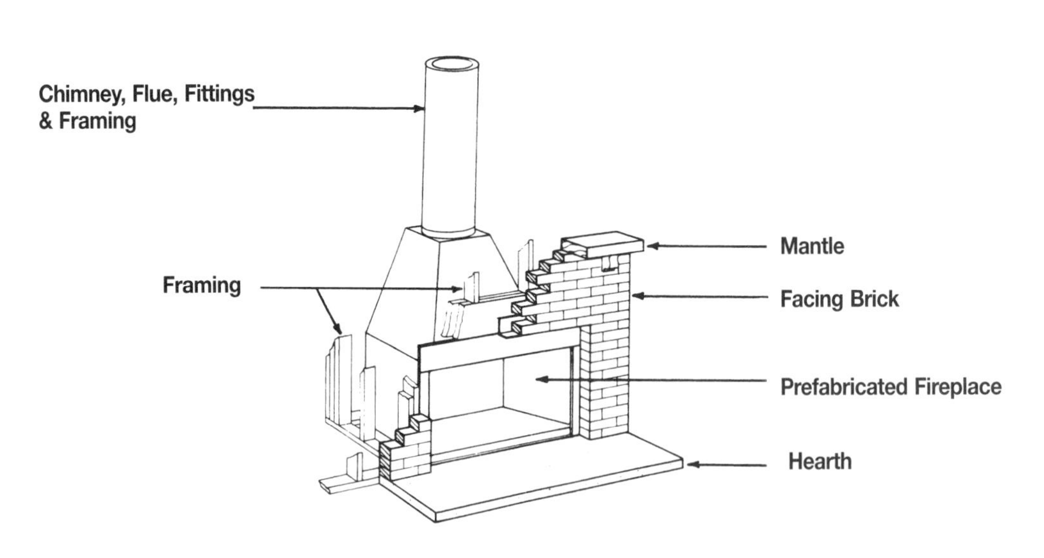

System Description	QUAN.	UNIT	LABOR HOURS	COST EACH		
				MAT.	INST.	TOTAL
PREFABRICATED FIREPLACE						
Prefabricated fireplace, metal, minimum	1.000	Ea.	6.154	1,075	260	1,335
Framing, 2″ x 4″ studs, 6′ x 5′	35.000	L.F.	.509	14.35	21.35	35.70
Fire resistant gypsum drywall, unfinished	40.000	S.F.	.320	13.20	13.60	26.80
Drywall finishing adder	40.000	S.F.	.320	1.60	13.60	15.20
Facing, brick, standard size brick, 6′ x 5′	30.000	S.F.	5.217	108.30	201	309.30
Hearth, standard size brick, 3′ x 6′	1.000	Ea.	8.000	168	300	468
Chimney, one story house, framing, 2″ x 4″ studs	80.000	L.F.	1.164	32.80	48.80	81.60
Sheathing, plywood, 5/8″ thick	32.000	S.F.	.758	83.84	32	115.84
Flue, 10″ metal, insulated pipe	12.000	V.L.F.	4.000	306	168.60	474.60
Fittings, ceiling support	1.000	Ea.	.667	142	28	170
Fittings, joist shield	1.000	Ea.	.667	81	28	109
Fittings, roof flashing	1.000	Ea.	.667	165	28	193
Mantle beam, wood, 4″ x 8″	6.000	L.F.	1.333	33.90	56.40	90.30
TOTAL		Ea.	29.776	2,224.99	1,199.35	3,424.34

The costs in this system are on a cost each basis.

Description	QUAN.	UNIT	LABOR HOURS	COST EACH		
				MAT.	INST.	TOTAL

Important: See the Reference Section for critical supporting data - Reference Nos., Crews & Location Factors

Prefabricated Fireplace Price Sheet	QUAN.	UNIT	LABOR HOURS	COST EACH		
				MAT.	INST.	TOTAL
Prefabricated fireplace, minimum	1.000	Ea.	6.154	1,075	260	1,335
Average	1.000	Ea.	8.000	1,500	340	1,840
Maximum	1.000	Ea.	8.889	3,700	375	4,075
Framing, 2″ x 4″ studs, fireplace, 6′ x 5′	35.000	L.F.	.509	14.35	21.50	35.85
7′ x 5′	40.000	L.F.	.582	16.40	24.50	40.90
8′ x 6′	45.000	L.F.	.655	18.45	27.50	45.95
Sheetrock, 1/2″ thick, fireplace, 6′ x 5′	40.000	S.F.	.640	14.80	27	41.80
7′ x 5′	45.000	S.F.	.720	16.65	30.50	47.15
8′ x 6′	50.000	S.F.	.800	18.50	34	52.50
Facing for fireplace, brick, 6′ x 5′	30.000	S.F.	5.217	108	201	309
7′ x 5′	35.000	S.F.	6.087	126	235	361
8′ x 6′	48.000	S.F.	8.348	173	320	493
Fieldstone, 6′ x 5′	30.000	S.F.	5.217	505	201	706
7′ x 5′	35.000	S.F.	6.087	590	235	825
8′ x 6′	48.000	S.F.	8.348	805	320	1,125
Hearth, standard size brick, 3′ x 6′	1.000	Ea.	8.000	168	300	468
3′ x 7′	1.000	Ea.	9.280	195	350	545
3′ x 8′	1.000	Ea.	10.640	223	400	623
Stone, 3′ x 6′	1.000	Ea.	8.000	181	300	481
3′ x 7′	1.000	Ea.	9.280	210	350	560
3′ x 8′	1.000	Ea.	10.640	241	400	641
Chimney, framing, 2″ x 4″, one story house	80.000	L.F.	1.164	33	49	82
Two story house	120.000	L.F.	1.746	49	73	122
Sheathing, plywood, 5/8″ thick	32.000	S.F.	.758	84	32	116
Stucco on plywood	32.000	S.F.	1.125	36	46	82
Flue, 10″ metal pipe, insulated, one story house	12.000	V.L.F.	4.000	305	169	474
Two story house	20.000	V.L.F.	6.667	510	281	791
Fittings, ceiling support	1.000	Ea.	.667	142	28	170
Fittings joist sheild, one story house	1.000	Ea.	.667	81	28	109
Two story house	2.000	Ea.	1.333	162	56	218
Fittings roof flashing	1.000	Ea.	.667	165	28	193
Mantle, wood beam, 4″ x 8″	6.000	L.F.	1.333	34	56.50	90.50
4″ x 10″	6.000	L.F.	1.371	41	58	99
Ornate prefabricated, 6′ x 3′-6″ opening, minimum	1.000	Ea.	1.600	153	67.50	220.50
Maximum	1.000	Ea.	1.600	190	67.50	257.50

System Description	QUAN.	UNIT	LABOR HOURS	COST EACH		
				MAT.	INST.	TOTAL
Economy, lean to, shell only, not including 2′ stub wall, fndtn, flrs, heat						
4′ x 16′	1.000	Ea.	26.212	2,375	1,100	3,475
4′ x 24′	1.000	Ea.	30.259	2,725	1,275	4,000
6′ x 10′	1.000	Ea.	16.552	1,975	700	2,675
6′ x 16′	1.000	Ea.	23.034	2,750	975	3,725
6′ x 24′	1.000	Ea.	29.793	3,575	1,250	4,825
8′ x 10′	1.000	Ea.	22.069	2,650	930	3,580
8′ x 16′	1.000	Ea.	38.400	4,600	1,625	6,225
8′ x 24′	1.000	Ea.	49.655	5,950	2,100	8,050
Free standing, 8′ x 8′	1.000	Ea.	17.356	3,075	735	3,810
8′ x 16′	1.000	Ea.	30.211	5,350	1,275	6,625
8′ x 24′	1.000	Ea.	39.051	6,900	1,650	8,550
10′ x 10′	1.000	Ea.	18.824	3,700	795	4,495
10′ x 16′	1.000	Ea.	24.095	4,725	1,025	5,750
10′ x 24′	1.000	Ea.	31.624	6,225	1,325	7,550
14′ x 10′	1.000	Ea.	20.741	4,625	875	5,500
14′ x 16′	1.000	Ea.	24.889	5,550	1,050	6,600
14′ x 24′	1.000	Ea.	33.349	7,425	1,400	8,825
Standard,lean to,shell only,not incl.2′stub wall, fndtn,flrs, heat 4′x10′	1.000	Ea.	28.235	2,550	1,200	3,750
4′ x 16′	1.000	Ea.	39.341	3,550	1,675	5,225
4′ x 24′	1.000	Ea.	45.412	4,100	1,925	6,025
6′ x 10′	1.000	Ea.	24.827	2,975	1,050	4,025
6′ x 16′	1.000	Ea.	34.538	4,125	1,450	5,575
6′ x 24′	1.000	Ea.	44.689	5,350	1,875	7,225
8′ x 10′	1.000	Ea.	33.103	3,950	1,400	5,350
8′ x 16′	1.000	Ea.	57.600	6,900	2,425	9,325
8′ x 24′	1.000	Ea.	74.482	8,900	3,150	12,050
Free standing, 8′ x 8′	1.000	Ea.	26.034	4,600	1,100	5,700
8′ x 16′	1.000	Ea.	45.316	8,025	1,925	9,950
8′ x 24′	1.000	Ea.	58.577	10,400	2,475	12,875
10′ x 10′	1.000	Ea.	28.236	5,550	1,200	6,750
10′ x 16′	1.000	Ea.	36.142	7,100	1,525	8,625
10′ x 24′	1.000	Ea.	47.436	9,325	2,000	11,325
14′ x 10′	1.000	Ea.	31.112	6,925	1,325	8,250
14′ x 16′	1.000	Ea.	37.334	8,325	1,575	9,900
14′ x 24′	1.000	Ea.	50.030	11,100	2,100	13,200
Deluxe,lean to,shell only,not incl.2′stub wall, fndtn, flrs or heat, 4′x10′	1.000	Ea.	20.645	4,400	880	5,280
4′ x 16′	1.000	Ea.	33.032	7,050	1,400	8,450
4′ x 24′	1.000	Ea.	49.548	10,600	2,100	12,700
6′ x 10′	1.000	Ea.	30.968	6,600	1,325	7,925
6′ x 16′	1.000	Ea.	49.548	10,600	2,100	12,700
6′ x 24′	1.000	Ea.	74.323	15,800	3,175	18,975
8′ x 10′	1.000	Ea.	41.290	8,800	1,750	10,550
8′ x 16′	1.000	Ea.	66.065	14,100	2,825	16,925
8′ x 24′	1.000	Ea.	99.097	21,100	4,225	25,325
Freestanding, 8′ x 8′	1.000	Ea.	18.618	6,050	785	6,835
8′ x 16′	1.000	Ea.	37.236	12,100	1,575	13,675
8′ x 24′	1.000	Ea.	55.855	18,100	2,350	20,450
10′ x 10′	1.000	Ea.	29.091	9,450	1,225	10,675
10′ x 16′	1.000	Ea.	46.546	15,100	1,975	17,075
10′ x 24′	1.000	Ea.	69.818	22,700	2,950	25,650
14′ x 10′	1.000	Ea.	40.727	13,200	1,725	14,925
14′ x 16′	1.000	Ea.	65.164	21,200	2,750	23,950
14′ x 24′	1.000	Ea.	97.746	31,800	4,125	35,925

System Description	QUAN.	UNIT	LABOR HOURS	COST EACH		
				MAT.	INST.	TOTAL
Swimming pools, vinyl lined, metal sides, sand bottom, 12′ x 28′	1.000	Ea.	50.177	3,450	2,125	5,575
12′ x 32′	1.000	Ea.	55.366	3,800	2,350	6,150
12′ x 36′	1.000	Ea.	60.061	4,125	2,550	6,675
16′ x 32′	1.000	Ea.	66.798	4,600	2,825	7,425
16′ x 36′	1.000	Ea.	71.190	4,900	3,000	7,900
16′ x 40′	1.000	Ea.	74.703	5,150	3,150	8,300
20′ x 36′	1.000	Ea.	77.860	5,350	3,300	8,650
20′ x 40′	1.000	Ea.	82.135	5,650	3,500	9,150
20′ x 44′	1.000	Ea.	90.348	6,225	3,825	10,050
24′ x 40′	1.000	Ea.	98.562	6,775	4,175	10,950
24′ x 44′	1.000	Ea.	108.418	7,475	4,600	12,075
24′ x 48′	1.000	Ea.	118.274	8,150	5,000	13,150
Vinyl lined, concrete sides, 12′ x 28′	1.000	Ea.	79.447	5,475	3,375	8,850
12′ x 32′	1.000	Ea.	88.818	6,125	3,775	9,900
12′ x 36′	1.000	Ea.	97.656	6,725	4,150	10,875
16′ x 32′	1.000	Ea.	111.393	7,675	4,725	12,400
16′ x 36′	1.000	Ea.	121.354	8,350	5,150	13,500
16′ x 40′	1.000	Ea.	130.445	8,975	5,525	14,500
28′ x 36′	1.000	Ea.	140.585	9,675	5,950	15,625
20′ x 40′	1.000	Ea.	149.336	10,300	6,325	16,625
20′ x 44′	1.000	Ea.	164.270	11,300	6,950	18,250
24′ x 40′	1.000	Ea.	179.203	12,300	7,600	19,900
24′ x 44′	1.000	Ea.	197.124	13,600	8,350	21,950
24′ x 48′	1.000	Ea.	215.044	14,800	9,100	23,900
Gunite, bottom and sides, 12′ x 28′	1.000	Ea.	129.767	7,225	5,475	12,700
12′ x 32′	1.000	Ea.	142.164	7,925	6,000	13,925
12′ x 36′	1.000	Ea.	153.028	8,525	6,475	15,000
16′ x 32′	1.000	Ea.	167.743	9,350	7,075	16,425
16′ x 36′	1.000	Ea.	176.421	9,825	7,450	17,275
16′ x 40′	1.000	Ea.	182.368	10,200	7,700	17,900
20′ x 36′	1.000	Ea.	187.949	10,500	7,950	18,450
20′ x 40′	1.000	Ea.	179.200	13,700	7,550	21,250
20′ x 44′	1.000	Ea.	197.120	15,000	8,325	23,325
24′ x 40′	1.000	Ea.	215.040	16,400	9,075	25,475
24′ x 44′	1.000	Ea.	273.244	15,200	11,500	26,700
24′ x 48′	1.000	Ea.	298.077	16,600	12,600	29,200

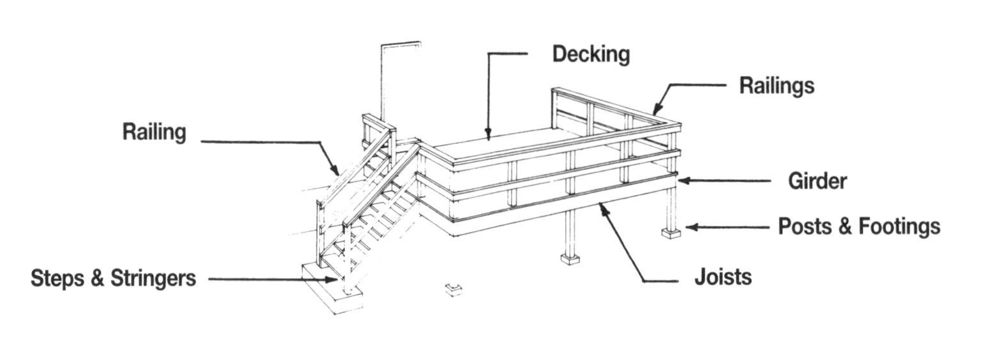

System Description	QUAN.	UNIT	LABOR HOURS	COST PER S.F.		
				MAT.	INST.	TOTAL
8′ X 12′ DECK, PRESSURE TREATED LUMBER, JOISTS 16″ O.C.						
Decking, 2″ x 6″ lumber	2.080	L.F.	.027	1.35	1.12	2.47
Lumber preservative	2.080	L.F.		.29		.29
Joists, 2″ x 8″, 16″ O.C.	1.000	L.F.	.015	.94	.61	1.55
Lumber preservative	1.000	L.F.		.19		.19
Girder, 2″ x 10″	.125	L.F.	.002	.18	.09	.27
Lumber preservative	.125	L.F.		.03		.03
Hand excavation for footings	.250	L.F.	.006		.18	.18
Concrete footings	.250	L.F.	.006	.27	.22	.49
4″ x 4″ Posts	.250	L.F.	.010	.35	.43	.78
Lumber preservative	.250	L.F.		.05		.05
Framing, pressure treated wood stairs, 3′ wide, 8 closed risers	1.000	Set	.080	1.32	3.40	4.72
Railings, 2″ x 4″	1.000	L.F.	.026	.41	1.09	1.50
Lumber preservative	1.000	L.F.		.09		.09
TOTAL		S.F.	.172	5.47	7.14	12.61
12′ X 16′ DECK, PRESSURE TREATED LUMBER, JOISTS 24″ O.C.						
Decking, 2″ x 6″	2.080	L.F.	.027	1.35	1.12	2.47
Lumber preservative	2.080	L.F.		.29		.29
Joists, 2″ x 10″, 24″ O.C.	.800	L.F.	.014	1.14	.60	1.74
Lumber preservative	.800	L.F.		.19		.19
Girder, 2″ x 10″	.083	L.F.	.001	.12	.06	.18
Lumber preservative	.083	L.F.		.02		.02
Hand excavation for footings	.122	L.F.	.006		.18	.18
Concrete footings	.122	L.F.	.006	.27	.22	.49
4″ x 4″ Posts	.122	L.F.	.005	.17	.21	.38
Lumber preservative	.122	L.F.		.02		.02
Framing, pressure treated wood stairs, 3′ wide, 8 closed risers	1.000	Set	.040	.66	1.70	2.36
Railings, 2″ x 4″	.670	L.F.	.017	.27	.73	1
Lumber preservative	.670	L.F.		.06		.06
TOTAL		S.F.	.116	4.56	4.82	9.38
12′ X 24′ DECK, REDWOOD OR CEDAR, JOISTS 16″ O.C.						
Decking, 2″ x 6″ redwood	2.080	L.F.	.027	7.55	1.12	8.67
Joists, 2″ x 10″, 16″ O.C.	1.000	L.F.	.018	7.85	.75	8.60
Girder, 2″ x 10″	.083	L.F.	.001	.65	.06	.71
Hand excavation for footings	.111	L.F.	.006		.18	.18
Concrete footings	.111	L.F.	.006	.27	.22	.49
Lumber preservative	.111	L.F.		.02		.02
Post, 4″ x 4″, including concrete footing	.111	L.F.	.009	3.50	.38	3.88
Framing, redwood or cedar stairs, 3′ wide, 8 closed risers	1.000	Set	.028	2.22	1.19	3.41
Railings, 2″ x 4″	.540	L.F.	.005	1.31	.19	1.50
TOTAL		S.F.	.100	23.37	4.09	27.46

The costs in this system are on a square foot basis.

Important: See the Reference Section for critical supporting data - Reference Nos., Crews & Location Factors

Wood Deck Price Sheet	QUAN.	UNIT	LABOR HOURS	COST PER S.F.		
				MAT.	INST.	TOTAL
Decking, treated lumber, 1″ x 4″	3.430	L.F.	.031	2.60	1.33	3.93
1″ x 6″	2.180	L.F.	.033	2.73	1.40	4.13
2″ x 4″	3.200	L.F.	.041	1.62	1.73	3.35
2″ x 6″	2.080	L.F.	.027	1.64	1.12	2.76
Redwood or cedar,, 1″ x 4″	3.430	L.F.	.035	3.04	1.46	4.50
1″ x 6″	2.180	L.F.	.036	3.17	1.53	4.70
2″ x 4″	3.200	L.F.	.028	8	1.19	9.19
2″ x 6″	2.080	L.F.	.027	7.55	1.12	8.67
Joists for deck, treated lumber, 2″ x 8″, 16″ O.C.	1.000	L.F.	.015	1.13	.61	1.74
24″ O.C.	.800	L.F.	.012	.90	.49	1.39
2″ x 10″, 16″ O.C.	1.000	L.F.	.018	1.65	.75	2.40
24″ O.C.	.800	L.F.	.014	1.33	.60	1.93
Redwood or cedar, 2″ x 8″, 16″ O.C.	1.000	L.F.	.015	4.83	.61	5.44
24″ O.C.	.800	L.F.	.012	3.86	.49	4.35
2″ x 10″, 16″ O.C.	1.000	L.F.	.018	7.85	.75	8.60
24″ O.C.	.800	L.F.	.014	6.30	.60	6.90
Girder for joists, treated lumber, 2″ x 10″, 8′ x 12′ deck	.125	L.F.	.002	.21	.09	.30
12′ x 16′ deck	.083	L.F.	.001	.14	.06	.20
12′ x 24′ deck	.083	L.F.	.001	.14	.06	.20
Redwood or cedar, 2″ x 10″, 8′ x 12′ deck	.125	L.F.	.002	.98	.09	1.07
12′ x 16′ deck	.083	L.F.	.001	.65	.06	.71
12′ x 24′ deck	.083	L.F.	.001	.65	.06	.71
Posts, 4″ x 4″, including concrete footing, 8′ x 12′ deck	.250	S.F.	.022	.67	.83	1.50
12′ x 16′ deck	.122	L.F.	.017	.46	.61	1.07
12′ x 24′ deck	.111	L.F.	.017	.44	.59	1.03
Stairs 2″ x 10″ stringers, treated lumber, 8′ x 12′ deck	1.000	Set	.020	1.32	3.40	4.72
12′ x 16′ deck	1.000	Set	.012	.66	1.70	2.36
12′ x 24′ deck	1.000	Set	.008	.46	1.19	1.65
Redwood or cedar, 8′ x 12′ deck	1.000	Set	.040	6.35	3.40	9.75
12′ x 16′ deck	1.000	Set	.020	3.18	1.70	4.88
12′ x 24′ deck	1.000	Set	.012	2.22	1.19	3.41
Railings 2″ x 4″, treated lumber, 8′ x 12′ deck	1.000	L.F.	.026	.50	1.09	1.59
12′ x 16′ deck	.670	L.F.	.017	.33	.73	1.06
12′ x 24′ deck	.540	L.F.	.014	.27	.59	.86
Redwood or cedar, 8′ x 12′ deck	1.000	L.F.	.009	2.43	.36	2.79
12′ x 16′ deck	.670	L.F.	.006	1.60	.24	1.84
12′ x 24′ deck	.540	L.F.	.005	1.31	.19	1.50

Division 8
Mechanical

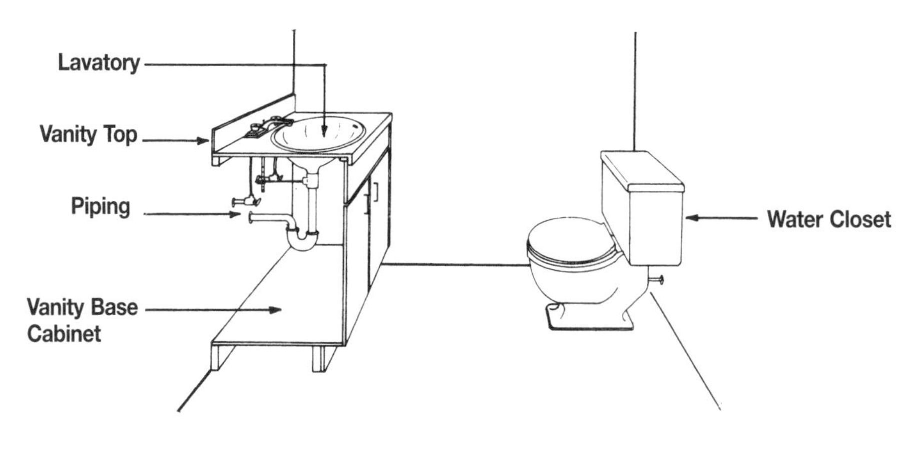

System Description	QUAN.	UNIT	LABOR HOURS	COST EACH		
				MAT.	INST.	TOTAL
LAVATORY INSTALLED WITH VANITY, PLUMBING IN 2 WALLS						
Water closet, floor mounted, 2 piece, close coupled, white	1.000	Ea.	3.019	186	126	312
Rough-in, vent, 2" diameter DWV piping	1.000	Ea.	.955	23.80	39.80	63.60
Waste, 4" diameter DWV piping	1.000	Ea.	.828	30.15	34.50	64.65
Supply, 1/2" diameter type "L" copper supply piping	1.000	Ea.	.593	11.70	27.48	39.18
Lavatory, 20" x 18", P.E. cast iron white	1.000	Ea.	2.500	234	104	338
Rough-in, vent, 1-1/2" diameter DWV piping	1.000	Ea.	.901	23	37.60	60.60
Waste, 2" diameter DWV piping	1.000	Ea.	.955	23.80	39.80	63.60
Supply, 1/2" diameter type "L" copper supply piping	1.000	Ea.	.988	19.50	45.80	65.30
Piping, supply, 1/2" diameter type "L" copper supply piping	10.000	L.F.	.988	19.50	45.80	65.30
Waste, 4" diameter DWV piping	7.000	L.F.	1.931	70.35	80.50	150.85
Vent, 2" diameter DWV piping	12.000	L.F.	2.866	71.40	119.40	190.80
Vanity base cabinet, 2 door, 30" wide	1.000	Ea.	1.000	232	42.50	274.50
Vanity top, plastic & laminated, square edge	2.670	L.F.	.712	82.77	30.04	112.81
TOTAL		Ea.	18.236	1,027.97	773.22	1,801.19
LAVATORY WITH WALL-HUNG LAVATORY, PLUMBING IN 2 WALLS						
Water closet, floor mounted, 2 piece close coupled, white	1.000	Ea.	3.019	186	126	312
Rough-in, vent, 2" diameter DWV piping	1.000	Ea.	.955	23.80	39.80	63.60
Waste, 4" diameter DWV piping	1.000	Ea.	.828	30.15	34.50	64.65
Supply, 1/2" diameter type "L" copper supply piping	1.000	Ea.	.593	11.70	27.48	39.18
Lavatory, 20" x 18", P.E. cast iron, wall hung, white	1.000	Ea.	2.000	265	83.50	348.50
Rough-in, vent, 1-1/2" diameter DWV piping	1.000	Ea.	.901	23	37.60	60.60
Waste, 2" diameter DWV piping	1.000	Ea.	.955	23.80	39.80	63.60
Supply, 1/2" diameter type "L" copper supply piping	1.000	Ea.	.988	19.50	45.80	65.30
Piping, supply, 1/2" diameter type "L" copper supply piping	10.000	L.F.	.988	19.50	45.80	65.30
Waste, 4" diameter DWV piping	7.000	L.F.	1.931	70.35	80.50	150.85
Vent, 2" diameter DWV piping	12.000	L.F.	2.866	71.40	119.40	190.80
Carrier, steel for studs, no arms	1.000	Ea.	1.143	51	53	104
TOTAL		Ea.	17.167	795.20	733.18	1,528.38

Description	QUAN.	UNIT	LABOR HOURS	COST EACH		
				MAT.	INST.	TOTAL

Important: See the Reference Section for critical supporting data - Reference Nos., Crews & Location Factors

Two Fixture Lavatory Price Sheet	QUAN.	UNIT	LABOR HOURS	COST EACH		
				MAT.	INST.	TOTAL
Water closet, close coupled standard 2 piece, white	1.000	Ea.	3.019	186	126	312
Color	1.000	Ea.	3.019	223	126	349
One piece elongated bowl, white	1.000	Ea.	3.019	545	126	671
Color	1.000	Ea.	3.019	680	126	806
Low profile, one piece elongated bowl, white	1.000	Ea.	3.019	775	126	901
Color	1.000	Ea.	3.019	1,025	126	1,151
Rough-in for water closet						
1/2" copper supply, 4" cast iron waste, 2" cast iron vent	1.000	Ea.	2.376	65.50	102	167.50
4" PVC waste, 2" PVC vent	1.000	Ea.	2.678	31.50	114	145.50
4" copper waste , 2" copper vent	1.000	Ea.	2.520	121	111	232
3" cast iron waste, 1-1/2" cast iron vent	1.000	Ea.	2.244	58.50	96.50	155
3" PVC waste, 1-1/2" PVC vent	1.000	Ea.	2.388	27.50	106	133.50
3" copper waste, 1-1/2" copper vent	1.000	Ea.	2.524	113	108	221
1/2" PVC supply, 4" PVC waste, 2" PVC vent	1.000	Ea.	2.974	36.50	128	164.50
3" PVC waste, 1-1/2" PVC vent	1.000	Ea.	2.684	32.50	120	152.50
1/2" steel supply, 4" cast iron waste, 2" cast iron vent	1.000	Ea.	2.545	72.50	110	182.50
4" cast iron waste, 2" steel vent	1.000	Ea.	2.590	90	112	202
4" PVC waste, 2" PVC vent	1.000	Ea.	2.847	38	122	160
Lavatory, vanity top mounted, P.E. on cast iron 20" x 18" white	1.000	Ea.	2.500	234	104	338
Color	1.000	Ea.	2.500	264	104	368
Steel, enameled 10" x 17" white	1.000	Ea.	2.759	152	115	267
Color	1.000	Ea.	2.500	152	104	256
Vitreous china 20" x 16", white	1.000	Ea.	2.963	244	124	368
Color	1.000	Ea.	2.963	244	124	368
Wall hung, P.E. on cast iron, 20" x 18", white	1.000	Ea.	2.000	265	83.50	348.50
Color	1.000	Ea.	2.000	305	83.50	388.50
Vitreous china 19" x 17", white	1.000	Ea.	2.286	187	95.50	282.50
Color	1.000	Ea.	2.286	187	95.50	282.50
Rough-in supply waste and vent for lavatory						
1/2" copper supply, 2" cast iron waste, 1-1/2" cast iron vent	1.000	Ea.	2.844	66.50	123	189.50
2" PVC waste, 1-1/2" PVC vent	1.000	Ea.	2.962	32.50	132	164.50
2" copper waste, 1-1/2" copper vent	1.000	Ea.	2.308	76.50	107	183.50
1-1/2" PVC waste, 1-1/4" PVC vent	1.000	Ea.	2.639	31.50	122	153.50
1-1/2" copper waste, 1-1/4" copper vent	1.000	Ea.	2.114	63.50	98	161.50
1/2" PVC supply, 2" PVC waste, 1-1/2" PVC vent	1.000	Ea.	3.456	41	155	196
1-1/2" PVC waste, 1-1/4" PVC vent	1.000	Ea.	3.133	40	145	185
1/2" steel supply, 2" cast iron waste, 1-1/2" cast iron vent	1.000	Ea.	3.126	77.50	136	213.50
2" cast iron waste, 2" steel vent	1.000	Ea.	3.225	96	141	237
2" PVC waste, 1-1/2" PVC vent	1.000	Ea.	3.244	43.50	145	188.50
1-1/2" PVC waste, 1-1/4" PVC vent	1.000	Ea.	2.921	43	136	179
Piping, supply, 1/2" copper, type "L"	10.000	L.F.	.988	19.50	46	65.50
1/2" steel	10.000	L.F.	1.270	30.50	59	89.50
1/2" PVC	10.000	L.F.	1.482	28	68.50	96.50
Waste, 4" cast iron	7.000	L.F.	1.931	70.50	80.50	151
4" copper	7.000	L.F.	2.800	179	117	296
4" PVC	7.000	L.F.	2.333	29.50	97.50	127
Vent, 2" cast iron	12.000	L.F.	2.866	71.50	119	190.50
2" copper	12.000	L.F.	2.182	98	101	199
2" PVC	12.000	L.F.	3.254	21	136	157
2" steel	12.000	Ea.	3.000	125	125	250
Vanity base cabinet, 2 door, 24" x 30"	1.000	Ea.	1.000	232	42.50	274.50
24" x 36"	1.000	Ea.	1.200	310	50.50	360.50
Vanity top, laminated plastic, square edge 25" x 32"	2.670	L.F.	.712	83	30	113
25" x 38"	3.170	L.F.	.845	98.50	35.50	134
Post formed, laminated plastic, 25" x 32"	2.670	L.F.	.712	26.50	30	56.50
25" x 38"	3.170	L.F.	.845	31.50	35.50	67
Cultured marble, 25" x 32" with bowl	1.000	Ea.	2.500	192	104	296
25" x 38" with bowl	1.000	Ea.	2.500	229	104	333
Carrier for lavatory, steel for studs	1.000	Ea.	1.143	51	53	104
Wood 2" x 8" blocking	1.330	L.F.	.053	1.25	2.25	3.50

12 Three Fixture Bathroom Systems

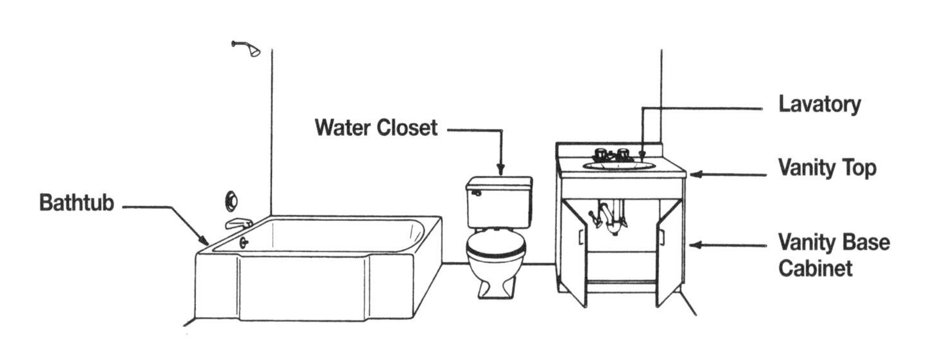

System Description	QUAN.	UNIT	LABOR HOURS	COST EACH MAT.	COST EACH INST.	COST EACH TOTAL
BATHROOM INSTALLED WITH VANITY						
Water closet, floor mounted, 2 piece, close coupled, white	1.000	Ea.	3.019	186	126	312
Rough-in, waste, 4" diameter DWV piping	1.000	Ea.	.828	30.15	34.50	64.65
Vent, 2" diameter DWV piping	1.000	Ea.	.955	23.80	39.80	63.60
Supply, 1/2" diameter type "L" copper supply piping	1.000	Ea.	.593	11.70	27.48	39.18
Lavatory, 20" x 18", P.E. cast iron with accessories, white	1.000	Ea.	2.500	234	104	338
Rough-in, supply, 1/2" diameter type "L" copper supply piping	1.000	Ea.	.988	19.50	45.80	65.30
Waste, 1-1/2" diameter DWV piping	1.000	Ea.	1.803	46	75.20	121.20
Bathtub, P.E. cast iron, 5' long with accessories, white	1.000	Ea.	3.636	805	152	957
Rough-in, waste, 4" diameter DWV piping	1.000	Ea.	.828	30.15	34.50	64.65
Vent, 1-1/2" diameter DWV piping	1.000	Ea.	.593	24.40	27.40	51.80
Supply, 1/2" diameter type "L" copper supply piping	1.000	Ea.	.988	19.50	45.80	65.30
Piping, supply, 1/2" diameter type "L" copper supply piping	20.000	L.F.	1.975	39	91.60	130.60
Waste, 4" diameter DWV piping	9.000	L.F.	2.483	90.45	103.50	193.95
Vent, 2" diameter DWV piping	6.000	L.F.	1.500	62.40	62.70	125.10
Vanity base cabinet, 2 door, 30" wide	1.000	Ea.	1.000	232	42.50	274.50
Vanity top, plastic laminated square edge	2.670	L.F.	.712	65.42	30.04	95.46
TOTAL		Ea.	24.401	1,919.47	1,042.82	2,962.29
BATHROOM WITH WALL HUNG LAVATORY						
Water closet, floor mounted, 2 piece, close coupled, white	1.000	Ea.	3.019	186	126	312
Rough-in, vent, 2" diameter DWV piping	1.000	Ea.	.955	23.80	39.80	63.60
Waste, 4" diameter DWV piping	1.000	Ea.	.828	30.15	34.50	64.65
Supply, 1/2" diameter type "L" copper supply piping	1.000	Ea.	.593	11.70	27.48	39.18
Lavatory, 20" x 18" P.E. cast iron, wall hung, white	1.000	Ea.	2.000	265	83.50	348.50
Rough-in, waste, 1-1/2" diameter DWV piping	1.000	Ea.	1.803	46	75.20	121.20
Supply, 1/2" diameter type "L" copper supply piping	1.000	Ea.	.988	19.50	45.80	65.30
Bathtub, P.E. cast iron, 5' long with accessories, white	1.000	Ea.	3.636	805	152	957
Rough-in, waste, 4" diameter DWV piping	1.000	Ea.	.828	30.15	34.50	64.65
Supply, 1/2" diameter type "L" copper supply piping	1.000	Ea.	.988	19.50	45.80	65.30
Vent, 1-1/2" diameter DWV piping	1.000	Ea.	1.482	61	68.50	129.50
Piping, supply, 1/2" diameter type "L" copper supply piping	20.000	L.F.	1.975	39	91.60	130.60
Waste, 4" diameter DWV piping	9.000	L.F.	2.483	90.45	103.50	193.95
Vent, 2" diameter DWV piping	6.000	L.F.	1.500	62.40	62.70	125.10
Carrier, steel, for studs, no arms	1.000	Ea.	1.143	51	53	104
TOTAL		Ea.	24.221	1,740.65	1,043.88	2,784.53

The costs in this system are a cost each basis. All necessary piping is included.

Three Fixture Bathroom Price Sheet	QUAN.	UNIT	LABOR HOURS	COST EACH		
				MAT.	INST.	TOTAL
Water closet, close coupled standard 2 piece, white	1.000	Ea.	3.019	186	126	312
Color	1.000	Ea.	3.019	223	126	349
One piece, elongated bowl, white	1.000	Ea.	3.019	545	126	671
Color	1.000	Ea.	3.019	680	126	806
Low profile, one piece elongated bowl, white	1.000	Ea.	3.019	775	126	901
Color	1.000	Ea.	3.019	1,025	126	1,151
Rough-in, for water closet						
1/2" copper supply, 4" cast iron waste, 2" cast iron vent	1.000	Ea.	2.376	65.50	102	167.50
4" PVC/DWV waste, 2" PVC vent	1.000	Ea.	2.678	31.50	114	145.50
4" copper waste, 2" copper vent	1.000	Ea.	2.520	121	111	232
3" cast iron waste, 1-1/2" cast iron vent	1.000	Ea.	2.244	58.50	96.50	155
3" PVC waste, 1-1/2" PVC vent	1.000	Ea.	2.388	27.50	106	133.50
3" copper waste, 1-1/2" copper vent	1.000	Ea.	2.014	79.50	89.50	169
1/2" PVC supply, 4" PVC waste, 2" PVC vent	1.000	Ea.	2.974	36.50	128	164.50
3" PVC waste, 1-1/2" PVC supply	1.000	Ea.	2.684	32.50	120	152.50
1/2" steel supply, 4" cast iron waste, 2" cast iron vent	1.000	Ea.	2.545	72.50	110	182.50
4" cast iron waste, 2" steel vent	1.000	Ea.	2.590	90	112	202
4" PVC waste, 2" PVC vent	1.000	Ea.	2.847	38	122	160
Lavatory, wall hung, P.E. cast iron 20" x 18", white	1.000	Ea.	2.000	265	83.50	348.50
Color	1.000	Ea.	2.000	305	83.50	388.50
Vitreous china 19" x 17", white	1.000	Ea.	2.286	187	95.50	282.50
Color	1.000	Ea.	2.286	187	95.50	282.50
Lavatory, for vanity top, P.E. cast iron 20" x 18"", white	1.000	Ea.	2.500	234	104	338
Color	1.000	Ea.	2.500	264	104	368
Steel, enameled 20" x 17", white	1.000	Ea.	2.759	152	115	267
Color	1.000	Ea.	2.500	152	104	256
Vitreous china 20" x 16", white	1.000	Ea.	2.963	244	124	368
Color	1.000	Ea.	2.963	244	124	368
Rough-in, for lavatory						
1/2" copper supply, 1-1/2" C.I. waste, 1-1/2" C.I. vent	1.000	Ea.	2.791	65.50	121	186.50
1-1/2" PVC waste, 1-1/4" PVC vent	1.000	Ea.	2.639	31.50	122	153.50
1/2" steel supply, 1-1/4" cast iron waste, 1-1/4" steel vent	1.000	Ea.	2.890	80	127	207
1-1/4" PVC@ waste, 1-1/4" PVC vent	1.000	Ea.	2.794	42.50	130	172.50
1/2" PVC supply, 1-1/2" PVC waste, 1-1/2" PVC vent	1.000	Ea.	3.260	40	151	191
Bathtub, P.E. cast iron, 5' long corner with fittings, white	1.000	Ea.	3.636	805	152	957
Color	1.000	Ea.	3.636	995	152	1,147
Rough-in, for bathtub						
1/2" copper supply, 4" cast iron waste, 1-1/2" copper vent	1.000	Ea.	2.409	74	108	182
4" PVC waste, 1-1/2" PVC vent	1.000	Ea.	2.877	38.50	129	167.50
1/2" steel supply, 4" cast iron waste, 1-1/2" steel vent	1.000	Ea.	2.898	92	127	219
4" PVC waste, 1-1/2" PVC vent	1.000	Ea.	3.159	49.50	142	191.50
1/2" PVC supply, 4" PVC waste, 1-1/2" PVC vent	1.000	Ea.	3.371	46.50	151	197.50
Piping, supply 1/2" copper	20.000	L.F.	1.975	39	91.50	130.50
1/2" steel	20.000	L.F.	2.540	61	118	179
1/2" PVC	20.000	L.F.	2.963	55.50	137	192.50
Piping, waste, 4" cast iron no hub	9.000	L.F.	2.483	90.50	104	194.50
4" PVC/DWV	9.000	L.F.	3.000	38	125	163
4" copper/DWV	9.000	L.F.	3.600	230	150	380
Piping, vent 2" cast iron no hub	6.000	L.F.	1.433	35.50	59.50	95
2" copper/DWV	6.000	L.F.	1.091	49	50.50	99.50
2" PVC/DWV	6.000	L.F.	1.627	10.40	68	78.40
2" steel, galvanized	6.000	L.F.	1.500	62.50	62.50	125
Vanity base cabinet, 2 door, 24" x 30"	1.000	Ea.	1.000	232	42.50	274.50
24" x 36"	1.000	Ea.	1.200	310	50.50	360.50
Vanity top, laminated plastic square edge 25" x 32"	2.670	L.F.	.712	65.50	30	95.50
25" x 38"	3.160	L.F.	.843	77.50	35.50	113
Cultured marble, 25" x 32", with bowl	1.000	Ea.	2.500	192	104	296
25" x 38", with bowl	1.000	Ea.	2.500	229	104	333
Carrier, for lavatory, steel for studs, no arms	1.000	Ea.	1.143	51	53	104
Wood, 2" x 8" blocking	1.300	L.F.	.052	1.22	2.20	3.42

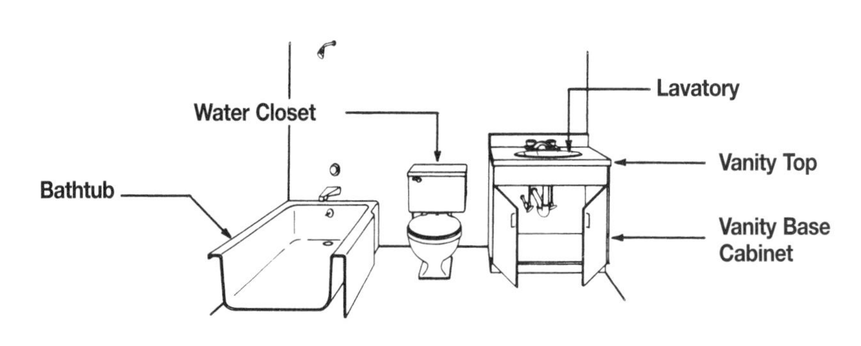

System Description	QUAN.	UNIT	LABOR HOURS	COST EACH		
				MAT.	INST.	TOTAL
BATHROOM WITH LAVATORY INSTALLED IN VANITY						
Water closet, floor mounted, 2 piece, close coupled, white	1.000	Ea.	3.019	186	126	312
Rough-in, waste, 4" diameter DWV piping	1.000	Ea.	.828	30.15	34.50	64.65
Vent, 2" diameter DWV piping	1.000	Ea.	.955	23.80	39.80	63.60
Supply, 1/2" diameter type "L" copper supply piping	1.000	Ea.	.593	11.70	27.48	39.18
Lavatory, 20" x 18", P.E. cast iron with accessories, white	1.000	Ea.	2.500	234	104	338
Rough-in, waste, 1-1/2" diameter DWV piping	1.000	Ea.	1.803	46	75.20	121.20
Supply, 1/2" diameter type "L" copper supply piping	1.000	Ea.	.988	19.50	45.80	65.30
Bathtub, P.E. cast iron 5' long with accessories, white	1.000	Ea.	3.636	805	152	957
Rough-in, waste, 4" diameter DWV piping	1.000	Ea.	.828	30.15	34.50	64.65
Vent, 1-1/2" diameter DWV piping	1.000	Ea.	.593	24.40	27.40	51.80
Supply, 1/2" diameter type "L" copper supply piping	1.000	Ea.	.988	19.50	45.80	65.30
Piping, supply, 1/2" diameter type "L" copper supply piping	10.000	L.F.	.988	19.50	45.80	65.30
Waste, 4" diameter DWV piping	6.000	L.F.	1.655	60.30	69	129.30
Vent, 2" diameter DWV piping	6.000	L.F.	1.500	62.40	62.70	125.10
Vanity base cabinet, 2 door, 30" wide	1.000	Ea.	1.000	232	42.50	274.50
Vanity top, plastic laminated square edge	2.670	L.F.	.712	65.42	30.04	95.46
TOTAL		Ea.	22.586	1,869.82	962.52	2,832.34
BATHROOM WITH WALL HUNG LAVATORY						
Water closet, floor mounted, 2 piece, close coupled, white	1.000	Ea.	3.019	186	126	312
Rough-in, vent, 2" diameter DWV piping	1.000	Ea.	.955	23.80	39.80	63.60
Waste, 4" diameter DWV piping	1.000	Ea.	.828	30.15	34.50	64.65
Supply, 1/2" diameter type "L" copper supply piping	1.000	Ea.	.593	11.70	27.48	39.18
Lavatory, 20" x 18" P.E. cast iron, wall hung, white	1.000	Ea.	2.000	265	83.50	348.50
Rough-in, waste, 1-1/2" diameter DWV piping	1.000	Ea.	1.803	46	75.20	121.20
Supply, 1/2" diameter type "L" copper supply piping	1.000	Ea.	.988	19.50	45.80	65.30
Bathtub, P.E. cast iron, 5' long with accessories, white	1.000	Ea.	3.636	805	152	957
Rough-in, waste, 4" diameter DWV piping	1.000	Ea.	.828	30.15	34.50	64.65
Supply, 1/2" diameter type "L" copper supply piping	1.000	Ea.	.988	19.50	45.80	65.30
Vent, 1-1/2" diameter DWV piping	1.000	Ea.	.593	24.40	27.40	51.80
Piping, supply, 1/2" diameter type "L" copper supply piping	10.000	L.F.	.988	19.50	45.80	65.30
Waste, 4" diameter DWV piping	6.000	L.F.	1.655	60.30	69	129.30
Vent, 2" diameter DWV piping	6.000	L.F.	1.500	62.40	62.70	125.10
Carrier, steel, for studs, no arms	1.000	Ea.	1.143	51	53	104
TOTAL		Ea.	21.517	1,654.40	922.48	2,576.88

The costs in this system are on a cost each basis. All necessary piping is included.

Three Fixture Bathroom Price Sheet	QUAN.	UNIT	LABOR HOURS	COST EACH		
				MAT.	INST.	TOTAL
Water closet, close coupled standard 2 piece, white	1.000	Ea.	3.019	186	126	312
Color	1.000	Ea.	3.019	223	126	349
One piece elongated bowl, white	1.000	Ea.	3.019	545	126	671
Color	1.000	Ea.	3.019	680	126	806
Low profile, one piece elongated bowl, white	1.000	Ea.	3.019	775	126	901
Color	1.000	Ea.	3.019	1,025	126	1,151
Rough-in for water closet						
1/2″ copper supply, 4″ cast iron waste, 2″ cast iron vent	1.000	Ea.	2.376	65.50	102	167.50
4″ PVC/DWV waste, 2″ PVC vent	1.000	Ea.	2.678	31.50	114	145.50
4″ carrier waste, 2″ copper vent	1.000	Ea.	2.520	121	111	232
3″ cast iron waste, 1-1/2″ cast iron vent	1.000	Ea.	2.244	58.50	96.50	155
3″ PVC waste, 1-1/2″ PVC vent	1.000	Ea.	2.388	27.50	106	133.50
3″ copper waste, 1-1/2″ copper vent	1.000	Ea.	2.014	79.50	89.50	169
1/2″ PVC supply, 4″ PVC waste, 2″ PVC vent	1.000	Ea.	2.974	36.50	128	164.50
3″ PVC waste, 1-1/2″ PVC supply	1.000	Ea.	2.684	32.50	120	152.50
1/2″ steel supply, 4″ cast iron waste, 2″ cast iron vent	1.000	Ea.	2.545	72.50	110	182.50
4″ cast iron waste, 2″ steel vent	1.000	Ea.	2.590	90	112	202
4″ PVC waste, 2″ PVC vent	1.000	Ea.	2.847	38	122	160
Lavatory, wall hung, PE cast iron 20″ x 18″, white	1.000	Ea.	2.000	265	83.50	348.50
Color	1.000	Ea.	2.000	305	83.50	388.50
Vitreous china 19″ x 17″, white	1.000	Ea.	2.286	187	95.50	282.50
Color	1.000	Ea.	2.286	187	95.50	282.50
Lavatory, for vanity top, PE cast iron 20″ x 18″, white	1.000	Ea.	2.500	234	104	338
Color	1.000	Ea.	2.500	264	104	368
Steel enameled 20″ x 17″, white	1.000	Ea.	2.759	152	115	267
Color	1.000	Ea.	2.500	152	104	256
Vitreous china 20″ x 16″, white	1.000	Ea.	2.963	244	124	368
Color	1.000	Ea.	2.963	244	124	368
Rough-in for lavatory						
1/2″ copper supply, 1-1/2″ cast iron waste, 1-1/2″ cast iron vent	1.000	Ea.	2.791	65.50	121	186.50
1-1/2″ PVC waste, 1-1/4″ PVC vent	1.000	Ea.	2.639	31.50	122	153.50
1/2″ steel supply, 1-1/4″ cast iron waste, 1-1/4″ steel vent	1.000	Ea.	2.890	80	127	207
1-1/4″ PVC waste, 1-1/4″ PVC vent	1.000	Ea.	2.794	42.50	130	172.50
1/2″ PVC supply, 1-1/2″ PVC waste, 1-1/2″ PVC vent	1.000	Ea.	3.260	40	151	191
Bathtub, PE cast iron, 5′ long corner with fittings, white	1.000	Ea.	3.636	805	152	957
Color	1.000	Ea.	3.636	995	152	1,147
Rough-in for bathtub						
1/2″ copper supply, 4″ cast iron waste, 1-1/2″ copper vent	1.000	Ea.	2.409	74	108	182
4″ PVC waste, 1/2″ PVC vent	1.000	Ea.	2.877	38.50	129	167.50
1/2″ steel supply, 4″ cast iron waste, 1-1/2″ steel vent	1.000	Ea.	2.898	92	127	219
4″ PVC waste, 1-1/2″ PVC vent	1.000	Ea.	3.159	49.50	142	191.50
1/2″ PVC supply, 4″ PVC waste, 1-1/2″ PVC vent	1.000	Ea.	3.371	46.50	151	197.50
Piping supply, 1/2″ copper	10.000	L.F.	.988	19.50	46	65.50
1/2″ steel	10.000	L.F.	1.270	30.50	59	89.50
1/2″ PVC	10.000	L.F.	1.482	28	68.50	96.50
Piping waste, 4″ cast iron no hub	6.000	L.F.	1.655	60.50	69	129.50
4″ PVC/DWV	6.000	L.F.	2.000	25.50	83.50	109
4″ copper/DWV	6.000	L.F.	2.400	153	100	253
Piping vent 2″ cast iron no hub	6.000	L.F.	1.433	35.50	59.50	95
2″ copper/DWV	6.000	L.F.	1.091	49	50.50	99.50
2″ PVC/DWV	6.000	L.F.	1.627	10.40	68	78.40
2″ steel, galvanized	6.000	L.F.	1.500	62.50	62.50	125
Vanity base cabinet, 2 door, 24″ x 30″	1.000	Ea.	1.000	232	42.50	274.50
24″ x 36″	1.000	Ea.	1.200	310	50.50	360.50
Vanity top, laminated plastic square edge 25″ x 32″	2.670	L.F.	.712	65.50	30	95.50
25″ x 38″	3.160	L.F.	.843	77.50	35.50	113
Cultured marble, 25″ x 32″, with bowl	1.000	Ea.	2.500	192	104	296
25″ x 38″, with bowl	1.000	Ea.	2.500	229	104	333
Carrier, for lavatory, steel for studs, no arms	1.000	Ea.	1.143	51	53	104
Wood, 2″ x 8″ blocking	1.300	L.F.	.052	1.22	2.20	3.42

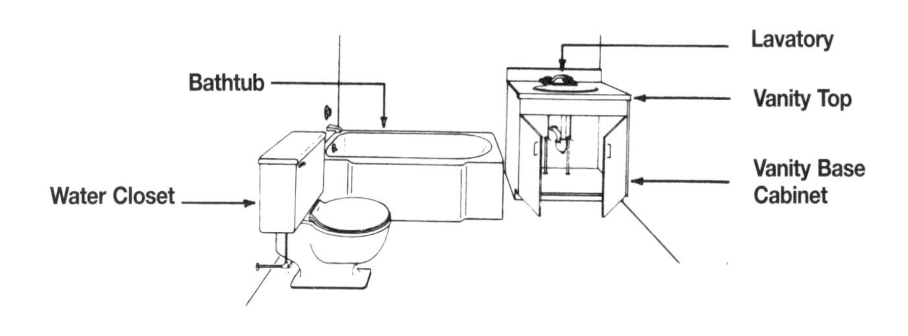

System Description	QUAN.	UNIT	LABOR HOURS	COST EACH		
				MAT.	INST.	TOTAL
BATHROOM WITH LAVATORY INSTALLED IN VANITY						
Water closet, floor mounted, 2 piece, close coupled, white	1.000	Ea.	3.019	186	126	312
Rough-in, vent, 2″ diameter DWV piping	1.000	Ea.	.955	23.80	39.80	63.60
Waste, 4″ diameter DWV piping	1.000	Ea.	.828	30.15	34.50	64.65
Supply, 1/2″ diameter type “L” copper supply piping	1.000	Ea.	.593	11.70	27.48	39.18
Lavatory, 20″ x 18″, PE cast iron with accessories, white	1.000	Ea.	2.500	234	104	338
Rough-in, vent, 1-1/2″ diameter DWV piping	1.000	Ea.	1.803	46	75.20	121.20
Supply, 1/2″ diameter type “L” copper supply piping	1.000	Ea.	.988	19.50	45.80	65.30
Bathtub, P.E. cast iron, 5′ long with accessories, white	1.000	Ea.	3.636	805	152	957
Rough-in, waste, 4″ diameter DWV piping	1.000	Ea.	.828	30.15	34.50	64.65
Supply, 1/2″ diameter type “L” copper supply piping	1.000	Ea.	.988	19.50	45.80	65.30
Vent, 1-1/2″ diameter DWV piping	1.000	Ea.	.593	24.40	27.40	51.80
Piping, supply, 1/2″ diameter type “L” copper supply piping	32.000	L.F.	3.161	62.40	146.56	208.96
Waste, 4″ diameter DWV piping	12.000	L.F.	3.310	120.60	138	258.60
Vent, 2″ diameter DWV piping	6.000	L.F.	1.500	62.40	62.70	125.10
Vanity base cabinet, 2 door, 30″ wide	1.000	Ea.	1.000	232	42.50	274.50
Vanity top, plastic laminated square edge	2.670	L.F.	.712	65.42	30.04	95.46
TOTAL		Ea.	26.414	1,973.02	1,132.28	3,105.30
BATHROOM WITH WALL HUNG LAVATORY						
Water closet, floor mounted, 2 piece, close coupled, white	1.000	Ea.	3.019	186	126	312
Rough-in, vent, 2″ diameter DWV piping	1.000	Ea.	.955	23.80	39.80	63.60
Waste, 4″ diameter DWV piping	1.000	Ea.	.828	30.15	34.50	64.65
Supply, 1/2″ diameter type “L” copper supply piping	1.000	Ea.	.593	11.70	27.48	39.18
Lavatory, 20″ x 18″ P.E. cast iron, wall hung, white	1.000	Ea.	2.000	265	83.50	348.50
Rough-in, waste, 1-1/2″ diameter DWV piping	1.000	Ea.	1.803	46	75.20	121.20
Supply, 1/2″ diameter type “L” copper supply piping	1.000	Ea.	.988	19.50	45.80	65.30
Bathtub, P.E. cast iron, 5′ long with accessories, white	1.000	Ea.	3.636	805	152	957
Rough-in, waste, 4″ diameter DWV piping	1.000	Ea.	.828	30.15	34.50	64.65
Supply, 1/2″ diameter type “L” copper supply piping	1.000	Ea.	.988	19.50	45.80	65.30
Vent, 1-1/2″ diameter DWV piping	1.000	Ea.	.593	24.40	27.40	51.80
Piping, supply, 1/2″ diameter type “L” copper supply piping	32.000	L.F.	3.161	62.40	146.56	208.96
Waste, 4″ diameter DWV piping	12.000	L.F.	3.310	120.60	138	258.60
Vent, 2″ diameter DWV piping	6.000	L.F.	1.500	62.40	62.70	125.10
Carrier steel, for studs, no arms	1.000	Ea.	1.143	51	53	104
TOTAL		Ea.	25.345	1,757.60	1,092.24	2,849.84

The costs in this system are on a cost each basis. All necessary piping is included.

Important: See the Reference Section for critical supporting data - Reference Nos., Crews & Location Factors

Three Fixture Bathroom Price Sheet	QUAN.	UNIT	LABOR HOURS	COST EACH		
				MAT.	INST.	TOTAL
Water closet, close coupled, standard 2 piece, white	1.000	Ea.	3.019	186	126	312
Color	1.000	Ea.	3.019	223	126	349
One piece, elongated bowl, white	1.000	Ea.	3.019	545	126	671
Color	1.000	Ea.	3.019	680	126	806
Low profile, one piece, elongated bowl, white	1.000	Ea.	3.019	775	126	901
Color	1.000	Ea.	3.019	1,025	126	1,151
Rough-in, for water closet						
1/2" copper supply, 4" cast iron waste, 2" cast iron vent	1.000	Ea.	2.376	65.50	102	167.50
4" PVC/DWV waste, 2" PVC vent	1.000	Ea.	2.678	31.50	114	145.50
4" copper waste, 2" copper vent	1.000	Ea.	2.520	121	111	232
3" cast iron waste, 1-1/2" cast iron vent	1.000	Ea.	2.244	58.50	96.50	155
3" PVC waste, 1-1/2" PVC vent	1.000	Ea.	2.388	27.50	106	133.50
3" copper waste, 1-1/2" copper vent	1.000	Ea.	2.014	79.50	89.50	169
1/2" PVC supply, 4" PVC waste, 2" PVC vent	1.000	Ea.	2.974	36.50	128	164.50
3" PVC waste, 1-1/2" PVC supply	1.000	Ea.	2.684	32.50	120	152.50
1/2" steel supply, 4" cast iron waste, 2" cast iron vent	1.000	Ea.	2.545	72.50	110	182.50
4" cast iron waste, 2" steel vent	1.000	Ea.	2.590	90	112	202
4" PVC waste, 2" PVC vent	1.000	Ea.	2.847	38	122	160
Lavatory wall hung, P.E. cast iron, 20" x 18", white	1.000	Ea.	2.000	265	83.50	348.50
Color	1.000	Ea.	2.000	305	83.50	388.50
Vitreous china, 19" x 17", white	1.000	Ea.	2.286	187	95.50	282.50
Color	1.000	Ea.	2.286	187	95.50	282.50
Lavatory, for vanity top, P.E., cast iron, 20" x 18", white	1.000	Ea.	2.500	234	104	338
Color	1.000	Ea.	2.500	264	104	368
Steel, enameled, 20" x 17", white	1.000	Ea.	2.759	152	115	267
Color	1.000	Ea.	2.500	152	104	256
Vitreous china, 20" x 16", white	1.000	Ea.	2.963	244	124	368
Color	1.000	Ea.	2.963	244	124	368
Rough-in, for lavatory						
1/2" copper supply, 1-1/2" C.I. waste, 1-1/2" C.I. vent	1.000	Ea.	2.791	65.50	121	186.50
1-1/2" PVC waste, 1-1/4" PVC vent	1.000	Ea.	2.639	31.50	122	153.50
1/2" steel supply, 1-1/4" cast iron waste, 1-1/4" steel vent	1.000	Ea.	2.890	80	127	207
1-1/4" PVC waste, 1-1/4" PVC vent	1.000	Ea.	2.794	42.50	130	172.50
1/2" PVC supply, 1-1/2" PVC waste, 1-1/2" PVC vent	1.000	Ea.	3.260	40	151	191
Bathtub, P.E. cast iron, 5' long corner with fittings, white	1.000	Ea.	3.636	805	152	957
Color	1.000	Ea.	3.636	995	152	1,147
Rough-in, for bathtub						
1/2" copper supply, 4" cast iron waste, 1-1/2" copper vent	1.000	Ea.	2.409	74	108	182
4" PVC waste, 1/2" PVC vent	1.000	Ea.	2.877	38.50	129	167.50
1/2" steel supply, 4" cast iron waste, 1-1/2" steel vent	1.000	Ea.	2.898	92	127	219
4" PVC waste, 1-1/2" PVC vent	1.000	Ea.	3.159	49.50	142	191.50
1/2" PVC supply, 4" PVC waste, 1-1/2" PVC vent	1.000	Ea.	3.371	46.50	151	197.50
Piping, supply, 1/2" copper	32.000	L.F.	3.161	62.50	147	209.50
1/2" steel	32.000	L.F.	4.063	98	189	287
1/2" PVC	32.000	L.F.	4.741	89	219	308
Piping, waste, 4" cast iron no hub	12.000	L.F.	3.310	121	138	259
4" PVC/DWV	12.000	L.F.	4.000	50.50	167	217.50
4" copper/DWV	12.000	L.F.	4.800	305	200	505
Piping, vent, 2" cast iron no hub	6.000	L.F.	1.433	35.50	59.50	95
2" copper/DWV	6.000	L.F.	1.091	49	50.50	99.50
2" PVC/DWV	6.000	L.F.	1.627	10.40	68	78.40
2" steel, galvanized	6.000	L.F.	1.500	62.50	62.50	125
Vanity base cabinet, 2 door, 24" x 30"	1.000	Ea.	1.000	232	42.50	274.50
24" x 36"	1.000	Ea.	1.200	310	50.50	360.50
Vanity top, laminated plastic square edge, 25" x 32"	2.670	L.F.	.712	65.50	30	95.50
25" x 38"	3.160	L.F.	.843	77.50	35.50	113
Cultured marble, 25" x 32", with bowl	1.000	Ea.	2.500	192	104	296
25" x 38", with bowl	1.000	Ea.	2.500	229	104	333
Carrier, for lavatory, steel for studs, no arms	1.000	Ea.	1.143	51	53	104
Wood, 2" x 8" blocking	1.300	L.F.	.052	1.22	2.20	3.42

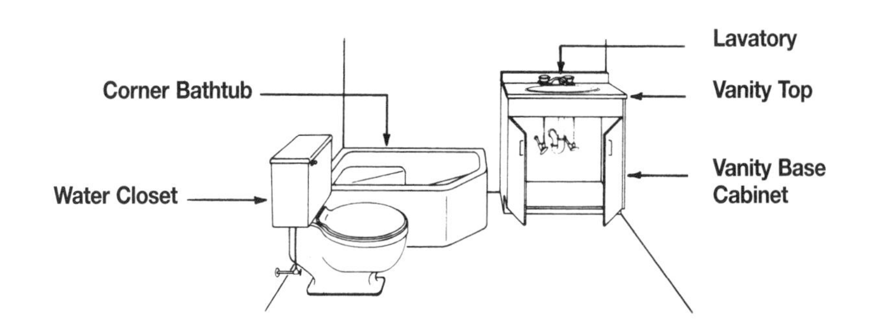

System Description	QUAN.	UNIT	LABOR HOURS	COST EACH		
				MAT.	INST.	TOTAL
BATHROOM WITH LAVATORY INSTALLED IN VANITY						
Water closet, floor mounted, 2 piece, close coupled, white	1.000	Ea.	3.019	186	126	312
Rough-in, vent, 2" diameter DWV piping	1.000	Ea.	.955	23.80	39.80	63.60
Waste, 4" diameter DWV piping	1.000	Ea.	.828	30.15	34.50	64.65
Supply, 1/2" diameter type "L" copper supply piping	1.000	Ea.	.593	11.70	27.48	39.18
Lavatory, 20" x 18", P.E. cast iron with fittings, white	1.000	Ea.	2.500	234	104	338
Rough-in, waste, 1-1/2" diameter DWV piping	1.000	Ea.	1.803	46	75.20	121.20
Supply, 1/2" diameter type "L" copper supply piping	1.000	Ea.	.988	19.50	45.80	65.30
Bathtub, P.E. cast iron, corner with fittings, white	1.000	Ea.	3.636	1,775	152	1,927
Rough-in, waste, 4" diameter DWV piping	1.000	Ea.	.828	30.15	34.50	64.65
Supply, 1/2" diameter type "L" copper supply piping	1.000	Ea.	.988	19.50	45.80	65.30
Vent, 1-1/2" diameter DWV piping	1.000	Ea.	.593	24.40	27.40	51.80
Piping, supply, 1/2" diameter type "L" copper supply piping	32.000	L.F.	3.161	62.40	146.56	208.96
Waste, 4" diameter DWV piping	12.000	L.F.	3.310	120.60	138	258.60
Vent, 2" diameter DWV piping	6.000	L.F.	1.500	62.40	62.70	125.10
Vanity base cabinet, 2 door, 30" wide	1.000	Ea.	1.000	232	42.50	274.50
Vanity top, plastic laminated, square edge	2.670	L.F.	.712	82.77	30.04	112.81
TOTAL		Ea.	26.414	2,960.37	1,132.28	4,092.65
BATHROOM WITH WALL HUNG LAVATORY						
Water closet, floor mounted, 2 piece, close coupled, white	1.000	Ea.	3.019	186	126	312
Rough-in, vent, 2" diameter DWV piping	1.000	Ea.	.955	23.80	39.80	63.60
Waste, 4" diameter DWV piping	1.000	Ea.	.828	30.15	34.50	64.65
Supply, 1/2" diameter type "L" copper supply piping	1.000	Ea.	.593	11.70	27.48	39.18
Lavatory, 20" x 18", P.E. cast iron, with fittings, white	1.000	Ea.	2.000	265	83.50	348.50
Rough-in, waste, 1-1/2" diameter DWV piping	1.000	Ea.	1.803	46	75.20	121.20
Supply, 1/2" diameter type "L" copper supply piping	1.000	Ea.	.988	19.50	45.80	65.30
Bathtub, P.E. cast iron, corner, with fittings, white	1.000	Ea.	3.636	1,775	152	1,927
Rough-in, waste, 4" diameter DWV piping	1.000	Ea.	.828	30.15	34.50	64.65
Supply, 1/2" diameter type "L" copper supply piping	1.000	Ea.	.988	19.50	45.80	65.30
Vent, 1-1/2" diameter DWV piping	1.000	Ea.	.593	24.40	27.40	51.80
Piping, supply, 1/2" diameter type "L" copper supply piping	32.000	L.F.	3.161	62.40	146.56	208.96
Waste, 4" diameter DWV piping	12.000	L.F.	3.310	120.60	138	258.60
Vent, 2" diameter DWV piping	6.000	L.F.	1.500	62.40	62.70	125.10
Carrier, steel, for studs, no arms	1.000	Ea.	1.143	51	53	104
TOTAL		Ea.	25.345	2,727.60	1,092.24	3,819.84

The costs in this system are on a cost each basis. All necessary piping is included.

Important: See the Reference Section for critical supporting data - Reference Nos., Crews & Location Factors

Three Fixture Bathroom Price Sheet	QUAN.	UNIT	LABOR HOURS	COST EACH		
				MAT.	INST.	TOTAL
Water closet, close coupled, standard 2 piece, white	1.000	Ea.	3.019	186	126	312
Color	1.000	Ea.	3.019	223	126	349
One piece elongated bowl, white	1.000	Ea.	3.019	545	126	671
Color	1.000	Ea.	3.019	680	126	806
Low profile, one piece elongated bowl, white	1.000	Ea.	3.019	775	126	901
Color	1.000	Ea.	3.019	1,025	126	1,151
Rough-in, for water closet						
1/2″ copper supply, 4″ cast iron waste, 2″ cast iron vent	1.000	Ea.	2.376	65.50	102	167.50
4″ PVC/DWV waste, 2″ PVC vent	1.000	Ea.	2.678	31.50	114	145.50
4″ copper waste, 2″ copper vent	1.000	Ea.	2.520	121	111	232
3″ cast iron waste, 1-1/2″ cast iron vent	1.000	Ea.	2.244	58.50	96.50	155
3″ PVC waste, 1-1/2″ PVC vent	1.000	Ea.	2.388	27.50	106	133.50
3″ copper waste, 1-1/2″ copper vent	1.000	Ea.	2.014	79.50	89.50	169
1/2″ PVC supply, 4″ PVC waste, 2″ PVC vent	1.000	Ea.	2.974	36.50	128	164.50
3″ PVC waste, 1-1/2″ PVC supply	1.000	Ea.	2.684	32.50	120	152.50
1/2″ steel supply, 4″ cast iron waste, 2″ cast iron vent	1.000	Ea.	2.545	72.50	110	182.50
4″ cast iron waste, 2″ steel vent	1.000	Ea.	2.590	90	112	202
4″ PVC waste, 2″ PVC vent	1.000	Ea.	2.847	38	122	160
Lavatory, wall hung P.E. cast iron 20″ x 18″, white	1.000	Ea.	2.000	265	83.50	348.50
Color	1.000	Ea.	2.000	305	83.50	388.50
Vitreous china 19″ x 17″, white	1.000	Ea.	2.286	187	95.50	282.50
Color	1.000	Ea.	2.286	187	95.50	282.50
Lavatory, for vanity top, P.E., cast iron, 20″ x 18″, white	1.000	Ea.	2.500	234	104	338
Color	1.000	Ea.	2.500	264	104	368
Steel enameled 20″ x 17″, white	1.000	Ea.	2.759	152	115	267
Color	1.000	Ea.	2.500	152	104	256
Vitreous china 20″ x 16″, white	1.000	Ea.	2.963	244	124	368
Color	1.000	Ea.	2.963	244	124	368
Rough-in, for lavatory						
1/2″ copper supply, 1-1/2″ cast iron waste, 1-1/2″ cast iron vent	1.000	Ea.	2.791	65.50	121	186.50
1-1/2″ PVC waste, 1-1/4″ PVC vent	1.000	Ea.	2.639	31.50	122	153.50
1/2″ steel supply, 1-1/4″ cast iron waste, 1-1/4″ steel vent	1.000	Ea.	2.890	80	127	207
1-1/4″ PVC waste, 1-1/4″ PVC vent	1.000	Ea.	2.794	42.50	130	172.50
1/2″ PVC supply, 1-1/2″ PVC waste, 1-1/2″ PVC vent	1.000	Ea.	3.260	40	151	191
Bathtub, P.E. cast iron, corner with fittings, white	1.000	Ea.	3.636	1,775	152	1,927
Color	1.000	Ea.	4.000	2,000	167	2,167
Rough-in, for bathtub						
1/2″ copper supply, 4″ cast iron waste, 1-1/2″ copper vent	1.000	Ea.	2.409	74	108	182
4″ PVC waste, 1-1/2″ PVC vent	1.000	Ea.	2.877	38.50	129	167.50
1/2″ steel supply, 4″ cast iron waste, 1-1/2″ steel vent	1.000	Ea.	2.898	92	127	219
4″ PVC waste, 1-1/2″ PVC vent	1.000	Ea.	3.159	49.50	142	191.50
1/2″ PVC supply, 4″ PVC waste, 1-1/2″ PVC vent	1.000	Ea.	3.371	46.50	151	197.50
Piping, supply, 1/2″ copper	32.000	L.F.	3.161	62.50	147	209.50
1/2″ steel	32.000	L.F.	4.063	98	189	287
1/2″ PVC	32.000	L.F.	4.741	89	219	308
Piping, waste, 4″ cast iron, no hub	12.000	L.F.	3.310	121	138	259
4″ PVC/DWV	12.000	L.F.	4.000	50.50	167	217.50
4″ copper/DWV	12.000	L.F.	4.800	305	200	505
Piping, vent 2″ cast iron, no hub	6.000	L.F.	1.433	35.50	59.50	95
2″ copper/DWV	6.000	L.F.	1.091	49	50.50	99.50
2″ PVC/DWV	6.000	L.F.	1.627	10.40	68	78.40
2″ steel, galvanized	6.000	L.F.	1.500	62.50	62.50	125
Vanity base cabinet, 2 door, 24″ x 30″	1.000	Ea.	1.000	232	42.50	274.50
24″ x 36″	1.000	Ea.	1.200	310	50.50	360.50
Vanity top, laminated plastic square edge 25″ x 32″	2.670	L.F.	.712	83	30	113
25″ x 38″	3.160	L.F.	.843	98	35.50	133.50
Cultured marble, 25″ x 32″, with bowl	1.000	Ea.	2.500	192	104	296
25″ x 38″, with bowl	1.000	Ea.	2.500	229	104	333
Carrier, for lavatory, steel for studs, no arms	1.000	Ea.	1.143	51	53	104
Wood, 2″ x 8″ blocking	1.300	L.F.	.053	1.25	2.25	3.50

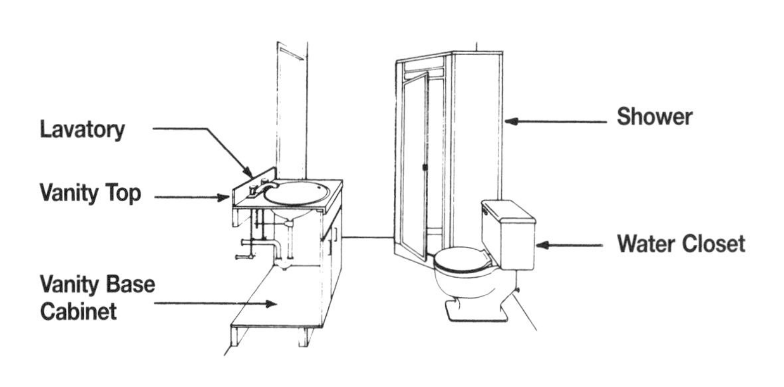

System Description	QUAN.	UNIT	LABOR HOURS	COST EACH		
				MAT.	INST.	TOTAL
BATHROOM WITH SHOWER, LAVATORY INSTALLED IN VANITY						
Water closet, floor mounted, 2 piece, close coupled, white	1.000	Ea.	3.019	186	126	312
Rough-in, vent, 2″ diameter DWV piping	1.000	Ea.	.955	23.80	39.80	63.60
Waste, 4″ diameter DWV piping	1.000	Ea.	.828	30.15	34.50	64.65
Supply, 1/2″ diameter type "L" copper supply piping	1.000	Ea.	.593	11.70	27.48	39.18
Lavatory, 20″ x 18″ P.E. cast iron with fittings, white	1.000	Ea.	2.500	234	104	338
Rough-in, waste, 1-1/2″ diameter DWV piping	1.000	Ea.	1.803	46	75.20	121.20
Supply, 1/2″ diameter type "L" copper supply piping	1.000	Ea.	.988	19.50	45.80	65.30
Shower, steel enameled, stone base, corner, white	1.000	Ea.	3.200	380	134	514
Rough-in, vent, 1-1/2″ diameter DWV piping	1.000	Ea.	.225	5.75	9.40	15.15
Waste, 2″ diameter DWV piping	1.000	Ea.	1.433	35.70	59.70	95.40
Supply, 1/2″ diameter type "L" copper supply piping	1.000	Ea.	1.580	31.20	73.28	104.48
Piping, supply, 1/2″ diameter type "L" copper supply piping	36.000	L.F.	4.148	81.90	192.36	274.26
Waste, 4″ diameter DWV piping	7.000	L.F.	2.759	100.50	115	215.50
Vent, 2″ diameter DWV piping	6.000	L.F.	2.250	93.60	94.05	187.65
Vanity base 2 door, 30″ wide	1.000	Ea.	1.000	232	42.50	274.50
Vanity top, plastic laminated, square edge	2.170	L.F.	.712	68.09	30.04	98.13
TOTAL		Ea.	27.993	1,579.89	1,203.11	2,783
BATHROOM WITH SHOWER, WALL HUNG LAVATORY						
Water closet, floor mounted, close coupled	1.000	Ea.	3.019	186	126	312
Rough-in, vent, 2″ diameter DWV piping	1.000	Ea.	.955	23.80	39.80	63.60
Waste, 4″ diameter DWV piping	1.000	Ea.	.828	30.15	34.50	64.65
Supply, 1/2″ diameter type "L" copper supply piping	1.000	Ea.	.593	11.70	27.48	39.18
Lavatory, 20″ x 18″ P.E. cast iron with fittings, white	1.000	Ea.	2.000	265	83.50	348.50
Rough-in, waste, 1-1/2″ diameter DWV piping	1.000	Ea.	1.803	46	75.20	121.20
Supply, 1/2″ diameter type "L" copper supply piping	1.000	Ea.	.988	19.50	45.80	65.30
Shower, steel enameled, stone base, white	1.000	Ea.	3.200	380	134	514
Rough-in, vent, 1-1/2″ diameter DWV piping	1.000	Ea.	.225	5.75	9.40	15.15
Waste, 2″ diameter DWV piping	1.000	Ea.	1.433	35.70	59.70	95.40
Supply, 1/2″ diameter type "L" copper supply piping	1.000	Ea.	1.580	31.20	73.28	104.48
Piping, supply, 1/2″ diameter type "L" copper supply piping	36.000	L.F.	4.148	81.90	192.36	274.26
Waste, 4″ diameter DWV piping	7.000	L.F.	2.759	100.50	115	215.50
Vent, 2″ diameter DWV piping	6.000	L.F.	2.250	93.60	94.05	187.65
Carrier, steel, for studs, no arms	1.000	Ea.	1.143	51	53	104
TOTAL		Ea.	26.924	1,361.80	1,163.07	2,524.87

The costs in this system are on a cost each basis. All necessary piping is included.

Important: See the Reference Section for critical supporting data - Reference Nos., Crews & Location Factors

Three Fixture Bathroom Price Sheet	QUAN.	UNIT	LABOR HOURS	COST EACH		
				MAT.	INST.	TOTAL
Water closet, close coupled, standard 2 piece, white	1.000	Ea.	3.019	186	126	312
Color	1.000	Ea.	3.019	223	126	349
One piece elongated bowl, white	1.000	Ea.	3.019	545	126	671
Color	1.000	Ea.	3.019	680	126	806
Low profile, one piece elongated bowl, white	1.000	Ea.	3.019	775	126	901
Color	1.000	Ea.	3.019	1,025	126	1,151
Rough-in, for water closet						
1/2" copper supply, 4" cast iron waste, 2" cast iron vent	1.000	Ea.	2.376	65.50	102	167.50
4" PVC/DWV waste, 2" PVC vent	1.000	Ea.	2.678	31.50	114	145.50
4" copper waste, 2" copper vent	1.000	Ea.	2.520	121	111	232
3" cast iron waste, 1-1/2" cast iron vent	1.000	Ea.	2.244	58.50	96.50	155
3" PVC waste, 1-1/2" PVC vent	1.000	Ea.	2.388	27.50	106	133.50
3" copper waste, 1-1/2" copper vent	1.000	Ea.	2.014	79.50	89.50	169
1/2" PVC supply, 4" PVC waste, 2" PVC vent	1.000	Ea.	2.974	36.50	128	164.50
3" PVC waste, 1-1/2" PVC supply	1.000	Ea.	2.684	32.50	120	152.50
1/2" steel supply, 4" cast iron waste, 2" cast iron vent	1.000	Ea.	2.545	72.50	110	182.50
4" cast iron waste, 2" steel vent	1.000	Ea.	2.590	90	112	202
4" PVC waste, 2" PVC vent	1.000	Ea.	2.847	38	122	160
Lavatory, wall hung, P.E. cast iron 20" x 18", white	1.000	Ea.	2.000	265	83.50	348.50
Color	1.000	Ea.	2.000	305	83.50	388.50
Vitreous china 19" x 17", white	1.000	Ea.	2.286	187	95.50	282.50
Color	1.000	Ea.	2.286	187	95.50	282.50
Lavatory, for vanity top, P.E. cast iron 20" x 18", white	1.000	Ea.	2.500	234	104	338
Color	1.000	Ea.	2.500	264	104	368
Steel enameled 20" x 17", white	1.000	Ea.	2.759	152	115	267
Color	1.000	Ea.	2.500	152	104	256
Vitreous china 20" x 16", white	1.000	Ea.	2.963	244	124	368
Color	1.000	Ea.	2.963	244	124	368
Rough-in, for lavatory						
1/2" copper supply, 1-1/2" cast iron waste, 1-1/2" cast iron vent	1.000	Ea.	2.791	65.50	121	186.50
1-1/2" PVC waste, 1-1/2" PVC vent	1.000	Ea.	2.639	31.50	122	153.50
1/2" steel supply, 1-1/4" cast iron waste, 1-1/4" steel vent	1.000	Ea.	2.890	80	127	207
1-1/4" PVC waste, 1-1/4" PVC vent	1.000	Ea.	2.921	43	136	179
1/2" PVC supply, 1-1/2" PVC waste, 1-1/2" PVC vent	1.000	Ea.	3.260	40	151	191
Shower, steel enameled stone base, 32" x 32", white	1.000	Ea.	8.000	380	134	514
Color	1.000	Ea.	7.822	845	122	967
36" x 36" white	1.000	Ea.	8.889	960	139	1,099
Color	1.000	Ea.	8.889	960	139	1,099
Rough-in, for shower						
1/2" copper supply, 4" cast iron waste, 1-1/2" copper vent	1.000	Ea.	3.238	72.50	142	214.50
4" PVC waste, 1-1/2" PVC vent	1.000	Ea.	3.429	43	151	194
1/2" steel supply, 4" cast iron waste, 1-1/2" steel vent	1.000	Ea.	3.665	92.50	162	254.50
4" PVC waste, 1-1/2" PVC vent	1.000	Ea.	3.881	61	173	234
1/2" PVC supply, 4" PVC waste, 1-1/2" PVC vent	1.000	Ea.	4.219	56.50	188	244.50
Piping, supply, 1/2" copper	36.000	L.F.	4.148	82	192	274
1/2" steel	36.000	L.F.	5.333	129	248	377
1/2" PVC	36.000	L.F.	6.222	117	288	405
Piping, waste, 4" cast iron no hub	7.000	L.F.	2.759	101	115	216
4" PVC/DWV	7.000	L.F.	3.333	42	139	181
4" copper/DWV	7.000	L.F.	4.000	255	167	422
Piping, vent, 2" cast iron no hub	6.000	L.F.	2.149	53.50	89.50	143
2" copper/DWV	6.000	L.F.	1.636	73.50	76	149.50
2" PVC/DWV	6.000	L.F.	2.441	15.55	102	117.55
2" steel, galvanized	6.000	L.F.	2.250	93.50	94	187.50
Vanity base cabinet, 2 door, 24" x 30"	1.000	Ea.	1.000	232	42.50	274.50
24" x 36"	1.000	Ea.	1.200	310	50.50	360.50
Vanity top, laminated plastic square edge, 25" x 32"	2.170	L.F.	.712	68	30	98
25" x 38"	2.670	L.F.	.845	81	35.50	116.50
Carrier, for lavatory, steel for studs, no arms	1.000	Ea.	1.143	51	53	104
Wood, 2" x 8" blocking	1.300	L.F.	.052	1.22	2.20	3.42

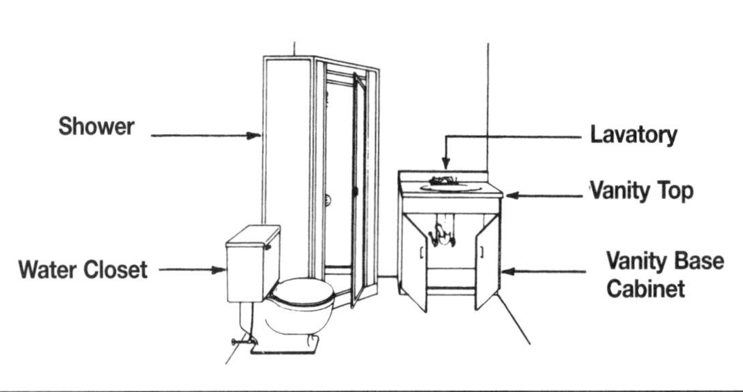

System Description	QUAN.	UNIT	LABOR HOURS	COST EACH		
				MAT.	INST.	TOTAL
BATHROOM WITH LAVATORY INSTALLED IN VANITY						
Water closet, floor mounted, 2 piece, close coupled, white	1.000	Ea.	3.019	186	126	312
Rough-in, vent, 2" diameter DWV piping	1.000	Ea.	.955	23.80	39.80	63.60
Waste, 4" diameter DWV piping	1.000	Ea.	.828	30.15	34.50	64.65
Supply, 1/2" diameter type "L" copper supply piping	1.000	Ea.	.593	11.70	27.48	39.18
Lavatory, 20" x 18", P.E. cast iron with fittings, white	1.000	Ea.	2.500	234	104	338
Rough-in, waste, 1-1/2" diameter DWV piping	1.000	Ea.	1.803	46	75.20	121.20
Supply, 1/2" diameter type "L" copper supply piping	1.000	Ea.	.988	19.50	45.80	65.30
Shower, steel enameled, stone base, corner, white	1.000	Ea.	3.200	380	134	514
Rough-in, vent, 1-1/2" diameter DWV piping	1.000	Ea.	.225	5.75	9.40	15.15
Waste, 2" diameter DWV piping	1.000	Ea.	1.433	35.70	59.70	95.40
Supply, 1/2" diameter type "L" copper supply piping	1.000	Ea.	1.580	31.20	73.28	104.48
Piping, supply, 1/2" diameter type "L" copper supply piping	36.000	L.F.	3.556	70.20	164.88	235.08
Waste, 4" diameter DWV piping	7.000	L.F.	1.931	70.35	80.50	150.85
Vent, 2" diameter DWV piping	6.000	L.F.	1.500	62.40	62.70	125.10
Vanity base, 2 door, 30" wide	1.000	Ea.	1.000	232	42.50	274.50
Vanity top, plastic laminated, square edge	2.670	L.F.	.712	65.42	30.04	95.46
TOTAL		Ea.	25.823	1,504.17	1,109.78	2,613.95
BATHROOM, WITH WALL HUNG LAVATORY						
Water closet, floor mounted, 2 piece, close coupled, white	1.000	Ea.	3.019	186	126	312
Rough-in, vent, 2" diameter DWV piping	1.000	Ea.	.955	23.80	39.80	63.60
Waste, 4" diameter DWV piping	1.000	Ea.	.828	30.15	34.50	64.65
Supply, 1/2" diameter type "L" copper supply piping	1.000	Ea.	.593	11.70	27.48	39.18
Lavatory, wall hung, 20" x 18" P.E. cast iron with fittings, white	1.000	Ea.	2.000	265	83.50	348.50
Rough-in, waste, 1-1/2" diameter DWV piping	1.000	Ea.	1.803	46	75.20	121.20
Supply, 1/2" diameter type "L" copper supply piping	1.000	Ea.	.988	19.50	45.80	65.30
Shower, steel enameled, stone base, corner, white	1.000	Ea.	3.200	380	134	514
Rough-in, waste, 1-1/2" diameter DWV piping	1.000	Ea.	.225	5.75	9.40	15.15
Waste, 2" diameter DWV piping	1.000	Ea.	1.433	35.70	59.70	95.40
Supply, 1/2" diameter type "L" copper supply piping	1.000	Ea.	1.580	31.20	73.28	104.48
Piping, supply, 1/2" diameter type "L" copper supply piping	36.000	L.F.	3.556	70.20	164.88	235.08
Waste, 4" diameter DWV piping	7.000	L.F.	1.931	70.35	80.50	150.85
Vent, 2" diameter DWV piping	6.000	L.F.	1.500	62.40	62.70	125.10
Carrier, steel, for studs, no arms	1.000	Ea.	1.143	51	53	104
TOTAL		Ea.	24.754	1,288.75	1,069.74	2,358.49

The costs in this system are on a cost each basis. All necessary piping is included.

Important: See the Reference Section for critical supporting data - Reference Nos., Crews & Location Factors

Three Fixture Bathroom Price Sheet	QUAN.	UNIT	LABOR HOURS	COST EACH		
				MAT.	INST.	TOTAL
Water closet, close coupled, standard 2 piece, white	1.000	Ea.	3.019	186	126	312
Color	1.000	Ea.	3.019	223	126	349
One piece elongated bowl, white	1.000	Ea.	3.019	545	126	671
Color	1.000	Ea.	3.019	680	126	806
Low profile one piece elongated bowl, white	1.000	Ea.	3.019	775	126	901
Color	1.000	Ea.	3.623	1,025	126	1,151
Rough-in, for water closet						
1/2" copper supply, 4" cast iron waste, 2" cast iron vent	1.000	Ea.	2.376	65.50	102	167.50
4" P.V.C./DWV waste, 2" PVC vent	1.000	Ea.	2.678	31.50	114	145.50
4" copper waste, 2" copper vent	1.000	Ea.	2.520	121	111	232
3" cast iron waste, 1-1/2" cast iron vent	1.000	Ea.	2.244	58.50	96.50	155
3" PVC waste, 1-1/2" PVC vent	1.000	Ea.	2.388	27.50	106	133.50
3" copper waste, 1-1/2" copper vent	1.000	Ea.	2.014	79.50	89.50	169
1/2" P.V.C. supply, 4" P.V.C. waste, 2" P.V.C. vent	1.000	Ea.	2.974	36.50	128	164.50
3" P.V.C. waste, 1-1/2" P.V.C. vent	1.000	Ea.	2.684	32.50	120	152.50
1/2" steel supply, 4" cast iron waste, 2" cast iron vent	1.000	Ea.	2.545	72.50	110	182.50
4" cast iron waste, 2" steel vent	1.000	Ea.	2.590	90	112	202
4" P.V.C. waste, 2" P.V.C. vent	1.000	Ea.	2.847	38	122	160
Lavatory, wall hung P.E. cast iron 20" x 18", white	1.000	Ea.	2.000	265	83.50	348.50
Color	1.000	Ea.	2.000	305	83.50	388.50
Vitreous china 19" x 17", white	1.000	Ea.	2.286	187	95.50	282.50
Color	1.000	Ea.	2.286	187	95.50	282.50
Lavatory, for vanity top P.E. cast iron 20" x 18", white	1.000	Ea.	2.500	234	104	338
Color	1.000	Ea.	2.500	264	104	368
Steel enameled 20" x 17", white	1.000	Ea.	2.759	152	115	267
Color	1.000	Ea.	2.500	152	104	256
Vitreous china 20" x 16", white	1.000	Ea.	2.963	244	124	368
Color	1.000	Ea.	2.963	244	124	368
Rough-in, for lavatory						
1/2" copper supply, 1-1/2" cast iron waste, 1-1/2" cast iron vent	1.000	Ea.	2.791	65.50	121	186.50
1-1/2" P.V.C. waste, 1-1/2" P.V.C. vent	1.000	Ea.	2.639	31.50	122	153.50
1/2" steel supply, 1-1/2" cast iron waste, 1-1/4" steel vent	1.000	Ea.	2.890	80	127	207
1-1/2" P.V.C. waste, 1-1/4" P.V.C. vent	1.000	Ea.	2.921	43	136	179
1/2" P.V.C. supply, 1-1/2" P.V.C. waste, 1-1/2" P.V.C. vent	1.000	Ea.	3.260	40	151	191
Shower, steel enameled stone base, 32" x 32", white	1.000	Ea.	8.000	380	134	514
Color	1.000	Ea.	7.822	845	122	967
36" x 36", white	1.000	Ea.	8.889	960	139	1,099
Color	1.000	Ea.	8.889	960	139	1,099
Rough-in, for shower						
1/2" copper supply, 2" cast iron waste, 1-1/2" copper vent	1.000	Ea.	3.161	73	140	213
2" P.V.C. waste, 1-1/2" P.V.C. vent	1.000	Ea.	3.429	43	151	194
1/2" steel supply, 2" cast iron waste, 1-1/2" steel vent	1.000	Ea.	3.887	117	172	289
2" P.V.C. waste, 1-1/2" P.V.C. vent	1.000	Ea.	3.881	61	173	234
1/2" P.V.C. supply, 2" P.V.C. waste, 1-1/2" P.V.C. vent	1.000	Ea.	4.219	56.50	188	244.50
Piping, supply, 1/2" copper	36.000	L.F.	3.556	70	165	235
1/2" steel	36.000	L.F.	4.571	110	212	322
1/2" P.V.C.	36.000	L.F.	5.333	100	247	347
Waste, 4" cast iron, no hub	7.000	L.F.	1.931	70.50	80.50	151
4" P.V.C./DWV	7.000	L.F.	2.333	29.50	97.50	127
4" copper/DWV	7.000	L.F.	2.800	179	117	296
Vent, 2" cast iron, no hub	6.000	L.F.	1.091	49	50.50	99.50
2" copper/DWV	6.000	L.F.	1.091	49	50.50	99.50
2" P.V.C./DWV	6.000	L.F.	1.627	10.40	68	78.40
2" steel, galvanized	6.000	L.F.	1.500	62.50	62.50	125
Vanity base cabinet, 2 door, 24" x 30"	1.000	Ea.	1.000	232	42.50	274.50
24" x 36"	1.000	Ea.	1.200	310	50.50	360.50
Vanity top, laminated plastic square edge, 25" x 32"	2.670	L.F.	.712	65.50	30	95.50
25" x 38"	3.170	L.F.	.845	77.50	35.50	113
Carrier , for lavatory, steel, for studs, no arms	1.000	Ea.	1.143	51	53	104
Wood, 2" x 8" blocking	1.300	L.F.	.052	1.22	2.20	3.42

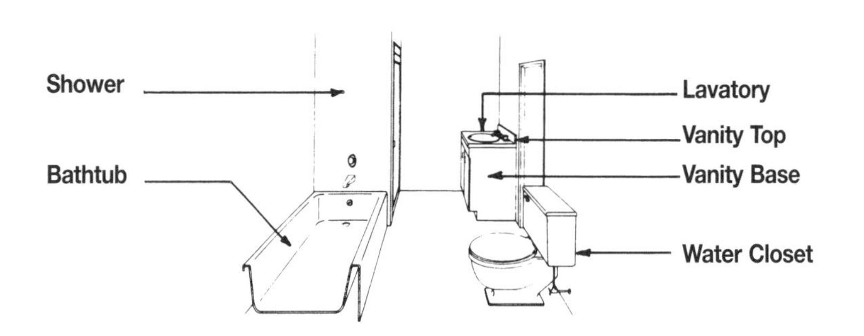

System Description	QUAN.	UNIT	LABOR HOURS	COST EACH		
				MAT.	INST.	TOTAL
BATHROOM WITH LAVATORY INSTALLED IN VANITY						
Water closet, floor mounted, 2 piece, close coupled, white	1.000	Ea.	3.019	186	126	312
Rough-in, vent, 2″ diameter DWV piping	1.000	Ea.	.955	23.80	39.80	63.60
Waste, 4″ diameter DWV piping	1.000	Ea.	.828	30.15	34.50	64.65
Supply, 1/2″ diameter type "L" copper supply piping	1.000	Ea.	.593	11.70	27.48	39.18
Lavatory, 20″ x 18″ P.E. cast iron with fittings, white	1.000	Ea.	2.500	234	104	338
Shower, steel, enameled, stone base, corner, white	1.000	Ea.	3.333	880	139	1,019
Rough-in, waste, 1-1/2″ diameter DWV piping	2.000	Ea.	4.507	115	188	303
Supply, 1/2″ diameter type "L" copper supply piping	2.000	Ea.	3.161	62.40	146.56	208.96
Bathtub, P.E. cast iron, 5′ long with fittings, white	1.000	Ea.	3.636	805	152	957
Rough-in, waste, 4″ diameter DWV piping	1.000	Ea.	.828	30.15	34.50	64.65
Supply, 1/2″ diameter type "L" copper supply piping	1.000	Ea.	.988	19.50	45.80	65.30
Vent, 1-1/2″ diameter DWV piping	1.000	Ea.	.593	24.40	27.40	51.80
Piping, supply, 1/2″ diameter type "L" copper supply piping	42.000	L.F.	4.148	81.90	192.36	274.26
Waste, 4″ diameter DWV piping	10.000	L.F.	2.759	100.50	115	215.50
Vent, 2″ diameter DWV piping	13.000	L.F.	3.250	135.20	135.85	271.05
Vanity base, 2 doors, 30″ wide	1.000	Ea.	1.000	232	42.50	274.50
Vanity top, plastic laminated, square edge	2.670	L.F.	.712	65.42	30.04	95.46
TOTAL		Ea.	36.810	3,037.12	1,580.79	4,617.91
BATHROOM WITH WALL HUNG LAVATORY						
Water closet, floor mounted, 2 piece, close coupled, white	1.000	Ea.	3.019	186	126	312
Rough-in, vent, 2″ diameter DWV piping	1.000	Ea.	.955	23.80	39.80	63.60
Waste, 4″ diameter DWV piping	1.000	Ea.	.828	30.15	34.50	64.65
Supply, 1/2″ diameter type "L" copper supply piping	1.000	Ea.	.593	11.70	27.48	39.18
Lavatory, 20″ x 18″ P.E. cast iron with fittings, white	1.000	Ea.	2.000	265	83.50	348.50
Shower, steel enameled, stone base, corner , white	1.000	Ea.	3.333	880	139	1,019
Rough-in, waste, 1-1/2″ diameter DWV piping	2.000	Ea.	4.507	115	188	303
Supply, 1/2″ diameter type "L" copper supply piping	2.000	Ea.	3.161	62.40	146.56	208.96
Bathtub, P.E. cast iron, 5′ long with fittings, white	1.000	Ea.	3.636	805	152	957
Rough-in, waste, 4″ diameter DWV piping	1.000	Ea.	.828	30.15	34.50	64.65
Supply, 1/2″ diameter type "L" copper supply piping	1.000	Ea.	.988	19.50	45.80	65.30
Vent, 1-1/2″ diameter copper DWV piping	1.000	Ea.	.593	24.40	27.40	51.80
Piping, supply, 1/2″ diameter type "L" copper supply piping	42.000	L.F.	4.148	81.90	192.36	274.26
Waste, 4″ diameter DWV piping	10.000	L.F.	2.759	100.50	115	215.50
Vent, 2″ diameter DWV piping	13.000	L.F.	3.250	135.20	135.85	271.05
Carrier, steel, for studs, no arms	1.000	Ea.	1.143	51	53	104
TOTAL		Ea.	35.741	2,821.70	1,540.75	4,362.45

The costs in this system are on a cost each basis. All necessary piping is included.

Important: See the Reference Section for critical supporting data - Reference Nos., Crews & Location Factors

Four Fixture Bathroom Price Sheet	QUAN.	UNIT	LABOR HOURS	COST EACH		
				MAT.	INST.	TOTAL
Water closet, close coupled, standard 2 piece, white	1.000	Ea.	3.019	186	126	312
Color	1.000	Ea.	3.019	223	126	349
One piece elongated bowl, white	1.000	Ea.	3.019	545	126	671
Color	1.000	Ea.	3.019	680	126	806
Low profile, one piece elongated bowl, white	1.000	Ea.	3.019	775	126	901
Color	1.000	Ea.	3.019	1,025	126	1,151
1/2" copper supply, 4" cast iron waste, 2" cast iron vent	1.000	Ea.	2.376	65.50	102	167.50
4" PVC/DWV waste, 2" PVC vent	1.000	Ea.	2.678	31.50	114	145.50
4" copper waste, 2" copper vent	1.000	Ea.	2.520	121	111	232
3" cast iron waste, 1-1/2" cast iron vent	1.000	Ea.	2.244	58.50	96.50	155
3" P.V.C. waste, 1-1/2" P.V.C. vent	1.000	Ea.	2.388	27.50	106	133.50
3" copper waste, 1-1/2" copper vent	1.000	Ea.	2.014	79.50	89.50	169
1/2" P.V.C. supply, 4" P.V.C. waste, 2" P.V.C. vent	1.000	Ea.	2.974	36.50	128	164.50
3" P.V.C. waste, 1-1/2" P.V.C. vent	1.000	Ea.	2.684	32.50	120	152.50
1/2" steel supply, 4" cast iron waste, 2" cast iron vent	1.000	Ea.	2.545	72.50	110	182.50
4" cast iron waste, 2" steel vent	1.000	Ea.	2.590	90	112	202
4" P.V.C. waste, 2" P.V.C. vent	1.000	Ea.	2.847	38	122	160
Lavatory, wall hung P.E. cast iron 20" x 18", white	1.000	Ea.	2.000	265	83.50	348.50
Color	1.000	Ea.	2.000	305	83.50	388.50
Vitreous china 19" x 17", white	1.000	Ea.	2.286	187	95.50	282.50
Color	1.000	Ea.	2.286	187	95.50	282.50
Lavatory for vanity top, P.E. cast iron 20" x 18", white	1.000	Ea.	2.500	234	104	338
Color	1.000	Ea.	2.500	264	104	368
Steel enameled, 20" x 17", white	1.000	Ea.	2.759	152	115	267
Color	1.000	Ea.	2.500	152	104	256
Vitreous china 20" x 16", white	1.000	Ea.	2.963	244	124	368
Color	1.000	Ea.	2.963	244	124	368
Shower, steel enameled stone base, 36" square, white	1.000	Ea.	8.889	880	139	1,019
Color	1.000	Ea.	8.889	895	139	1,034
Rough-in, for lavatory or shower						
1/2" copper supply, 1-1/2" cast iron waste, 1-1/2" cast iron vent	1.000	Ea.	3.834	88.50	167	255.50
1-1/2" P.V.C. waste, 1-1/4" P.V.C. vent	1.000	Ea.	3.675	46.50	170	216.50
1/2" steel supply, 1-1/4" cast iron waste, 1-1/4" steel vent	1.000	Ea.	4.103	110	181	291
1-1/4" P.V.C. waste, 1-1/4" P.V.C. vent	1.000	Ea.	3.937	64	183	247
1/2" P.V.C. supply, 1-1/2" P.V.C. waste, 1-1/2" P.V.C. vent	1.000	Ea.	4.592	60	213	273
Bathtub, P.E. cast iron, 5' long with fittings, white	1.000	Ea.	3.636	805	152	957
Color	1.000	Ea.	3.636	995	152	1,147
Steel, enameled 5' long with fittings, white	1.000	Ea.	2.909	330	121	451
Color	1.000	Ea.	2.909	330	121	451
Rough-in, for bathtub						
1/2" copper supply, 4" cast iron waste, 1-1/2" copper vent	1.000	Ea.	2.409	74	108	182
4" P.V.C. waste, 1-1/2" P.V.C. vent	1.000	Ea.	2.877	38.50	129	167.50
1/2" steel supply, 4" cast iron waste, 1-1/2" steel vent	1.000	Ea.	2.898	92	127	219
4" P.V.C. waste, 1-1/2" P.V.C. vent	1.000	Ea.	3.159	49.50	142	191.50
1/2" P.V.C. supply, 4" P.V.C. waste, 1-1/2" P.V.C. vent	1.000	Ea.	3.371	46.50	151	197.50
Piping, supply, 1/2" copper	42.000	L.F.	4.148	82	192	274
1/2" steel	42.000	L.F.	5.333	129	248	377
1/2" P.V.C.	42.000	L.F.	6.222	117	288	405
Waste, 4" cast iron, no hub	10.000	L.F.	2.759	101	115	216
4" P.V.C./DWV	10.000	L.F.	3.333	42	139	181
4" copper/DWV	10.000	Ea.	4.000	255	167	422
Vent 2" cast iron, no hub	13.000	L.F.	3.105	77.50	129	206.50
2" copper/DWV	13.000	L.F.	2.364	106	110	216
2" P.V.C./DWV	13.000	L.F.	3.525	22.50	147	169.50
2" steel, galvanized	13.000	L.F.	3.250	135	136	271
Vanity base cabinet, 2 doors, 30" wide	1.000	Ea.	1.000	232	42.50	274.50
Vanity top, plastic laminated, square edge	2.670	L.F.	.712	65.50	30	95.50
Carrier, steel for studs, no arms	1.000	Ea.	1.143	51	53	104
Wood, 2" x 8" blocking	1.300	L.F.	.052	1.22	2.20	3.42

8 MECHANICAL

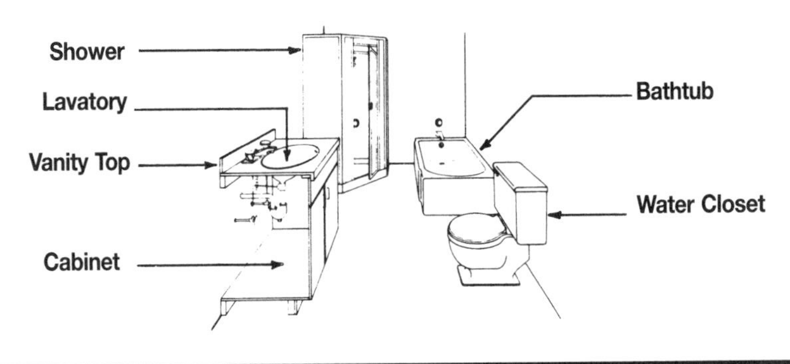

System Description	QUAN.	UNIT	LABOR HOURS	COST EACH MAT.	COST EACH INST.	COST EACH TOTAL
BATHROOM WITH LAVATORY INSTALLED IN VANITY						
Water closet, floor mounted, 2 piece, close coupled, white	1.000	Ea.	3.019	186	126	312
Rough-in, vent, 2″ diameter DWV piping	1.000	Ea.	.955	23.80	39.80	63.60
Waste, 4″ diameter DWV piping	1.000	Ea.	.828	30.15	34.50	64.65
Supply, 1/2″ diameter type "L" copper supply piping	1.000	Ea.	.593	11.70	27.48	39.18
Lavatory, 20″ x 18″ P.E. cast iron with fittings, white	1.000	Ea.	2.500	234	104	338
Shower, steel, enameled, stone base, corner, white	1.000	Ea.	3.333	880	139	1,019
Rough-in, waste, 1-1/2″ diameter DWV piping	2.000	Ea.	4.507	115	188	303
Supply, 1/2″ diameter type "L" copper supply piping	2.000	Ea.	3.161	62.40	146.56	208.96
Bathtub, P.E. cast iron, 5′ long with fittings, white	1.000	Ea.	3.636	805	152	957
Rough-in, waste, 4″ diameter DWV piping	1.000	Ea.	.828	30.15	34.50	64.65
Supply, 1/2″ diameter type "L" copper supply piping	1.000	Ea.	.988	19.50	45.80	65.30
Vent, 1-1/2″ diameter DWV piping	1.000	Ea.	.593	24.40	27.40	51.80
Piping, supply, 1/2″ diameter type "L" copper supply piping	42.000	L.F.	4.939	97.50	229	326.50
Waste, 4″ diameter DWV piping	10.000	L.F.	4.138	150.75	172.50	323.25
Vent, 2″ diameter DWV piping	13.000	L.F.	4.500	187.20	188.10	375.30
Vanity base, 2 doors, 30″ wide	1.000	Ea.	1.000	232	42.50	274.50
Vanity top, plastic laminated, square edge	2.670	L.F.	.712	68.09	30.04	98.13
TOTAL		Ea.	40.230	3,157.64	1,727.18	4,884.82
BATHROOM WITH WALL HUNG LAVATORY						
Water closet, floor mounted, 2 piece, close coupled, white	1.000	Ea.	3.019	186	126	312
Rough-in, vent, 2″ diameter DWV piping	1.000	Ea.	.955	23.80	39.80	63.60
Waste, 4″ diameter DWV piping	1.000	Ea.	.828	30.15	34.50	64.65
Supply, 1/2″ diameter type "L" copper supply piping	1.000	Ea.	.593	11.70	27.48	39.18
Lavatory, 20″ x 18″ P.E. cast iron with fittings, white	1.000	Ea.	2.000	265	83.50	348.50
Shower, steel enameled, stone base, corner, white	1.000	Ea.	3.333	880	139	1,019
Rough-in, waste, 1-1/2″ diameter DWV piping	2.000	Ea.	4.507	115	188	303
Supply, 1/2″ diameter type "L" copper supply piping	2.000	Ea.	3.161	62.40	146.56	208.96
Bathtub, P.E. cast iron, 5″ long with fittings, white	1.000	Ea.	3.636	805	152	957
Rough-in, waste, 4″ diameter DWV piping	1.000	Ea.	.828	30.15	34.50	64.65
Supply, 1/2″ diameter type "L" copper supply piping	1.000	Ea.	.988	19.50	45.80	65.30
Vent, 1-1/2″ diameter DWV piping	1.000	Ea.	.593	24.40	27.40	51.80
Piping, supply, 1/2″ diameter type "L" copper supply piping	42.000	L.F.	4.939	97.50	229	326.50
Waste, 4″ diameter DWV piping	10.000	L.F.	4.138	150.75	172.50	323.25
Vent, 2″ diameter DWV piping	13.000	L.F.	4.500	187.20	188.10	375.30
Carrier, steel for studs, no arms	1.000	Ea.	1.143	51	53	104
TOTAL		Ea.	39.161	2,939.55	1,687.14	4,626.69

The costs in this system are on a cost each basis. All necessary piping is included

Four Fixture Bathroom Price Sheet	QUAN.	UNIT	LABOR HOURS	COST EACH		
				MAT.	INST.	TOTAL
Water closet, close coupled, standard 2 piece, white	1.000	Ea.	3.019	186	126	312
Color	1.000	Ea.	3.019	223	126	349
One piece, elongated bowl, white	1.000	Ea.	3.019	545	126	671
Color	1.000	Ea.	3.019	680	126	806
Low profile, one piece elongated bowl, white	1.000	Ea.	3.019	775	126	901
Color	1.000	Ea.	3.019	1,025	126	1,151
Rough-in, for water closet						
1/2" copper supply, 4" cast iron waste, 2" cast iron vent	1.000	Ea.	2.376	65.50	102	167.50
4" PVC/DWV waste, 2" PVC vent	1.000	Ea.	2.678	31.50	114	145.50
4" copper waste, 2" copper vent	1.000	Ea.	2.520	121	111	232
3" cast iron waste, 1-1/2" cast iron vent	1.000	Ea.	2.244	58.50	96.50	155
3" PVC waste, 1-1/2" PVC vent	1.000	Ea.	2.388	27.50	106	133.50
3" PVC waste, 1-1/2" PVC vent	1.000	Ea.	2.014	79.50	89.50	169
1/2" PVC supply, 4" PVC waste, 2" PVC vent	1.000	Ea.	2.974	36.50	128	164.50
3" PVC waste, 1-1/2" PVC vent	1.000	Ea.	2.684	32.50	120	152.50
1/2" steel supply, 4" cast iron waste, 2" cast iron vent	1.000	Ea.	2.545	72.50	110	182.50
4" cast iron waste, 2" steel vent	1.000	Ea.	2.590	90	112	202
4" PVC waste, 2" PVC vent	1.000	Ea.	2.847	38	122	160
Lavatory wall hung, P.E. cast iron 20" x 18", white	1.000	Ea.	2.000	265	83.50	348.50
Color	1.000	Ea.	2.000	305	83.50	388.50
Vitreous china 19" x 17", white	1.000	Ea.	2.286	187	95.50	282.50
Color	1.000	Ea.	2.286	187	95.50	282.50
Lavatory for vanity top, P.E. cast iron, 20" x 18", white	1.000	Ea.	2.500	234	104	338
Color	1.000	Ea.	2.500	264	104	368
Steel, enameled 20" x 17", white	1.000	Ea.	2.759	152	115	267
Color	1.000	Ea.	2.500	152	104	256
Vitreous china 20" x 16", white	1.000	Ea.	2.963	244	124	368
Color	1.000	Ea.	2.963	244	124	368
Shower, steel enameled, stone base 36" square, white	1.000	Ea.	8.889	880	139	1,019
Color	1.000	Ea.	8.889	895	139	1,034
Rough-in, for lavatory and shower						
1/2" copper supply, 1-1/2" cast iron waste, 1-1/2" cast iron vent	1.000	Ea.	7.668	177	335	512
1-1/2" PVC waste, 1-1/4" PVC vent	1.000	Ea.	7.352	93	340	433
1/2" steel supply, 1-1/4" cast iron waste, 1-1/4" steel vent	1.000	Ea.	8.205	220	360	580
1-1/4" PVC waste, 1-1/4" PVC vent	1.000	Ea.	7.873	128	365	493
1/2" PVC supply, 1-1/2" PVC waste, 1-1/2" PVC vent	1.000	Ea.	9.185	120	425	545
Bathtub, P.E. cast iron, 5' long with fittings, white	1.000	Ea.	3.636	805	152	957
Color	1.000	Ea.	3.636	995	152	1,147
Steel enameled, 5' long with fittings, white	1.000	Ea.	2.909	330	121	451
Color	1.000	Ea.	2.909	330	121	451
Rough-in, for bathtub						
1/2" copper supply, 4" cast iron waste, 1-1/2" copper vent	1.000	Ea.	2.409	74	108	182
4" PVC waste, 1-1/2" PVC vent	1.000	Ea.	2.877	38.50	129	167.50
1/2" steel supply, 4" cast iron waste, 1-1/2" steel vent	1.000	Ea.	2.898	92	127	219
4" PVC waste, 1-1/2" PVC vent	1.000	Ea.	3.159	49.50	142	191.50
1/2" PVC supply, 4" PVC waste, 1-1/2" PVC vent	1.000	Ea.	3.371	46.50	151	197.50
Piping supply, 1/2" copper	42.000	L.F.	4.148	82	192	274
1/2" steel	42.000	L.F.	5.333	129	248	377
1/2" PVC	42.000	L.F.	6.222	117	288	405
Piping, waste, 4" cast iron, no hub	10.000	L.F.	3.586	131	150	281
4" PVC/DWV	10.000	L.F.	4.333	55	181	236
4" copper/DWV	10.000	L.F.	5.200	330	217	547
Piping, vent, 2" cast iron, no hub	13.000	L.F.	3.105	77.50	129	206.50
2" copper/DWV	13.000	L.F.	2.364	106	110	216
2" PVC/DWV	13.000	L.F.	3.525	22.50	147	169.50
2" steel, galvanized	13.000	L.F.	3.250	135	136	271
Vanity base cabinet, 2 doors, 30" wide	1.000	Ea.	1.000	232	42.50	274.50
Vanity top, plastic laminated, square edge	3.160	L.F.	.843	77.50	35.50	113
Carrier, steel, for studs, no arms	1.000	Ea.	1.143	51	53	104
Wood, 2" x 8" blocking	1.300	L.F.	.052	1.22	2.20	3.42

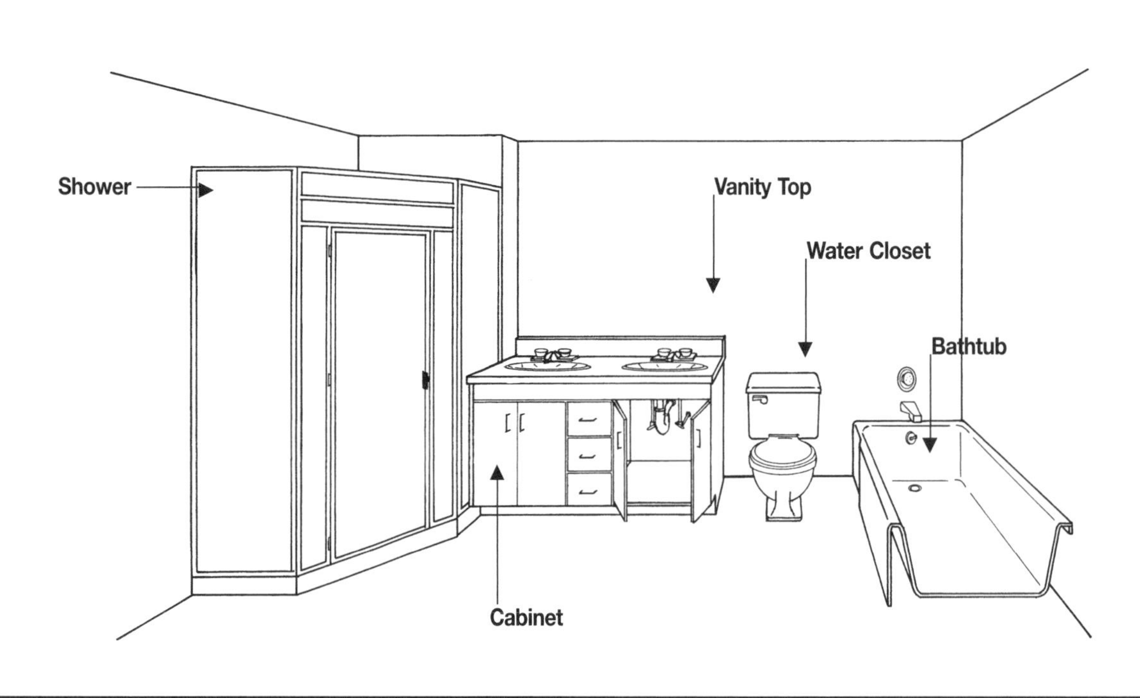

System Description	QUAN.	UNIT	LABOR HOURS	COST EACH		
				MAT.	INST.	TOTAL
BATHROOM WITH SHOWER, BATHTUB, LAVATORIES IN VANITY						
Water closet, floor mounted, 1 piece combination, white	1.000	Ea.	3.019	775	126	901
Rough-in, vent, 2″ diameter DWV piping	1.000	Ea.	.955	23.80	39.80	63.60
Waste, 4″ diameter DWV piping	1.000	Ea.	.828	30.15	34.50	64.65
Supply, 1/2″ diameter type "L" copper supply piping	1.000	Ea.	.593	11.70	27.48	39.18
Lavatory, 20″ x 16″, vitreous china oval, with fittings, white	2.000	Ea.	5.926	488	248	736
Shower, steel enameled, stone base, corner, white	1.000	Ea.	3.333	880	139	1,019
Rough-in, waste, 1-1/2″ diameter DWV piping	3.000	Ea.	5.408	138	225.60	363.60
Supply, 1/2″ diameter type "L" copper supply piping	3.000	Ea.	2.963	58.50	137.40	195.90
Bathtub, P.E. cast iron, 5′ long with fittings, white	1.000	Ea.	3.636	805	152	957
Rough-in, waste, 4″ diameter DWV piping	1.000	Ea.	1.103	40.20	46	86.20
Supply, 1/2″ diameter type "L" copper supply piping	1.000	Ea.	.988	19.50	45.80	65.30
Vent, 1-1/2″ diameter copper DWV piping	1.000	Ea.	.593	24.40	27.40	51.80
Piping, supply, 1/2″ diameter type "L" copper supply piping	42.000	L.F.	4.148	81.90	192.36	274.26
Waste, 4″ diameter DWV piping	10.000	L.F.	2.759	100.50	115	215.50
Vent, 2″ diameter DWV piping	13.000	L.F.	3.250	135.20	135.85	271.05
Vanity base, 2 door, 24″ x 48″	1.000	Ea.	1.400	370	59	429
Vanity top, plastic laminated, square edge	4.170	L.F.	1.112	102.17	46.91	149.08
TOTAL		Ea.	42.014	4,084.02	1,798.10	5,882.12

The costs in this system are on a cost each basis. All necessary piping is included.

Description	QUAN.	UNIT	LABOR HOURS	COST EACH		
				MAT.	INST.	TOTAL

 Important: See the Reference Section for critical supporting data - Reference Nos., Crews & Location Factors

Five Fixture Bathroom Price Sheet	QUAN.	UNIT	LABOR HOURS	COST EACH		
				MAT.	INST.	TOTAL
Water closet, close coupled, standard 2 piece, white	1.000	Ea.	3.019	186	126	312
Color	1.000	Ea.	3.019	223	126	349
One piece elongated bowl, white	1.000	Ea.	3.019	545	126	671
Color	1.000	Ea.	3.019	680	126	806
Low profile, one piece elongated bowl, white	1.000	Ea.	3.019	775	126	901
Color	1.000	Ea.	3.019	1,025	126	1,151
Rough-in, supply, waste and vent for water closet						
1/2″ copper supply, 4″ cast iron waste, 2″ cast iron vent	1.000	Ea.	2.376	65.50	102	167.50
4″ P.V.C./DWV waste, 2″ P.V.C. vent	1.000	Ea.	2.678	31.50	114	145.50
4″ copper waste, 2″ copper vent	1.000	Ea.	2.520	121	111	232
3″ cast iron waste, 1-1/2″ cast iron vent	1.000	Ea.	2.244	58.50	96.50	155
3″ P.V.C. waste, 1-1/2″ P.V.C. vent	1.000	Ea.	2.388	27.50	106	133.50
3″ copper waste, 1-1/2″ copper vent	1.000	Ea.	2.014	79.50	89.50	169
1/2″ P.V.C. supply, 4″ P.V.C. waste, 2″ P.V.C. vent	1.000	Ea.	2.974	36.50	128	164.50
3″ P.V.C. waste, 1-1/2″ P.V.C. supply	1.000	Ea.	2.684	32.50	120	152.50
1/2″ steel supply, 4″ cast iron waste, 2″ cast iron vent	1.000	Ea.	2.545	72.50	110	182.50
4″ cast iron waste, 2″ steel vent	1.000	Ea.	2.590	90	112	202
4″ P.V.C. waste, 2″ P.V.C. vent	1.000	Ea.	2.847	38	122	160
Lavatory, wall hung, P.E. cast iron 20″ x 18″, white	2.000	Ea.	4.000	530	167	697
Color	2.000	Ea.	4.000	610	167	777
Vitreous china, 19″ x 17″, white	2.000	Ea.	4.571	375	191	566
Color	2.000	Ea.	4.571	375	191	566
Lavatory, for vanity top, P.E. cast iron, 20″ x 18″, white	2.000	Ea.	5.000	470	208	678
Color	2.000	Ea.	5.000	530	208	738
Steel enameled 20″ x 17″, white	2.000	Ea.	5.517	305	230	535
Color	2.000	Ea.	5.000	305	208	513
Vitreous china 20″ x 16″, white	2.000	Ea.	5.926	490	248	738
Color	2.000	Ea.	5.926	490	248	738
Shower, steel enameled, stone base 36″ square, white	1.000	Ea.	8.889	880	139	1,019
Color	1.000	Ea.	8.889	895	139	1,034
Rough-in, for lavatory or shower						
1/2″ copper supply, 1-1/2″ cast iron waste, 1-1/2″ cast iron vent	3.000	Ea.	8.371	197	365	562
1-1/2″ P.V.C. waste, 1-1/4″ P.V.C. vent	3.000	Ea.	7.916	95	365	460
1/2″ steel supply, 1-1/4″ cast iron waste, 1-1/4″ steel vent	3.000	Ea.	8.670	240	380	620
1-1/4″ P.V.C. waste, 1-1/4″ P.V.C. vent	3.000	Ea.	8.381	128	390	518
1/2″ P.V.C. supply, 1-1/2″ P.V.C. waste, 1-1/2″ P.V.C. vent	3.000	Ea.	9.778	120	455	575
Bathtub, P.E. cast iron 5′ long with fittings, white	1.000	Ea.	3.636	805	152	957
Color	1.000	Ea.	3.636	995	152	1,147
Steel, enameled 5′ long with fittings, white	1.000	Ea.	2.909	330	121	451
Color	1.000	Ea.	2.909	330	121	451
Rough-in, for bathtub						
1/2″ copper supply, 4″ cast iron waste, 1-1/2″ copper vent	1.000	Ea.	2.684	84	119	203
4″ P.V.C. waste, 1-1/2″ P.V.C. vent	1.000	Ea.	3.210	42.50	143	185.50
1/2″ steel supply, 4″ cast iron waste, 1-1/2″ steel vent	1.000	Ea.	3.173	102	138	240
4″ P.V.C. waste, 1-1/2″ P.V.C. vent	1.000	Ea.	3.492	53.50	156	209.50
1/2″ P.V.C. supply, 4″ P.V.C. waste, 1-1/2″ P.V.C. vent	1.000	Ea.	3.704	51	165	216
Piping, supply, 1/2″ copper	42.000	L.F.	4.148	82	192	274
1/2″ steel	42.000	L.F.	5.333	129	248	377
1/2″ P.V.C.	42.000	L.F.	6.222	117	288	405
Piping, waste, 4″ cast iron, no hub	10.000	L.F.	2.759	101	115	216
4″ P.V.C./DWV	10.000	L.F.	3.333	42	139	181
4″ copper/DWV	10.000	L.F.	4.000	255	167	422
Piping, vent, 2″ cast iron, no hub	13.000	L.F.	3.105	77.50	129	206.50
2″ copper/DWV	13.000	L.F.	2.364	106	110	216
2″ P.V.C./DWV	13.000	L.F.	3.525	22.50	147	169.50
2″ steel, galvanized	13.000	L.F.	3.250	135	136	271
Vanity base cabinet, 2 doors, 24″ x 48″	1.000	Ea.	1.400	370	59	429
Vanity top, plastic laminated, square edge	4.170	L.F.	1.112	102	47	149
Carrier, steel, for studs, no arms	1.000	Ea.	1.143	51	53	104
Wood, 2″ x 8″ blocking	1.300	L.F.	.052	1.22	2.20	3.42

60 Gas Heating/Cooling Systems

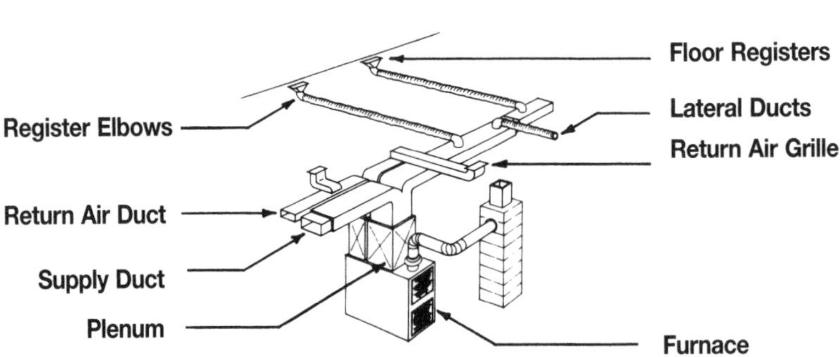

System Description	QUAN.	UNIT	LABOR HOURS	COST PER SYSTEM		
				MAT.	INST.	TOTAL
HEATING ONLY, GAS FIRED HOT AIR, ONE ZONE, 1200 S.F. BUILDING						
Furnace, gas, up flow	1.000	Ea.	5.000	745	210	955
Intermittent pilot	1.000	Ea.		151		151
Supply duct, rigid fiberglass	176.000	S.F.	12.068	119.68	526.24	645.92
Return duct, sheet metal, galvanized	158.000	Lb.	16.137	165.90	704.68	870.58
Lateral ducts, 6″ flexible fiberglass	144.000	L.F.	8.862	401.76	372.96	774.72
Register, elbows	12.000	Ea.	3.200	456	134.40	590.40
Floor registers, enameled steel	12.000	Ea.	3.000	246	140.40	386.40
Floor grille, return air	2.000	Ea.	.727	56	34	90
Thermostat	1.000	Ea.	1.000	29.50	47	76.50
Plenum	1.000	Ea.	1.000	77	42	119
TOTAL		System	50.994	2,447.84	2,211.68	4,659.52
HEATING/COOLING, GAS FIRED FORCED AIR, ONE ZONE, 1200 S.F. BUILDING						
Furnace, including plenum, compressor, coil	1.000	Ea.	14.720	4,140	621	4,761
Intermittent pilot	1.000	Ea.		151		151
Supply duct, rigid fiberglass	176.000	S.F.	12.068	119.68	526.24	645.92
Return duct, sheet metal, galvanized	158.000	Lb.	16.137	165.90	704.68	870.58
Lateral duct, 6″ flexible fiberglass	144.000	L.F.	8.862	401.76	372.96	774.72
Register elbows	12.000	Ea.	3.200	456	134.40	590.40
Floor registers, enameled steel	12.000	Ea.	3.000	246	140.40	386.40
Floor grille return air	2.000	Ea.	.727	56	34	90
Thermostat	1.000	Ea.	1.000	29.50	47	76.50
Refrigeration piping, 25 ft. (pre-charged)	1.000	Ea.		213		213
TOTAL		System	59.714	5,978.84	2,580.68	8,559.52

The costs in these systems are based on complete system basis. For larger buildings use the price sheet on the opposite page.

Description	QUAN.	UNIT	LABOR HOURS	COST PER SYSTEM		
				MAT.	INST.	TOTAL

Important: See the Reference Section for critical supporting data - Reference Nos., Crews & Location Factors

Gas Heating/Cooling Price Sheet	QUAN.	UNIT	LABOR HOURS	COST EACH		
				MAT.	INST.	TOTAL
Furnace, heating only, 100 MBH, area to 1200 S.F.	1.000	Ea.	5.000	745	210	955
120 MBH, area to 1500 S.F.	1.000	Ea.	5.000	745	210	955
160 MBH, area to 2000 S.F.	1.000	Ea.	5.714	1,000	240	1,240
200 MBH, area to 2400 S.F.	1.000	Ea.	6.154	2,300	259	2,559
Heating/cooling, 100 MBH heat, 36 MBH cool, to 1200 S.F.	1.000	Ea.	16.000	4,500	675	5,175
120 MBH heat, 42 MBH cool, to 1500 S.F.	1.000	Ea.	18.462	4,800	805	5,605
144 MBH heat, 47 MBH cool, to 2000 S.F.	1.000	Ea.	20.000	5,550	875	6,425
200 MBH heat, 60 MBH cool, to 2400 S.F.	1.000	Ea.	34.286	5,850	1,500	7,350
Intermittent pilot, 100 MBH furnace	1.000	Ea.		151		151
200 MBH furnace	1.000	Ea.		151		151
Supply duct, rectangular, area to 1200 S.F., rigid fiberglass	176.000	S.F.	12.068	120	525	645
Sheet metal insulated	228.000	Lb.	31.331	350	1,325	1,675
Area to 1500 S.F., rigid fiberglass	176.000	S.F.	12.068	120	525	645
Sheet metal insulated	228.000	Lb.	31.331	350	1,325	1,675
Area to 2400 S.F., rigid fiberglass	205.000	S.F.	14.057	139	615	754
Sheet metal insulated	271.000	Lb.	37.048	410	1,575	1,985
Round flexible, insulated 6″ diameter, to 1200 S.F.	156.000	L.F.	9.600	435	405	840
To 1500 S.F.	184.000	L.F.	11.323	515	475	990
8″ diameter, to 2000 S.F.	269.000	L.F.	23.911	935	1,000	1,935
To 2400 S.F.	248.000	L.F.	22.045	860	930	1,790
Return duct, sheet metal galvanized, to 1500 S.F.	158.000	Lb.	16.137	166	705	871
To 2400 S.F.	191.000	Lb.	19.507	201	850	1,051
Lateral ducts, flexible round 6″ insulated, to 1200 S.F.	144.000	L.F.	8.862	400	375	775
To 1500 S.F.	172.000	L.F.	10.585	480	445	925
To 2000 S.F.	261.000	L.F.	16.062	730	675	1,405
To 2400 S.F.	300.000	L.F.	18.462	835	775	1,610
Spiral steel insulated, to 1200 S.F.	144.000	L.F.	20.067	440	815	1,255
To 1500 S.F.	172.000	L.F.	23.952	525	975	1,500
To 2000 S.F.	261.000	L.F.	36.352	795	1,475	2,270
To 2400 S.F.	300.000	L.F.	41.825	915	1,700	2,615
Rectangular sheet metal galvanized insulated, to 1200 S.F.	228.000	Lb.	39.056	455	1,650	2,105
To 1500 S.F.	344.000	Lb.	53.966	615	2,275	2,890
To 2000 S.F.	522.000	Lb.	81.926	935	3,475	4,410
To 2400 S.F.	600.000	Lb.	94.189	1,075	3,975	5,050
Register elbows, to 1500 S.F.	12.000	Ea.	3.200	455	134	589
To 2400 S.F.	14.000	Ea.	3.733	530	157	687
Floor registers, enameled steel w/damper, to 1500 S.F.	12.000	Ea.	3.000	246	140	386
To 2400 S.F.	14.000	Ea.	4.308	335	202	537
Return air grille, area to 1500 S.F. 12″ x 12″	2.000	Ea.	.727	56	34	90
Area to 2400 S.F. 8″ x 16″	2.000	Ea.	.444	49.50	21	70.50
Area to 2400 S.F. 8″ x 16″	2.000	Ea.	.727	56	34	90
16″ x 16″	1.000	Ea.	.364	40	17	57
Thermostat, manual, 1 set back	1.000	Ea.	1.000	29.50	47	76.50
Electric, timed, 1 set back	1.000	Ea.	1.000	89	47	136
2 set back	1.000	Ea.	1.000	196	47	243
Plenum, heating only, 100 M.B.H.	1.000	Ea.	1.000	77	42	119
120 MBH	1.000	Ea.	1.000	77	42	119
160 MBH	1.000	Ea.	1.000	77	42	119
200 MBH	1.000	Ea.	1.000	77	42	119
Refrigeration piping, 3/8″	25.000	L.F.		21.50		21.50
3/4″	25.000	L.F.		43.50		43.50
7/8″	25.000	L.F.		50.50		50.50
Refrigerant piping, 25 ft. (precharged)	1.000	Ea.		213		213
Diffusers, ceiling, 6″ diameter, to 1500 S.F.	10.000	Ea.	4.444	197	210	407
To 2400 S.F.	12.000	Ea.	6.000	258	282	540
Floor, aluminum, adjustable, 2-1/4″ x 12″ to 1500 S.F.	12.000	Ea.	3.000	179	140	319
To 2400 S.F.	14.000	Ea.	3.500	209	164	373
Side wall, aluminum, adjustable, 8″ x 4″, to 1500 S.F.	12.000	Ea.	3.000	400	140	540
5″ x 10″ to 2400 S.F.	12.000	Ea.	3.692	520	173	693

64 Oil Fired Heating/Cooling Systems

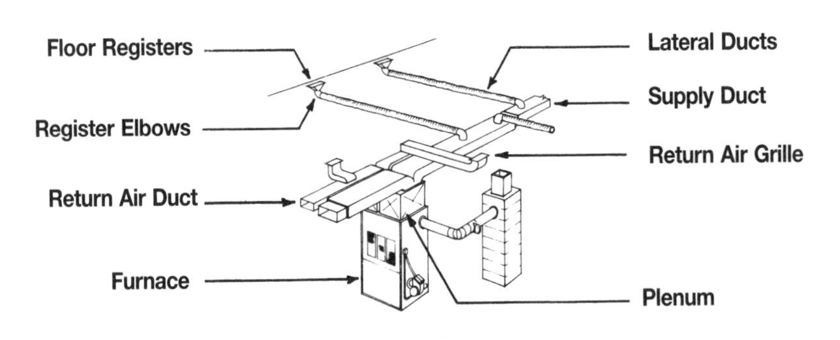

System Description	QUAN.	UNIT	LABOR HOURS	COST PER SYSTEM		
				MAT.	INST.	TOTAL
HEATING ONLY, OIL FIRED HOT AIR, ONE ZONE, 1200 S.F. BUILDING						
Furnace, oil fired, atomizing gun type burner	1.000	Ea.	4.571	850	192	1,042
3/8″ diameter copper supply pipe	1.000	Ea.	2.759	41.10	128.10	169.20
Shut off valve	1.000	Ea.	.333	8.90	15.45	24.35
Oil tank, 275 gallon, on legs	1.000	Ea.	3.200	355	135	490
Supply duct, rigid fiberglass	176.000	S.F.	12.068	119.68	526.24	645.92
Return duct, sheet metal, galvanized	158.000	Lb.	16.137	165.90	704.68	870.58
Lateral ducts, 6″ flexible fiberglass	144.000	L.F.	8.862	401.76	372.96	774.72
Register elbows	12.000	Ea.	3.200	456	134.40	590.40
Floor register, enameled steel	12.000	Ea.	3.000	246	140.40	386.40
Floor grille, return air	2.000	Ea.	.727	56	34	90
Thermostat	1.000	Ea.	1.000	29.50	47	76.50
TOTAL		System	55.857	2,729.84	2,430.23	5,160.07
HEATING/COOLING, OIL FIRED, FORCED AIR, ONE ZONE, 1200 S.F. BUILDING						
Furnace, including plenum, compressor, coil	1.000	Ea.	16.000	4,800	675	5,475
3/8″ diameter copper supply pipe	1.000	Ea.	2.759	41.10	128.10	169.20
Shut off valve	1.000	Ea.	.333	8.90	15.45	24.35
Oil tank, 275 gallon on legs	1.000	Ea.	3.200	355	135	490
Supply duct, rigid fiberglass	176.000	S.F.	12.068	119.68	526.24	645.92
Return duct, sheet metal, galvanized	158.000	Lb.	16.137	165.90	704.68	870.58
Lateral ducts, 6″ flexible fiberglass	144.000	L.F.	8.862	401.76	372.96	774.72
Register elbows	12.000	Ea.	3.200	456	134.40	590.40
Floor registers, enameled steel	12.000	Ea.	3.000	246	140.40	386.40
Floor grille, return air	2.000	Ea.	.727	56	34	90
Refrigeration piping (precharged)	25.000	L.F.		213		213
TOTAL		System	66.286	6,863.34	2,866.23	9,729.57

Description	QUAN.	UNIT	LABOR HOURS	COST EACH		
				MAT.	INST.	TOTAL

Important: See the Reference Section for critical supporting data - Reference Nos., Crews & Location Factors

Oil Fired Heating/Cooling	QUAN.	UNIT	LABOR HOURS	COST EACH		
				MAT.	INST.	TOTAL
Furnace, heating, 95.2 MBH,area to 1200 S.F.	1.000	Ea.	4.706	870	198	1,068
123.2 MBH, area to 1500 S.F.	1.000	Ea.	5.000	1,200	210	1,410
151.2 MBH, area to 2000 S.F.	1.000	Ea.	5.333	1,325	224	1,549
200 MBH, area to 2400 S.F.	1.000	Ea.	6.154	2,450	259	2,709
Heating/cooling, 95.2 MBH heat, 36 MBH cool, to 1200 S.F.	1.000	Ea.	16.000	4,800	675	5,475
112 MBH heat, 42 MBH cool, to 1500 S.F.	1.000	Ea.	24.000	7,200	1,025	8,225
151 MBH heat, 47 MBH cool, to 2000 S.F.	1.000	Ea.	20.800	6,250	880	7,130
184.8 MBH heat, 60 MBH cool, to 2400 S.F.	1.000	Ea.	24.000	6,600	1,050	7,650
Oil piping to furnace, 3/8″ dia., copper	1.000	Ea.	3.412	146	157	303
Oil tank, on legs above ground, 275 gallons	1.000	Ea.	3.200	355	135	490
550 gallons	1.000	Ea.	5.926	1,750	249	1,999
Below ground, 275 gallons	1.000	Ea.	3.200	355	135	490
550 gallons	1.000	Ea.	5.926	1,750	249	1,999
1000 gallons	1.000	Ea.	6.400	2,750	269	3,019
Supply duct, rectangular, area to 1200 S.F., rigid fiberglass	176.000	S.F.	12.068	120	525	645
Sheet metal, insulated	228.000	Lb.	31.331	350	1,325	1,675
Area to 1500 S.F., rigid fiberglass	176.000	S.F.	12.068	120	525	645
Sheet metal, insulated	228.000	Lb.	31.331	350	1,325	1,675
Area to 2400 S.F., rigid fiberglass	205.000	S.F.	14.057	139	615	754
Sheet metal, insulated	271.000	Lb.	37.048	410	1,575	1,985
Round flexible, insulated, 6″ diameter to 1200 S.F.	156.000	L.F.	9.600	435	405	840
To 1500 S.F.	184.000	L.F.	11.323	515	475	990
8″ diameter to 2000 S.F.	269.000	L.F.	23.911	935	1,000	1,935
To 2400 S.F.	269.000	L.F.	22.045	860	930	1,790
Return duct, sheet metal galvanized, to 1500 S.F.	158.000	Lb.	16.137	166	705	871
To 2400 S.F.	191.000	Lb.	19.507	201	850	1,051
Lateral ducts, flexible round, 6″, insulated to 1200 S.F.	144.000	L.F.	8.862	400	375	775
To 1500 S.F.	172.000	L.F.	10.585	480	445	925
To 2000 S.F.	261.000	L.F.	16.062	730	675	1,405
To 2400 S.F.	300.000	L.F.	18.462	835	775	1,610
Spiral steel, insulated to 1200 S.F.	144.000	L.F.	20.067	440	815	1,255
To 1500 S.F.	172.000	L.F.	23.952	525	975	1,500
To 2000 S.F.	261.000	L.F.	36.352	795	1,475	2,270
To 2400 S.F.	300.000	L.F.	41.825	915	1,700	2,615
Rectangular sheet metal galvanized insulated, to 1200 S.F.	288.000	Lb.	45.183	515	1,900	2,415
To 1500 S.F.	344.000	Lb.	53.966	615	2,275	2,890
To 2000 S.F.	522.000	Lb.	81.926	935	3,475	4,410
To 2400 S.F.	600.000	Lb.	94.189	1,075	3,975	5,050
Register elbows, to 1500 S.F.	12.000	Ea.	3.200	455	134	589
To 2400 S.F.	14.000	Ea.	3.733	530	157	687
Floor registers, enameled steel w/damper, to 1500 S.F.	12.000	Ea.	3.000	246	140	386
To 2400 S.F.	14.000	Ea.	4.308	335	202	537
Return air grille, area to 1500 S.F., 12″ x 12″	2.000	Ea.	.727	56	34	90
12″ x 24″	1.000	Ea.	.444	49.50	21	70.50
Area to 2400 S.F., 8″ x 16″	2.000	Ea.	.727	56	34	90
16″ x 16″	1.000	Ea.	.364	40	17	57
Thermostat, manual, 1 set back	1.000	Ea.	1.000	29.50	47	76.50
Electric, timed, 1 set back	1.000	Ea.	1.000	89	47	136
2 set back	1.000	Ea.	1.000	196	47	243
Refrigeration piping, 3/8″	25.000	L.F.		21.50		21.50
3/4″	25.000	L.F.		43.50		43.50
Diffusers, ceiling, 6″ diameter, to 1500 S.F.	10.000	Ea.	4.444	197	210	407
To 2400 S.F.	12.000	Ea.	6.000	258	282	540
Floor, aluminum, adjustable, 2-1/4″ x 12″ to 1500 S.F.	12.000	Ea.	3.000	179	140	319
To 2400 S.F.	14.000	Ea.	3.500	209	164	373
Side wall, aluminum, adjustable, 8″ x 4″, to 1500 S.F.	12.000	Ea.	3.000	400	140	540
5″ x 10″ to 2400 S.F.	12.000	Ea.	3.692	520	173	693

Hot Water Heating Systems

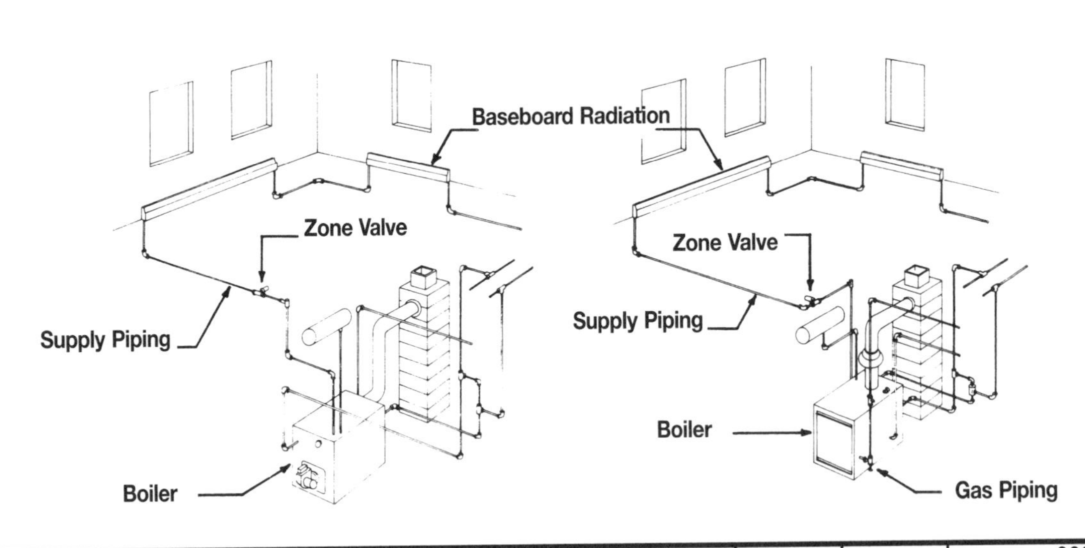

System Description	QUAN.	UNIT	LABOR HOURS	COST EACH MAT.	COST EACH INST.	COST EACH TOTAL
OIL FIRED HOT WATER HEATING SYSTEM, AREA TO 1200 S.F.						
Boiler package, oil fired, 97 MBH, area to 1200 S.F. building	1.000	Ea.	15.000	1,525	605	2,130
3/8″ diameter copper supply pipe	1.000	Ea.	2.759	41.10	128.10	169.20
Shut off valve	1.000	Ea.	.333	8.90	15.45	24.35
Oil tank, 275 gallon, with black iron filler pipe	1.000	Ea.	3.200	355	135	490
Supply piping, 3/4″ copper tubing	176.000	L.F.	18.526	503.36	858.88	1,362.24
Supply fittings, copper 3/4″	36.000	Ea.	15.158	45	703.80	748.80
Supply valves, 3/4″	2.000	Ea.	.800	125	37.10	162.10
Baseboard radiation, 3/4″	106.000	L.F.	35.333	433.54	1,484	1,917.54
Zone valve	1.000	Ea.	.400	149	18.70	167.70
TOTAL		Ea.	91.509	3,185.90	3,986.03	7,171.93
OIL FIRED HOT WATER HEATING SYSTEM, AREA TO 2400 S.F.						
Boiler package, oil fired, 225 MBH, area to 2400 S.F. building	1.000	Ea.	19.704	3,725	795	4,520
3/8″ diameter copper supply pipe	1.000	Ea.	2.759	41.10	128.10	169.20
Shut off valve	1.000	Ea.	.333	8.90	15.45	24.35
Oil tank, 550 gallon, with black iron pipe filler pipe	1.000	Ea.	5.926	1,750	249	1,999
Supply piping, 3/4″ copper tubing	228.000	L.F.	23.999	652.08	1,112.64	1,764.72
Supply fittings, copper	46.000	Ea.	19.368	57.50	899.30	956.80
Supply valves	2.000	Ea.	.800	125	37.10	162.10
Baseboard radiation	212.000	L.F.	70.666	867.08	2,968	3,835.08
Zone valve	1.000	Ea.	.400	149	18.70	167.70
TOTAL		Ea.	143.955	7,375.66	6,223.29	13,598.95

The costs in this system are on a cost each basis. The costs represent total cost for the system based on a gross square foot of plan area.

Description	QUAN.	UNIT	LABOR HOURS	COST EACH MAT.	COST EACH INST.	COST EACH TOTAL

Important: See the Reference Section for critical supporting data - Reference Nos., Crews & Location Factors

Hot Water Heating Price Sheet

Hot Water Heating Price Sheet	QUAN.	UNIT	LABOR HOURS	COST EACH		
				MAT.	INST.	TOTAL
Boiler, oil fired, 97 MBH, area to 1200 S.F.	1.000	Ea.	15.000	1,525	605	2,130
118 MBH, area to 1500 S.F.	1.000	Ea.	16.506	2,875	670	3,545
161 MBH, area to 2000 S.F.	1.000	Ea.	18.405	3,625	745	4,370
215 MBH, area to 2400 S.F.	1.000	Ea.	19.704	3,725	795	4,520
Oil piping, (valve & filter), 3/8″ copper	1.000	Ea.	3.289	87.50	152	239.50
1/4″ copper	1.000	Ea.	3.242	65.50	150	215.50
Oil tank, filler pipe and cap on legs, 275 gallon	1.000	Ea.	3.200	355	135	490
550 gallon	1.000	Ea.	5.926	1,750	249	1,999
Buried underground, 275 gallon	1.000	Ea.	3.200	355	135	490
550 gallon	1.000	Ea.	5.926	1,750	249	1,999
1000 gallon	1.000	Ea.	6.400	2,750	269	3,019
Supply piping copper, area to 1200 S.F., 1/2″ tubing	176.000	L.F.	17.384	345	805	1,150
3/4″ tubing	176.000	L.F.	18.526	505	860	1,365
Area to 1500 S.F., 1/2″ tubing	186.000	L.F.	18.371	365	850	1,215
3/4″ tubing	186.000	L.F.	19.578	530	910	1,440
Area to 2000 S.F., 1/2″ tubing	204.000	L.F.	20.149	400	935	1,335
3/4″ tubing	204.000	L.F.	21.473	585	995	1,580
Area to 2400 S.F., 1/2″ tubing	228.000	L.F.	22.520	445	1,050	1,495
3/4″ tubing	228.000	L.F.	23.999	650	1,125	1,775
Supply pipe fittings copper, area to 1200 S.F., 1/2″	36.000	Ea.	14.400	20	670	690
3/4″	36.000	Ea.	15.158	45	705	750
Area to 1500 S.F., 1/2″	40.000	Ea.	16.000	22.50	740	762.50
3/4″	40.000	Ea.	16.842	50	780	830
Area to 2000 S.F., 1/2″	44.000	Ea.	17.600	24.50	815	839.50
3/4″	44.000	Ea.	18.526	55	860	915
Area to 2400, S.F., 1/2″	46.000	Ea.	18.400	26	855	881
3/4″	46.000	Ea.	19.368	57.50	900	957.50
Supply valves, 1/2″ pipe size	2.000	Ea.	.667	92	31	123
3/4″	2.000	Ea.	.800	125	37	162
Baseboard radiation, area to 1200 S.F., 1/2″ tubing	106.000	L.F.	28.267	725	1,175	1,900
3/4″ tubing	106.000	L.F.	35.333	435	1,475	1,910
Area to 1500 S.F., 1/2″ tubing	134.000	L.F.	35.734	920	1,500	2,420
3/4″ tubing	134.000	L.F.	44.666	550	1,875	2,425
Area to 2000 S.F., 1/2″ tubing	178.000	L.F.	47.467	1,225	2,000	3,225
3/4″ tubing	178.000	L.F.	59.333	730	2,500	3,230
Area to 2400 S.F., 1/2″ tubing	212.000	L.F.	56.534	1,450	2,375	3,825
3/4″ tubing	212.000	L.F.	70.666	865	2,975	3,840
Zone valves, 1/2″ tubing	1.000	Ea.	.400	149	18.70	167.70
3/4″ tubing	1.000	Ea.	.400	160	18.70	178.70

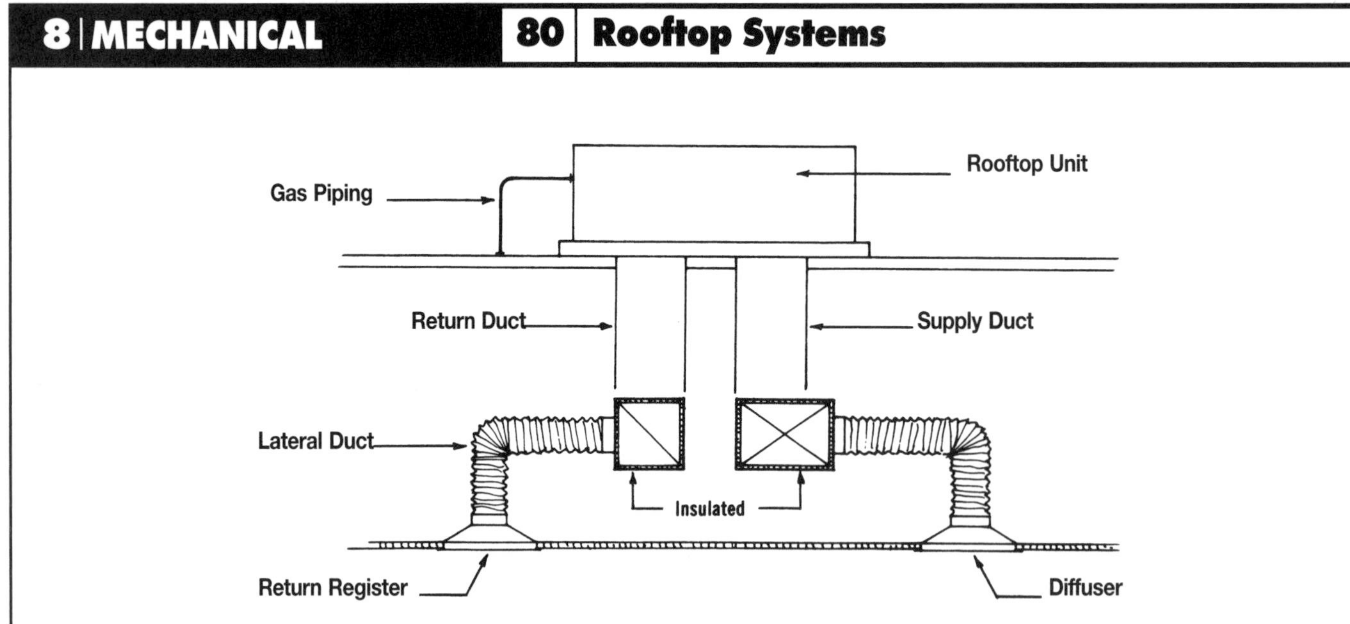

System Description	QUAN.	UNIT	LABOR HOURS	COST EACH MAT.	COST EACH INST.	COST EACH TOTAL
ROOFTOP HEATING/COOLING UNIT, AREA TO 2000 S.F.						
Rooftop unit, single zone, electric cool, gas heat, to 2000 s.f.	1.000	Ea.	28.521	5,125	1,200	6,325
Gas piping	34.500	L.F.	5.207	131.79	241.50	373.29
Duct, supply and return, galvanized steel	38.000	Lb.	3.881	39.90	169.48	209.38
Insulation, ductwork	33.000	S.F.	1.508	20.46	60.06	80.52
Lateral duct, flexible duct 12″ diameter, insulated	72.000	L.F.	11.520	367.20	486	853.20
Diffusers	4.000	Ea.	4.571	1,320	214	1,534
Return registers	1.000	Ea.	.727	132	34	166
TOTAL		Ea.	55.935	7,136.35	2,405.04	9,541.39
ROOFTOP HEATING/COOLING UNIT, AREA TO 5000 S.F.						
Rooftop unit, single zone, electric cool, gas heat, to 5000 s.f.	1.000	Ea.	42.032	15,100	1,700	16,800
Gas piping	86.250	L.F.	13.019	329.48	603.75	933.23
Duct supply and return, galvanized steel	95.000	Lb.	9.702	99.75	423.70	523.45
Insulation, ductwork	82.000	S.F.	3.748	50.84	149.24	200.08
Lateral duct, flexible duct, 12″ diameter, insulated	180.000	L.F.	28.800	918	1,215	2,133
Diffusers	10.000	Ea.	11.429	3,300	535	3,835
Return registers	3.000	Ea.	2.182	396	102	498
TOTAL		Ea.	110.912	20,194.07	4,728.69	24,922.76

Description	QUAN.	UNIT	LABOR HOURS	COST EACH MAT.	COST EACH INST.	COST EACH TOTAL

Important: See the Reference Section for critical supporting data - Reference Nos., Crews & Location Factors

Rooftop Price Sheet	QUAN.	UNIT	LABOR HOURS	COST EACH		
				MAT.	INST.	TOTAL
Rooftop unit, single zone, electric cool, gas heat to 2000 S.F.	1.000	Ea.	28.521	5,125	1,200	6,325
Area to 3000 S.F.	1.000	Ea.	35.982	9,600	1,450	11,050
Area to 5000 S.F.	1.000	Ea.	42.032	15,100	1,700	16,800
Area to 10000 S.F.	1.000	Ea.	68.376	31,100	2,875	33,975
Gas piping, area 2000 through 4000 S.F.	34.500	L.F.	5.207	132	242	374
Area 5000 to 10000 S.F.	86.250	L.F.	13.019	330	605	935
Duct, supply and return, galvanized steel, to 2000 S.F.	38.000	Lb.	3.881	40	169	209
Area to 3000 S.F.	57.000	Lb.	5.821	60	254	314
Area to 5000 S.F.	95.000	Lb.	9.702	100	425	525
Area to 10000 S.F.	190.000	Lb.	19.405	200	845	1,045
Rigid fiberglass, area to 2000 S.F.	33.000	S.F.	2.263	22.50	98.50	121
Area to 3000 S.F.	49.000	S.F.	3.360	33.50	147	180.50
Area to 5000 S.F.	82.000	S.F.	5.623	56	245	301
Area to 10000 S.F.	164.000	S.F.	11.245	112	490	602
Insulation, supply and return, blanket type, area to 2000 S.F.	33.000	S.F.	1.508	20.50	60	80.50
Area to 3000 S.F.	49.000	S.F.	2.240	30.50	89	119.50
Area to 5000 S.F.	82.000	S.F.	3.748	51	149	200
Area to 10000 S.F.	164.000	S.F.	7.496	102	298	400
Lateral ducts, flexible round, 12″ insulated, to 2000 S.F.	72.000	L.F.	11.520	365	485	850
Area to 3000 S.F.	108.000	L.F.	17.280	550	730	1,280
Area to 5000 S.F.	180.000	L.F.	28.800	920	1,225	2,145
Area to 10000 S.F.	360.000	L.F.	57.600	1,825	2,425	4,250
Rectangular, galvanized steel, to 2000 S.F.	239.000	Lb.	24.409	251	1,075	1,326
Area to 3000 S.F.	360.000	Lb.	36.767	380	1,600	1,980
Area to 5000 S.F.	599.000	Lb.	61.176	630	2,675	3,305
Area to 10000 S.F.	998.000	Lb.	101.926	1,050	4,450	5,500
Diffusers, ceiling, 1 to 4 way blow, 24″ x 24″, to 2000 S.F.	4.000	Ea.	4.571	1,325	214	1,539
Area to 3000 S.F.	6.000	Ea.	6.857	1,975	320	2,295
Area to 5000 S.F.	10.000	Ea.	11.429	3,300	535	3,835
Area to 10000 S.F.	20.000	Ea.	22.857	6,600	1,075	7,675
Return grilles, 24″ x 24″, to 2000 S.F.	1.000	Ea.	.727	132	34	166
Area to 3000 S.F.	2.000	Ea.	1.455	264	68	332
Area to 5000 S.F.	3.000	Ea.	2.182	395	102	497
Area to 10000 S.F.	5.000	Ea.	3.636	660	170	830

Division 9
Electrical

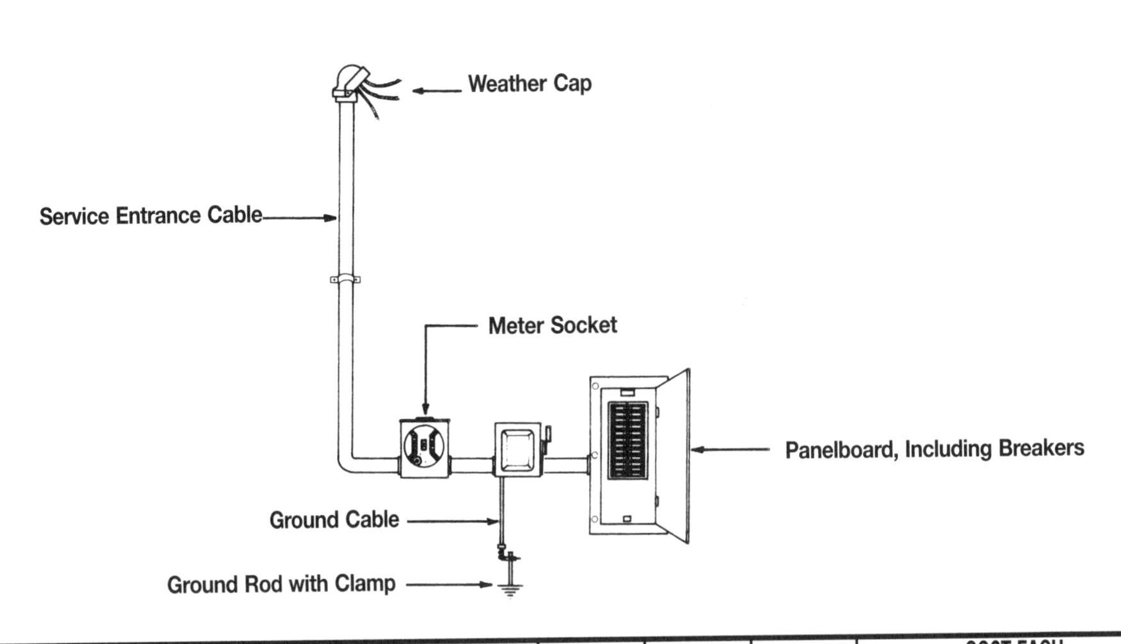

System Description	QUAN.	UNIT	LABOR HOURS	COST EACH MAT.	COST EACH INST.	COST EACH TOTAL
100 AMP SERVICE						
Weather cap	1.000	Ea.	.667	10.45	30.50	40.95
Service entrance cable	10.000	L.F.	.762	32.20	35	67.20
Meter socket	1.000	Ea.	2.500	39.50	115	154.50
Ground rod with clamp	1.000	Ea.	1.455	15.05	67	82.05
Ground cable	5.000	L.F.	.250	8.20	11.50	19.70
Panel board, 12 circuit	1.000	Ea.	6.667	238	305	543
TOTAL		Ea.	12.301	343.40	564	907.40
200 AMP SERVICE						
Weather cap	1.000	Ea.	1.000	32	46	78
Service entrance cable	10.000	L.F.	1.143	69	52.50	121.50
Meter socket	1.000	Ea.	4.211	59	193	252
Ground rod with clamp	1.000	Ea.	1.818	35	83.50	118.50
Ground cable	10.000	L.F.	.500	16.40	23	39.40
3/4″ EMT	5.000	L.F.	.308	5.40	14.10	19.50
Panel board, 24 circuit	1.000	Ea.	12.308	545	470	1,015
TOTAL		Ea.	21.288	761.80	882.10	1,643.90
400 AMP SERVICE						
Weather cap	1.000	Ea.	2.963	380	136	516
Service entrance cable	180.000	L.F.	5.760	327.60	264.60	592.20
Meter socket	1.000	Ea.	4.211	59	193	252
Ground rod with clamp	1.000	Ea.	2.000	95.50	92	187.50
Ground cable	20.000	L.F.	.485	24.40	22.20	46.60
3/4″ greenfield	20.000	L.F.	1.000	11.20	46	57.20
Current transformer cabinet	1.000	Ea.	6.154	168	282	450
Panel board, 42 circuit	1.000	Ea.	33.333	2,725	1,525	4,250
TOTAL		Ea.	55.906	3,790.70	2,560.80	6,351.50

Important: See the Reference Section for critical supporting data - Reference Nos., Crews & Location Factors

20 | Electric Perimeter Heating Systems

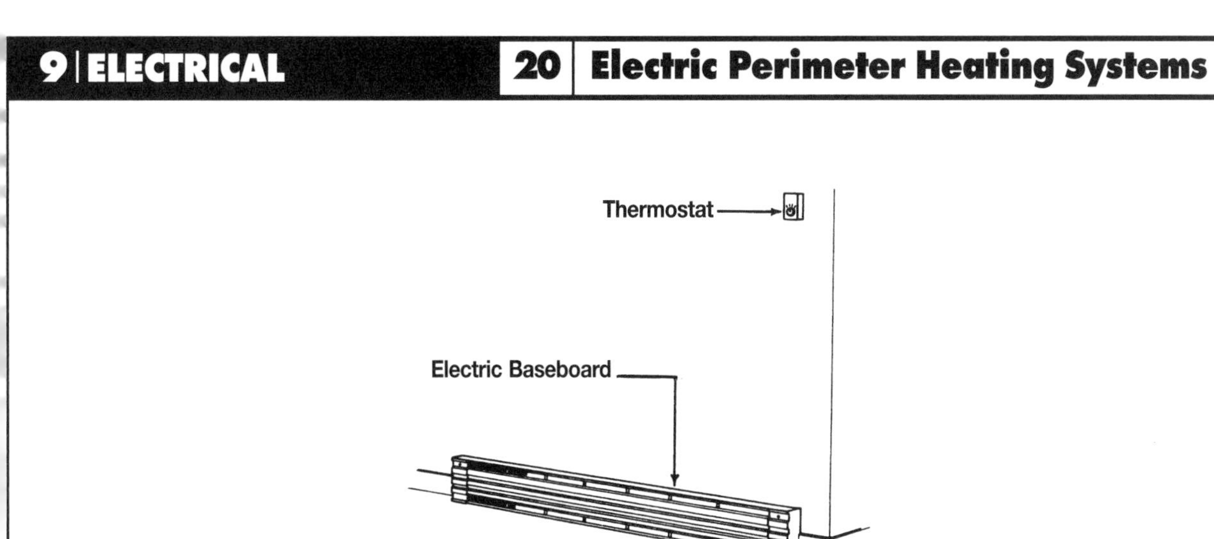

System Description	QUAN.	UNIT	LABOR HOURS	COST EACH		
				MAT.	INST.	TOTAL
4′ BASEBOARD HEATER						
Electric baseboard heater, 4′ long	1.000	Ea.	1.194	51	55	106
Thermostat, integral	1.000	Ea.	.500	19.80	23	42.80
Romex, 12-3 with ground	40.000	L.F.	1.600	21.20	73.60	94.80
Panel board breaker, 20 Amp	1.000	Ea.	.300	9.45	13.80	23.25
TOTAL		Ea.	3.594	101.45	165.40	266.85
6′ BASEBOARD HEATER						
Electric baseboard heater, 6′ long	1.000	Ea.	1.600	67	73.50	140.50
Thermostat, integral	1.000	Ea.	.500	19.80	23	42.80
Romex, 12-3 with ground	40.000	L.F.	1.600	21.20	73.60	94.80
Panel board breaker, 20 Amp	1.000	Ea.	.400	12.60	18.40	31
TOTAL		Ea.	4.100	120.60	188.50	309.10
8′ BASEBOARD HEATER						
Electric baseboard heater, 8′ long	1.000	Ea.	2.000	84.50	92	176.50
Thermostat, integral	1.000	Ea.	.500	19.80	23	42.80
Romex, 12-3 with ground	40.000	L.F.	1.600	21.20	73.60	94.80
Panel board breaker, 20 Amp	1.000	Ea.	.500	15.75	23	38.75
TOTAL		Ea.	4.600	141.25	211.60	352.85
10′ BASEBOARD HEATER						
Electric baseboard heater, 10′ long	1.000	Ea.	2.424	105	111	216
Thermostat, integral	1.000	Ea.	.500	19.80	23	42.80
Romex, 12-3 with ground	40.000	L.F.	1.600	21.20	73.60	94.80
Panel board breaker, 20 Amp	1.000	Ea.	.750	23.63	34.50	58.13
TOTAL		Ea.	5.274	169.63	242.10	411.73

The costs in this system are on a cost each basis and include all necessary conduit fittings.

Description	QUAN.	UNIT	LABOR HOURS	COST EACH		
				MAT.	INST.	TOTAL

System Description	QUAN.	UNIT	LABOR HOURS	COST EACH		
				MAT.	INST.	TOTAL
Air conditioning receptacles						
Using non-metallic sheathed cable	1.000	Ea.	.800	17.85	36.50	54.35
Using BX cable	1.000	Ea.	.964	31	44	75
Using EMT conduit	1.000	Ea.	1.194	40	55	95
Disposal wiring						
Using non-metallic sheathed cable	1.000	Ea.	.889	13.45	41	54.45
Using BX cable	1.000	Ea.	1.067	26	49	75
Using EMT conduit	1.000	Ea.	1.333	36	61	97
Dryer circuit						
Using non-metallic sheathed cable	1.000	Ea.	1.455	38.50	67	105.50
Using BX cable	1.000	Ea.	1.739	52.50	80	132.50
Using EMT conduit	1.000	Ea.	2.162	54	99	153
Duplex receptacles						
Using non-metallic sheathed cable	1.000	Ea.	.615	17.85	28.50	46.35
Using BX cable	1.000	Ea.	.741	31	34	65
Using EMT conduit	1.000	Ea.	.920	40	42	82
Exhaust fan wiring						
Using non-metallic sheathed cable	1.000	Ea.	.800	15.95	36.50	52.45
Using BX cable	1.000	Ea.	.964	29	44	73
Using EMT conduit	1.000	Ea.	1.194	38	55	93
Furnace circuit & switch						
Using non-metallic sheathed cable	1.000	Ea.	1.333	22	61	83
Using BX cable	1.000	Ea.	1.600	36.50	73.50	110
Using EMT conduit	1.000	Ea.	2.000	44	92	136
Ground fault						
Using non-metallic sheathed cable	1.000	Ea.	1.000	47.50	46	93.50
Using BX cable	1.000	Ea.	1.212	59	55.50	114.50
Using EMT conduit	1.000	Ea.	1.481	77	68	145
Heater circuits						
Using non-metallic sheathed cable	1.000	Ea.	1.000	19.45	46	65.45
Using BX cable	1.000	Ea.	1.212	28	55.50	83.50
Using EMT conduit	1.000	Ea.	1.481	36.50	68	104.50
Lighting wiring						
Using non-metallic sheathed cable	1.000	Ea.	.500	19.40	23	42.40
Using BX cable	1.000	Ea.	.602	28	27.50	55.50
Using EMT conduit	1.000	Ea.	.748	35	34.50	69.50
Range circuits						
Using non-metallic sheathed cable	1.000	Ea.	2.000	83	92	175
Using BX cable	1.000	Ea.	2.424	117	111	228
Using EMT conduit	1.000	Ea.	2.963	85	136	221
Switches, single pole						
Using non-metallic sheathed cable	1.000	Ea.	.500	15.95	23	38.95
Using BX cable	1.000	Ea.	.602	29	27.50	56.50
Using EMT conduit	1.000	Ea.	.748	38	34.50	72.50
Switches, 3-way						
Using non-metallic sheathed cable	1.000	Ea.	.667	21.50	30.50	52
Using BX cable	1.000	Ea.	.800	32	36.50	68.50
Using EMT conduit	1.000	Ea.	1.333	52.50	61	113.50
Water heater						
Using non-metallic sheathed cable	1.000	Ea.	1.600	25.50	73.50	99
Using BX cable	1.000	Ea.	1.905	44.50	87.50	132
Using EMT conduit	1.000	Ea.	2.353	42.50	108	150.50
Weatherproof receptacle						
Using non-metallic sheathed cable	1.000	Ea.	1.333	126	61	187
Using BX cable	1.000	Ea.	1.600	134	73.50	207.50
Using EMT conduit	1.000	Ea.	2.000	143	92	235

40 Light Fixture Systems

DESCRIPTION	QUAN.	UNIT	LABOR HOURS	COST EACH		
				MAT.	INST.	TOTAL
Fluorescent strip, 4′ long, 1 light, average	1.000	Ea.	.941	31	43	74
Deluxe	1.000	Ea.	1.129	37	51.50	88.50
2 lights, average	1.000	Ea.	1.000	33.50	46	79.50
Deluxe	1.000	Ea.	1.200	40	55	95
8′ long, 1 light, average	1.000	Ea.	1.194	47	55	102
Deluxe	1.000	Ea.	1.433	56.50	66	122.50
2 lights, average	1.000	Ea.	1.290	56.50	59	115.50
Deluxe	1.000	Ea.	1.548	68	71	139
Surface mounted, 4′ x 1′, economy	1.000	Ea.	.914	65	42	107
Average	1.000	Ea.	1.143	81	52.50	133.50
Deluxe	1.000	Ea.	1.371	97	63	160
4′ x 2′, economy	1.000	Ea.	1.208	82.50	55.50	138
Average	1.000	Ea.	1.509	103	69.50	172.50
Deluxe	1.000	Ea.	1.811	124	83.50	207.50
Recessed, 4′x 1′, 2 lamps, economy	1.000	Ea.	1.123	43	51.50	94.50
Average	1.000	Ea.	1.404	53.50	64.50	118
Deluxe	1.000	Ea.	1.684	64	77.50	141.50
4′ x 2′, 4′ lamps, economy	1.000	Ea.	1.362	52	62.50	114.50
Average	1.000	Ea.	1.702	65	78	143
Deluxe	1.000	Ea.	2.043	78	93.50	171.50
Incandescent, exterior, 150W, single spot	1.000	Ea.	.500	21.50	23	44.50
Double spot	1.000	Ea.	1.167	81	53.50	134.50
Recessed, 100W, economy	1.000	Ea.	.800	53	37	90
Average	1.000	Ea.	1.000	66.50	46	112.50
Deluxe	1.000	Ea.	1.200	80	55	135
150W, economy	1.000	Ea.	.800	77	37	114
Average	1.000	Ea.	1.000	96	46	142
Deluxe	1.000	Ea.	1.200	115	55	170
Surface mounted, 60W, economy	1.000	Ea.	.800	41.50	36.50	78
Average	1.000	Ea.	1.000	46.50	46	92.50
Deluxe	1.000	Ea.	1.194	66	55	121
Metal halide, recessed 2′ x 2′ 250W	1.000	Ea.	2.500	320	115	435
2′ x 2′, 400W	1.000	Ea.	2.759	360	127	487
Surface mounted, 2′ x 2′, 250W	1.000	Ea.	2.963	315	136	451
2′ x 2′, 400W	1.000	Ea.	3.333	375	153	528
High bay, single, unit, 400W	1.000	Ea.	3.478	380	160	540
Twin unit, 400W	1.000	Ea.	5.000	760	230	990
Low bay, 250W	1.000	Ea.	2.500	370	115	485

Location Factors

Costs shown in *Means cost data publications* are based on National Averages for materials and installation. To adjust these costs to a specific location, simply multiply the base cost by the factor for that city. The data is arranged alphabetically by state and postal zip code numbers. For a city not listed, use the factor for a nearby city with similar economic characteristics.

STATE	CITY	Residential
ALABAMA		
350-352	Birmingham	.86
354	Tuscaloosa	.73
355	Jasper	.71
356	Decatur	.76
357-358	Huntsville	.84
359	Gadsden	.73
360-361	Montgomery	.75
362	Anniston	.68
363	Dothan	.74
364	Evergreen	.70
365-366	Mobile	.79
367	Selma	.72
368	Phenix City	.72
369	Butler	.71
ALASKA		
995-996	Anchorage	1.27
997	Fairbanks	1.29
998	Juneau	1.27
999	Ketchikan	1.28
ARIZONA		
850,853	Phoenix	.86
852	Mesa/Tempe	.84
855	Globe	.80
856-857	Tucson	.84
859	Show Low	.82
860	Flagstaff	.87
863	Prescott	.82
864	Kingman	.83
865	Chambers	.81
ARKANSAS		
716	Pine Bluff	.75
717	Camden	.65
718	Texarkana	.69
719	Hot Springs	.63
720-722	Little Rock	.80
723	West Memphis	.73
724	Jonesboro	.72
725	Batesville	.70
726	Harrison	.71
727	Fayetteville	.67
728	Russellville	.69
729	Fort Smith	.76
CALIFORNIA		
900-902	Los Angeles	1.06
903-905	Inglewood	1.05
906-908	Long Beach	1.03
910-912	Pasadena	1.05
913-916	Van Nuys	1.08
917-918	Alhambra	1.09
919-921	San Diego	1.04
922	Palm Springs	1.03
923-924	San Bernardino	1.05
925	Riverside	1.07
926-927	Santa Ana	1.06
928	Anaheim	1.07
930	Oxnard	1.07
931	Santa Barbara	1.07
932-933	Bakersfield	1.04
934	San Luis Obispo	1.09
935	Mojave	1.05
936-938	Fresno	1.09
939	Salinas	1.12
940-941	San Francisco	1.22
942,956-958	Sacramento	1.11
943	Palo Alto	1.17
944	San Mateo	1.21
945	Vallejo	1.14
946	Oakland	1.20
947	Berkeley	1.23
948	Richmond	1.24
949	San Rafael	1.21
950	Santa Cruz	1.14
951	San Jose	1.19
952	Stockton	1.09
953	Modesto	1.08
954	Santa Rosa	1.14
955	Eureka	1.08
959	Marysville	1.10
960	Redding	1.09
961	Susanville	1.10
COLORADO		
800-802	Denver	.95
803	Boulder	.94
804	Golden	.92
805	Fort Collins	.91
806	Greeley	.79
807	Fort Morgan	.94
808-809	Colorado Springs	.91
810	Pueblo	.92
811	Alamosa	.89
812	Salida	.91
813	Durango	.93
814	Montrose	.88
815	Grand Junction	.93
816	Glenwood Springs	.91
CONNECTICUT		
060	New Britain	1.10
061	Hartford	1.10
062	Willimantic	1.11
063	New London	1.10
064	Meriden	1.10
065	New Haven	1.10
066	Bridgeport	1.11
067	Waterbury	1.11
068	Norwalk	1.11
069	Stamford	1.12
D.C.		
200-205	Washington	.93
DELAWARE		
197	Newark	1.00
198	Wilmington	1.00
199	Dover	1.00
FLORIDA		
320,322	Jacksonville	.78
321	Daytona Beach	.85
323	Tallahassee	.72
324	Panama City	.67
325	Pensacola	.75
326,344	Gainesville	.77
327-328,347	Orlando	.85
329	Melbourne	.87
330-332,340	Miami	.83
333	Fort Lauderdale	.83
334,349	West Palm Beach	.83
335-336,346	Tampa	.86
337	St. Petersburg	.76
338	Lakeland	.82
339,341	Fort Myers	.80
342	Sarasota	.84
GEORGIA		
300-303,399	Atlanta	.88
304	Statesboro	.66
305	Gainesville	.74
306	Athens	.74
307	Dalton	.70
308-309	Augusta	.76
310-312	Macon	.77
313-314	Savannah	.79
315	Waycross	.71
316	Valdosta	.70
317,398	Albany	.75
318-319	Columbus	.79
HAWAII		
967	Hilo	1.22

Location Factors

STATE	CITY	Residential
968	Honolulu	1.23
STATES & POSS.		
969	Guam	1.73
IDAHO		
832	Pocatello	.87
833	Twin Falls	.74
834	Idaho Falls	.72
835	Lewiston	.98
836-837	Boise	.89
838	Coeur d'Alene	.94
ILLINOIS		
600-603	North Suburban	1.13
604	Joliet	1.13
605	South Suburban	1.13
606-608	Chicago	1.14
609	Kankakee	.99
610-611	Rockford	1.04
612	Rock Island	.97
613	La Salle	1.03
614	Galesburg	.98
615-616	Peoria	1.00
617	Bloomington	.96
618-619	Champaign	.98
620-622	East St. Louis	1.00
623	Quincy	.97
624	Effingham	.98
625	Decatur	.99
626-627	Springfield	.99
628	Centralia	1.01
629	Carbondale	.96
INDIANA		
460	Anderson	.93
461-462	Indianapolis	.95
463-464	Gary	1.03
465-466	South Bend	.94
467-468	Fort Wayne	.92
469	Kokomo	.94
470	Lawrenceburg	.88
471	New Albany	.87
472	Columbus	.93
473	Muncie	.92
474	Bloomington	.96
475	Washington	.92
476-477	Evansville	.92
478	Terre Haute	.92
479	Lafayette	.93
IOWA		
500-503,509	Des Moines	.93
504	Mason City	.77
505	Fort Dodge	.75
506-507	Waterloo	.80
508	Creston	.82
510-511	Sioux City	.88
512	Sibley	.73
513	Spencer	.75
514	Carroll	.74
515	Council Bluffs	.81
516	Shenandoah	.75
520	Dubuque	.84
521	Decorah	.76
522-524	Cedar Rapids	.94
525	Ottumwa	.84
526	Burlington	.88
527-528	Davenport	.98
KANSAS		
660-662	Kansas City	.97
664-666	Topeka	.79
667	Fort Scott	.86
668	Emporia	.72
669	Belleville	.78
670-672	Wichita	.81
673	Independence	.86
674	Salina	.76
675	Hutchinson	.68
676	Hays	.83
677	Colby	.76
678	Dodge City	.82
679	Liberal	.68

STATE	CITY	Residential
KENTUCKY		
400-402	Louisville	.92
403-405	Lexington	.84
406	Frankfort	.81
407-409	Corbin	.66
410	Covington	.92
411-412	Ashland	.93
413-414	Campton	.68
415-416	Pikeville	.76
417-418	Hazard	.66
420	Paducah	.89
421-422	Bowling Green	.89
423	Owensboro	.81
424	Henderson	.92
425-426	Somerset	.66
427	Elizabethtown	.87
LOUISIANA		
700-701	New Orleans	.85
703	Thibodaux	.83
704	Hammond	.78
705	Lafayette	.81
706	Lake Charles	.82
707-708	Baton Rouge	.83
710-711	Shreveport	.78
712	Monroe	.73
713-714	Alexandria	.74
MAINE		
039	Kittery	.80
040-041	Portland	.91
042	Lewiston	.91
043	Augusta	.83
044	Bangor	.89
045	Bath	.81
046	Machias	.82
047	Houlton	.86
048	Rockland	.82
049	Waterville	.81
MARYLAND		
206	Waldorf	.84
207-208	College Park	.86
209	Silver Spring	.85
210-212	Baltimore	.90
214	Annapolis	.86
215	Cumberland	.87
216	Easton	.68
217	Hagerstown	.86
218	Salisbury	.75
219	Elkton	.82
MASSACHUSETTS		
010-011	Springfield	1.05
012	Pittsfield	1.03
013	Greenfield	1.02
014	Fitchburg	1.11
015-016	Worcester	1.12
017	Framingham	1.13
018	Lowell	1.14
019	Lawrence	1.14
020-022, 024	Boston	1.20
023	Brockton	1.13
025	Buzzards Bay	1.11
026	Hyannis	1.10
027	New Bedford	1.13
MICHIGAN		
480,483	Royal Oak	1.03
481	Ann Arbor	1.04
482	Detroit	1.10
484-485	Flint	.97
486	Saginaw	.94
487	Bay City	.95
488-489	Lansing	.97
490	Battle Creek	.92
491	Kalamazoo	.91
492	Jackson	.92
493,495	Grand Rapids	.82
494	Muskegon	.88
496	Traverse City	.80
497	Gaylord	.83
498-499	Iron Mountain	.90

Location Factors

STATE	CITY	Residential
MINNESOTA		
550-551	Saint Paul	1.14
553-555	Minneapolis	1.17
556-558	Duluth	1.11
559	Rochester	1.06
560	Mankato	1.02
561	Windom	.83
562	Willmar	.85
563	St. Cloud	1.09
564	Brainerd	.99
565	Detroit Lakes	.97
566	Bemidji	.97
567	Thief River Falls	.95
MISSISSIPPI		
386	Clarksdale	.61
387	Greenville	.68
388	Tupelo	.63
389	Greenwood	.65
390-392	Jackson	.73
393	Meridian	.66
394	Laurel	.62
395	Biloxi	.75
396	McComb	.73
397	Columbus	.64
MISSOURI		
630-631	St. Louis	1.01
633	Bowling Green	.95
634	Hannibal	.87
635	Kirksville	.80
636	Flat River	.95
637	Cape Girardeau	.87
638	Sikeston	.82
639	Poplar Bluff	.83
640-641	Kansas City	1.04
644-645	St. Joseph	.95
646	Chillicothe	.86
647	Harrisonville	.96
648	Joplin	.85
650-651	Jefferson City	.89
652	Columbia	.88
653	Sedalia	.85
654-655	Rolla	.89
656-658	Springfield	.87
MONTANA		
590-591	Billings	.87
592	Wolf Point	.83
593	Miles City	.86
594	Great Falls	.88
595	Havre	.80
596	Helena	.87
597	Butte	.82
598	Missoula	.83
599	Kalispell	.82
NEBRASKA		
680-681	Omaha	.90
683-685	Lincoln	.78
686	Columbus	.69
687	Norfolk	.77
688	Grand Island	.77
689	Hastings	.76
690	Mccook	.70
691	North Platte	.75
692	Valentine	.66
693	Alliance	.65
NEVADA		
889-891	Las Vegas	1.01
893	Ely	.89
894-895	Reno	.96
897	Carson City	.96
898	Elko	.95
NEW HAMPSHIRE		
030	Nashua	.90
031	Manchester	.90
032-033	Concord	.86
034	Keene	.73
035	Littleton	.81
036	Charleston	.70
037	Claremont	.72
038	Portsmouth	.84
NEW JERSEY		
070-071	Newark	1.15
072	Elizabeth	1.18
073	Jersey City	1.14
074-075	Paterson	1.15
076	Hackensack	1.14
077	Long Branch	1.15
078	Dover	1.14
079	Summit	1.15
080,083	Vineland	1.09
081	Camden	1.10
082,084	Atlantic City	1.13
085-086	Trenton	1.13
087	Point Pleasant	1.11
088-089	New Brunswick	1.15
NEW MEXICO		
870-872	Albuquerque	.85
873	Gallup	.85
874	Farmington	.85
875	Santa Fe	.85
877	Las Vegas	.85
878	Socorro	.85
879	Truth/Consequences	.84
880	Las Cruces	.82
881	Clovis	.85
882	Roswell	.85
883	Carrizozo	.85
884	Tucumcari	.86
NEW YORK		
100-102	New York	1.36
103	Staten Island	1.28
104	Bronx	1.30
105	Mount Vernon	1.17
106	White Plains	1.20
107	Yonkers	1.22
108	New Rochelle	1.21
109	Suffern	1.14
110	Queens	1.28
111	Long Island City	1.31
112	Brooklyn	1.33
113	Flushing	1.30
114	Jamaica	1.30
115,117,118	Hicksville	1.21
116	Far Rockaway	1.29
119	Riverhead	1.22
120-122	Albany	.96
123	Schenectady	.96
124	Kingston	1.04
125-126	Poughkeepsie	1.06
127	Monticello	1.06
128	Glens Falls	.89
129	Plattsburgh	.94
130-132	Syracuse	.97
133-135	Utica	.94
136	Watertown	.90
137-139	Binghamton	.94
140-142	Buffalo	1.06
143	Niagara Falls	1.02
144-146	Rochester	1.00
147	Jamestown	.89
148-149	Elmira	.86
NORTH CAROLINA		
270,272-274	Greensboro	.74
271	Winston-Salem	.74
275-276	Raleigh	.75
277	Durham	.74
278	Rocky Mount	.64
279	Elizabeth City	.62
280	Gastonia	.75
281-282	Charlotte	.75
283	Fayetteville	.72
284	Wilmington	.73
285	Kinston	.62
286	Hickory	.62
287-288	Asheville	.73
289	Murphy	.66

STATE	CITY	Residential
NORTH DAKOTA		
580-581	Fargo	.80
582	Grand Forks	.76
583	Devils Lake	.79
584	Jamestown	.74
585	Bismarck	.80
586	Dickinson	.76
587	Minot	.80
588	Williston	.77
OHIO		
430-432	Columbus	.95
433	Marion	.93
434-436	Toledo	1.00
437-438	Zanesville	.91
439	Steubenville	.96
440	Lorain	1.00
441	Cleveland	1.01
442-443	Akron	.99
444-445	Youngstown	.95
446-447	Canton	.94
448-449	Mansfield	.95
450	Hamilton	.95
451-452	Cincinnati	.95
453-454	Dayton	.92
455	Springfield	.94
456	Chillicothe	.97
457	Athens	.89
458	Lima	.91
OKLAHOMA		
730-731	Oklahoma City	.80
734	Ardmore	.78
735	Lawton	.81
736	Clinton	.77
737	Enid	.77
738	Woodward	.76
739	Guymon	.66
740-741	Tulsa	.78
743	Miami	.82
744	Muskogee	.72
745	Mcalester	.74
746	Ponca City	.77
747	Durant	.76
748	Shawnee	.75
749	Poteau	.78
OREGON		
970-972	Portland	1.02
973	Salem	1.02
974	Eugene	1.01
975	Medford	1.00
976	Klamath Falls	1.00
977	Bend	1.03
978	Pendleton	1.00
979	Vale	.99
PENNSYLVANIA		
150-152	Pittsburgh	.99
153	Washington	.94
154	Uniontown	.91
155	Bedford	.88
156	Greensburg	.96
157	Indiana	.92
158	Dubois	.90
159	Johnstown	.91
160	Butler	.93
161	New Castle	.93
162	Kittanning	.94
163	Oil City	.90
164-165	Erie	.96
166	Altoona	.89
167	Bradford	.91
168	State College	.92
169	Wellsboro	.90
170-171	Harrisburg	.95
172	Chambersburg	.89
173-174	York	.92
175-176	Lancaster	.92
177	Williamsport	.85
178	Sunbury	.92
179	Pottsville	.91
180	Lehigh Valley	1.02
181	Allentown	1.04

STATE	CITY	Residential
182	Hazleton	.91
183	Stroudsburg	.92
184-185	Scranton	.96
186-187	Wilkes-Barre	.93
188	Montrose	.90
189	Doylestown	1.06
190-191	Philadelphia	1.15
193	Westchester	1.10
194	Norristown	1.08
195-196	Reading	.97
PUERTO RICO		
009	San Juan	.84
RHODE ISLAND		
028	Newport	1.09
029	Providence	1.09
SOUTH CAROLINA		
290-292	Columbia	.73
293	Spartanburg	.71
294	Charleston	.72
295	Florence	.66
296	Greenville	.70
297	Rock Hill	.65
298	Aiken	.84
299	Beaufort	.67
SOUTH DAKOTA		
570-571	Sioux Falls	.77
572	Watertown	.72
573	Mitchell	.75
574	Aberdeen	.76
575	Pierre	.75
576	Mobridge	.72
577	Rapid City	.76
TENNESSEE		
370-372	Nashville	.84
373-374	Chattanooga	.76
375,380-381	Memphis	.85
376	Johnson City	.71
377-379	Knoxville	.74
382	Mckenzie	.73
383	Jackson	.70
384	Columbia	.71
385	Cookeville	.68
TEXAS		
750	Mckinney	.74
751	Waxahachie	.75
752-753	Dallas	.83
754	Greenville	.68
755	Texarkana	.73
756	Longview	.67
757	Tyler	.74
758	Palestine	.66
759	Lufkin	.71
760-761	Fort Worth	.82
762	Denton	.76
763	Wichita Falls	.79
764	Eastland	.72
765	Temple	.74
766-767	Waco	.77
768	Brownwood	.68
769	San Angelo	.71
770-772	Houston	.85
773	Huntsville	.68
774	Wharton	.70
775	Galveston	.84
776-777	Beaumont	.82
778	Bryan	.73
779	Victoria	.73
780	Laredo	.72
781-782	San Antonio	.80
783-784	Corpus Christi	.76
785	McAllen	.75
786-787	Austin	.79
788	Del Rio	.65
789	Giddings	.69
790-791	Amarillo	.77
792	Childress	.75
793-794	Lubbock	.75
795-796	Abilene	.74

STATE	CITY	Residential
797	Midland	.76
798-799,885	El Paso	.75
UTAH		
840-841	Salt Lake City	.82
842,844	Ogden	.81
843	Logan	.81
845	Price	.72
846-847	Provo	.82
VERMONT		
050	White River Jct.	.73
051	Bellows Falls	.75
052	Bennington	.74
053	Brattleboro	.74
054	Burlington	.79
056	Montpelier	.81
057	Rutland	.81
058	St. Johnsbury	.75
059	Guildhall	.74
VIRGINIA		
220-221	Fairfax	.85
222	Arlington	.87
223	Alexandria	.90
224-225	Fredericksburg	.75
226	Winchester	.70
227	Culpeper	.77
228	Harrisonburg	.67
229	Charlottesville	.73
230-232	Richmond	.81
233-235	Norfolk	.82
236	Newport News	.81
237	Portsmouth	.78
238	Petersburg	.79
239	Farmville	.69
240-241	Roanoke	.72
242	Bristol	.68
243	Pulaski	.66
244	Staunton	.69
245	Lynchburg	.70
246	Grundy	.67
WASHINGTON		
980-981,987	Seattle	1.01
982	Everett	1.04
983-984	Tacoma	1.00
985	Olympia	1.00
986	Vancouver	.97
988	Wenatchee	.92
989	Yakima	.96
990-992	Spokane	1.00
993	Richland	.98
994	Clarkston	.97
WEST VIRGINIA		
247-248	Bluefield	.88
249	Lewisburg	.89
250-253	Charleston	.97
254	Martinsburg	.86
255-257	Huntington	.97
258-259	Beckley	.90
260	Wheeling	.94
261	Parkersburg	.92
262	Buckhannon	.93
263-264	Clarksburg	.92
265	Morgantown	.93
266	Gassaway	.92
267	Romney	.88
268	Petersburg	.90
WISCONSIN		
530,532	Milwaukee	1.07
531	Kenosha	1.06
534	Racine	1.04
535	Beloit	1.01
537	Madison	1.01
538	Lancaster	.99
539	Portage	.98
540	New Richmond	1.01
541-543	Green Bay	1.03
544	Wausau	.96
545	Rhinelander	.96
546	La Crosse	.95

STATE	CITY	Residential
547	Eau Claire	1.00
548	Superior	1.01
549	Oshkosh	.97
WYOMING		
820	Cheyenne	.76
821	Yellowstone Nat. Pk.	.72
822	Wheatland	.73
823	Rawlins	.71
824	Worland	.70
825	Riverton	.71
826	Casper	.75
827	Newcastle	.70
828	Sheridan	.74
829-831	Rock Springs	.75
CANADIAN FACTORS (reflect Canadian Currency)		
ALBERTA		
	Calgary	1.05
	Edmonton	1.04
	Fort McMurray	1.02
	Lethbridge	1.03
	Lloydminster	1.02
	Medicine Hat	1.03
	Red Deer	1.03
BRITISH COLUMBIA		
	Kamloops	1.00
	Prince George	1.00
	Vancouver	1.07
	Victoria	1.00
MANITOBA		
	Brandon	.99
	Portage la Prairie	.99
	Winnipeg	1.00
NEW BRUNSWICK		
	Bathurst	.90
	Dalhousie	.90
	Fredericton	.98
	Moncton	.90
	Newcastle	.90
	Saint John	.99
NEWFOUNDLAND		
	Corner Brook	.92
	St. John's	.92
NORTHWEST TERRITORIES		
	Yellowknife	1.01
NOVA SCOTIA		
	Dartmouth	.93
	Halifax	.93
	New Glasgow	.92
	Sydney	.91
	Yarmouth	.92
ONTARIO		
	Barrie	1.09
	Brantford	1.11
	Cornwall	1.10
	Hamilton	1.11
	Kingston	1.11
	Kitchener	1.05
	London	1.09
	North Bay	1.08
	Oshawa	1.09
	Ottawa	1.11
	Owen Sound	1.08
	Peterborough	1.08
	Sarnia	1.11
	St. Catharines	1.04
	Sudbury	1.02
	Thunder Bay	1.07
	Toronto	1.14
	Windsor	1.08
PRINCE EDWARD ISLAND		
	Charlottetown	.87
	Summerside	.87

Location Factors

STATE	CITY	Residential
QUEBEC		
	Cap-de-la-Madeleine	1.10
	Charlesbourg	1.10
	Chicoutimi	1.10
	Gatineau	1.09
	Laval	1.09
	Montreal	1.10
	Quebec	1.12
	Sherbrooke	1.09
	Trois Rivieres	1.10
SASKATCHEWAN		
	Moose Jaw	.90
	Prince Albert	.90
	Regina	.92
	Saskatoon	.90
YUKON		
	Whitehorse	.89

Abbreviations

A	Area Square Feet; Ampere
ABS	Acrylonitrile Butadiene Stryrene; Asbestos Bonded Steel
A.C.	Alternating Current; Air-Conditioning; Asbestos Cement; Plywood Grade A & C
A.C.I.	American Concrete Institute
AD	Plywood, Grade A & D
Addit.	Additional
Adj.	Adjustable
af	Audio-frequency
A.G.A.	American Gas Association
Agg.	Aggregate
A.H.	Ampere Hours
A hr.	Ampere-hour
A.H.U.	Air Handling Unit
A.I.A.	American Institute of Architects
AIC	Ampere Interrupting Capacity
Allow.	Allowance
alt.	Altitude
Alum.	Aluminum
a.m.	Ante Meridiem
Amp.	Ampere
Anod.	Anodized
Approx.	Approximate
Apt.	Apartment
Asb.	Asbestos
A.S.B.C.	American Standard Building Code
Asbe.	Asbestos Worker
A.S.H.R.A.E.	American Society of Heating, Refrig. & AC Engineers
A.S.M.E.	American Society of Mechanical Engineers
A.S.T.M.	American Society for Testing and Materials
Attchmt.	Attachment
Avg.	Average
A.W.G.	American Wire Gauge
AWWA	American Water Works Assoc.
Bbl.	Barrel
B&B	Grade B and Better; Balled & Burlapped
B.&S.	Bell and Spigot
B.&W.	Black and White
b.c.c.	Body-centered Cubic
B.C.Y.	Bank Cubic Yards
BE	Bevel End
B.F.	Board Feet
Bg. cem.	Bag of Cement
BHP	Boiler Horsepower; Brake Horsepower
B.I.	Black Iron
Bit.; Bitum.	Bituminous
Bk.	Backed
Bkrs.	Breakers
Bldg.	Building
Blk.	Block
Bm.	Beam
Boil.	Boilermaker
B.P.M.	Blows per Minute
BR	Bedroom
Brg.	Bearing
Brhe.	Bricklayer Helper
Bric.	Bricklayer
Brk.	Brick
Brng.	Bearing
Brs.	Brass
Brz.	Bronze
Bsn.	Basin
Btr.	Better
BTU	British Thermal Unit
BTUH	BTU per Hour
B.U.R.	Built-up Roofing
BX	Interlocked Armored Cable
c	Conductivity, Copper Sweat
C	Hundred; Centigrade
C/C	Center to Center, Cedar on Cedar
Cab.	Cabinet
Cair.	Air Tool Laborer
Calc	Calculated
Cap.	Capacity
Carp.	Carpenter
C.B.	Circuit Breaker
C.C.A.	Chromate Copper Arsenate
C.C.F.	Hundred Cubic Feet
cd	Candela
cd/sf	Candela per Square Foot
CD	Grade of Plywood Face & Back
CDX	Plywood, Grade C & D, exterior glue
Cefi.	Cement Finisher
Cem.	Cement
CF	Hundred Feet
C.F.	Cubic Feet
CFM	Cubic Feet per Minute
c.g.	Center of Gravity
CHW	Chilled Water; Commercial Hot Water
C.I.	Cast Iron
C.I.P.	Cast in Place
Circ.	Circuit
C.L.	Carload Lot
Clab.	Common Laborer
Clam	Common maintenance laborer
C.L.F.	Hundred Linear Feet
CLF	Current Limiting Fuse
CLP	Cross Linked Polyethylene
cm	Centimeter
CMP	Corr. Metal Pipe
C.M.U.	Concrete Masonry Unit
CN	Change Notice
Col.	Column
CO_2	Carbon Dioxide
Comb.	Combination
Compr.	Compressor
Conc.	Concrete
Cont.	Continuous; Continued
Corr.	Corrugated
Cos	Cosine
Cot	Cotangent
Cov.	Cover
C/P	Cedar on Paneling
CPA	Control Point Adjustment
Cplg.	Coupling
C.P.M.	Critical Path Method
CPVC	Chlorinated Polyvinyl Chloride
C.Pr.	Hundred Pair
CRC	Cold Rolled Channel
Creos.	Creosote
Crpt.	Carpet & Linoleum Layer
CRT	Cathode-ray Tube
CS	Carbon Steel, Constant Shear Bar Joist
Csc	Cosecant
C.S.F.	Hundred Square Feet
CSI	Construction Specifications Institute
C.T.	Current Transformer
CTS	Copper Tube Size
Cu	Copper, Cubic
Cu. Ft.	Cubic Foot
cw	Continuous Wave
C.W.	Cool White; Cold Water
Cwt.	100 Pounds
C.W.X.	Cool White Deluxe
C.Y.	Cubic Yard (27 cubic feet)
C.Y./Hr.	Cubic Yard per Hour
Cyl.	Cylinder
d	Penny (nail size)
D	Deep; Depth; Discharge
Dis.;Disch.	Discharge
Db.	Decibel
Dbl.	Double
DC	Direct Current
DDC	Direct Digital Control
Demob.	Demobilization
d.f.u.	Drainage Fixture Units
D.H.	Double Hung
DHW	Domestic Hot Water
Diag.	Diagonal
Diam.	Diameter
Distrib.	Distribution
Dk.	Deck
D.L.	Dead Load; Diesel
DLH	Deep Long Span Bar Joist
Do.	Ditto
Dp.	Depth
D.P.S.T.	Double Pole, Single Throw
Dr.	Driver
Drink.	Drinking
D.S.	Double Strength
D.S.A.	Double Strength A Grade
D.S.B.	Double Strength B Grade
Dty.	Duty
DWV	Drain Waste Vent
DX	Deluxe White, Direct Expansion
dyn	Dyne
e	Eccentricity
E	Equipment Only; East
Ea.	Each
E.B.	Encased Burial
Econ.	Economy
E.C.Y	Embankment Cubic Yards
EDP	Electronic Data Processing
EIFS	Exterior Insulation Finish System
E.D.R.	Equiv. Direct Radiation
Eq.	Equation
Elec.	Electrician; Electrical
Elev.	Elevator; Elevating
EMT	Electrical Metallic Conduit; Thin Wall Conduit
Eng.	Engine, Engineered
EPDM	Ethylene Propylene Diene Monomer
EPS	Expanded Polystyrene
Eqhv.	Equip. Oper., Heavy
Eqlt.	Equip. Oper., Light
Eqmd.	Equip. Oper., Medium
Eqmm.	Equip. Oper., Master Mechanic
Eqol.	Equip. Oper., Oilers
Equip.	Equipment
ERW	Electric Resistance Welded
E.S.	Energy Saver
Est.	Estimated
esu	Electrostatic Units
E.W.	Each Way
EWT	Entering Water Temperature
Excav.	Excavation
Exp.	Expansion, Exposure
Ext.	Exterior
Extru.	Extrusion
f.	Fiber stress
F	Fahrenheit; Female; Fill
Fab.	Fabricated
FBGS	Fiberglass
F.C.	Footcandles
f.c.c.	Face-centered Cubic
f'c.	Compressive Stress in Concrete; Extreme Compressive Stress
F.E.	Front End
FEP	Fluorinated Ethylene Propylene (Teflon)
F.G.	Flat Grain
F.H.A.	Federal Housing Administration
Fig.	Figure
Fin.	Finished
Fixt.	Fixture
Fl. Oz.	Fluid Ounces
Flr.	Floor
F.M.	Frequency Modulation; Factory Mutual
Fmg.	Framing
Fndtn.	Foundation

Abbreviations

Fori.	Foreman, Inside
Foro.	Foreman, Outside
Fount.	Fountain
FPM	Feet per Minute
FPT	Female Pipe Thread
Fr.	Frame
F.R.	Fire Rating
FRK	Foil Reinforced Kraft
FRP	Fiberglass Reinforced Plastic
FS	Forged Steel
FSC	Cast Body; Cast Switch Box
Ft.	Foot; Feet
Ftng.	Fitting
Ftg.	Footing
Ft. Lb.	Foot Pound
Furn.	Furniture
FVNR	Full Voltage Non-Reversing
FXM	Female by Male
Fy.	Minimum Yield Stress of Steel
g	Gram
G	Gauss
Ga.	Gauge
Gal.	Gallon
Gal./Min.	Gallon per Minute
Galv.	Galvanized
Gen.	General
G.F.I.	Ground Fault Interrupter
Glaz.	Glazier
GPD	Gallons per Day
GPH	Gallons per Hour
GPM	Gallons per Minute
GR	Grade
Gran.	Granular
Grnd.	Ground
H	High; High Strength Bar Joist; Henry
H.C.	High Capacity
H.D.	Heavy Duty; High Density
H.D.O.	High Density Overlaid
Hdr.	Header
Hdwe.	Hardware
Help.	Helper Average
HEPA	High Efficiency Particulate Air Filter
Hg	Mercury
HIC	High Interrupting Capacity
HM	Hollow Metal
H.O.	High Output
Horiz.	Horizontal
H.P.	Horsepower; High Pressure
H.P.F.	High Power Factor
Hr.	Hour
Hrs./Day	Hours per Day
HSC	High Short Circuit
Ht.	Height
Htg.	Heating
Htrs.	Heaters
HVAC	Heating, Ventilation & Air-Conditioning
Hvy.	Heavy
HW	Hot Water
Hyd.;Hydr.	Hydraulic
Hz.	Hertz (cycles)
I.	Moment of Inertia
I.C.	Interrupting Capacity
ID	Inside Diameter
I.D.	Inside Dimension; Identification
I.F.	Inside Frosted
I.M.C.	Intermediate Metal Conduit
In.	Inch
Incan.	Incandescent
Incl.	Included; Including
Int.	Interior
Inst.	Installation
Insul.	Insulation/Insulated
I.P.	Iron Pipe
I.P.S.	Iron Pipe Size
I.P.T.	Iron Pipe Threaded
I.W.	Indirect Waste
J	Joule
J.I.C.	Joint Industrial Council
K	Thousand; Thousand Pounds; Heavy Wall Copper Tubing, Kelvin
K.A.H.	Thousand Amp. Hours
KCMIL	Thousand Circular Mils
KD	Knock Down
K.D.A.T.	Kiln Dried After Treatment
kg	Kilogram
kG	Kilogauss
kgf	Kilogram Force
kHz	Kilohertz
Kip.	1000 Pounds
KJ	Kiljoule
K.L.	Effective Length Factor
K.L.F.	Kips per Linear Foot
Km	Kilometer
K.S.F.	Kips per Square Foot
K.S.I.	Kips per Square Inch
kV	Kilovolt
kVA	Kilovolt Ampere
K.V.A.R.	Kilovar (Reactance)
KW	Kilowatt
KWh	Kilowatt-hour
L	Labor Only; Length; Long; Medium Wall Copper Tubing
Lab.	Labor
lat	Latitude
Lath.	Lather
Lav.	Lavatory
lb.; #	Pound
L.B.	Load Bearing; L Conduit Body
L. & E.	Labor & Equipment
lb./hr.	Pounds per Hour
lb./L.F.	Pounds per Linear Foot
lbf/sq.in.	Pound-force per Square Inch
L.C.L.	Less than Carload Lot
L.C.Y.	Loose Cubic Yard
Ld.	Load
LE	Lead Equivalent
LED	Light Emitting Diode
L.F.	Linear Foot
Lg.	Long; Length; Large
L & H	Light and Heat
LH	Long Span Bar Joist
L.H.	Labor Hours
L.L.	Live Load
L.L.D.	Lamp Lumen Depreciation
lm	Lumen
lm/sf	Lumen per Square Foot
lm/W	Lumen per Watt
L.O.A.	Length Over All
log	Logarithm
L-O-L	Lateralolet
L.P.	Liquefied Petroleum; Low Pressure
L.P.F.	Low Power Factor
LR	Long Radius
L.S.	Lump Sum
Lt.	Light
Lt. Ga.	Light Gauge
L.T.L.	Less than Truckload Lot
Lt. Wt.	Lightweight
L.V.	Low Voltage
M	Thousand; Material; Male; Light Wall Copper Tubing
M^2CA	Meters Squared Contact Area
m/hr; M.H.	Man-hour
mA	Milliampere
Mach.	Machine
Mag. Str.	Magnetic Starter
Maint.	Maintenance
Marb.	Marble Setter
Mat; Mat'l.	Material
Max.	Maximum
MBF	Thousand Board Feet
MBH	Thousand BTU's per hr.
MC	Metal Clad Cable
M.C.F.	Thousand Cubic Feet
M.C.F.M.	Thousand Cubic Feet per Minute
M.C.M.	Thousand Circular Mils
M.C.P.	Motor Circuit Protector
MD	Medium Duty
M.D.O.	Medium Density Overlaid
Med.	Medium
MF	Thousand Feet
M.F.B.M.	Thousand Feet Board Measure
Mfg.	Manufacturing
Mfrs.	Manufacturers
mg	Milligram
MGD	Million Gallons per Day
MGPH	Thousand Gallons per Hour
MH, M.H.	Manhole; Metal Halide; Man-Hour
MHz	Megahertz
Mi.	Mile
MI	Malleable Iron; Mineral Insulated
mm	Millimeter
Mill.	Millwright
Min., min.	Minimum, minute
Misc.	Miscellaneous
ml	Milliliter, Mainline
M.L.F.	Thousand Linear Feet
Mo.	Month
Mobil.	Mobilization
Mog.	Mogul Base
MPH	Miles per Hour
MPT	Male Pipe Thread
MRT	Mile Round Trip
ms	Millisecond
M.S.F.	Thousand Square Feet
Mstz.	Mosaic & Terrazzo Worker
M.S.Y.	Thousand Square Yards
Mtd.	Mounted
Mthe.	Mosaic & Terrazzo Helper
Mtng.	Mounting
Mult.	Multi; Multiply
M.V.A.	Million Volt Amperes
M.V.A.R.	Million Volt Amperes Reactance
MV	Megavolt
MW	Megawatt
MXM	Male by Male
MYD	Thousand Yards
N	Natural; North
nA	Nanoampere
NA	Not Available; Not Applicable
N.B.C.	National Building Code
NC	Normally Closed
N.E.M.A.	National Electrical Manufacturers Assoc.
NEHB	Bolted Circuit Breaker to 600V.
N.L.B.	Non-Load-Bearing
NM	Non-Metallic Cable
nm	Nanometer
No.	Number
NO	Normally Open
N.O.C.	Not Otherwise Classified
Nose.	Nosing
N.P.T.	National Pipe Thread
NQOD	Combination Plug-on/Bolt on Circuit Breaker to 240V.
N.R.C.	Noise Reduction Coefficient
N.R.S.	Non Rising Stem
ns	Nanosecond
nW	Nanowatt
OB	Opposing Blade
OC	On Center
OD	Outside Diameter
O.D.	Outside Dimension
ODS	Overhead Distribution System
O.G.	Ogee
O.H.	Overhead
O&P	Overhead and Profit
Oper.	Operator
Opng.	Opening
Orna.	Ornamental
OSB	Oriented Strand Board

Abbreviations

O.S.&Y.	Outside Screw and Yoke
Ovhd.	Overhead
OWG	Oil, Water or Gas
Oz.	Ounce
P.	Pole; Applied Load; Projection
p.	Page
Pape.	Paperhanger
P.A.P.R.	Powered Air Purifying Respirator
PAR	Parabolic Reflector
Pc., Pcs.	Piece, Pieces
P.C.	Portland Cement; Power Connector
P.C.F.	Pounds per Cubic Foot
P.C.M.	Phase Contrast Microscopy
P.E.	Professional Engineer; Porcelain Enamel; Polyethylene; Plain End
Perf.	Perforated
Ph.	Phase
P.I.	Pressure Injected
Pile.	Pile Driver
Pkg.	Package
Pl.	Plate
Plah.	Plasterer Helper
Plas.	Plasterer
Pluh.	Plumbers Helper
Plum.	Plumber
Ply.	Plywood
p.m.	Post Meridiem
Pntd.	Painted
Pord.	Painter, Ordinary
pp	Pages
PP; PPL	Polypropylene
P.P.M.	Parts per Million
Pr.	Pair
P.E.S.B.	Pre-engineered Steel Building
Prefab.	Prefabricated
Prefin.	Prefinished
Prop.	Propelled
PSF; psf	Pounds per Square Foot
PSI; psi	Pounds per Square Inch
PSIG	Pounds per Square Inch Gauge
PSP	Plastic Sewer Pipe
Pspr.	Painter, Spray
Psst.	Painter, Structural Steel
P.T.	Potential Transformer
P. & T.	Pressure & Temperature
Ptd.	Painted
Ptns.	Partitions
Pu	Ultimate Load
PVC	Polyvinyl Chloride
Pvmt.	Pavement
Pwr.	Power
Q	Quantity Heat Flow
Quan.;Qty.	Quantity
Q.C.	Quick Coupling
r	Radius of Gyration
R	Resistance
R.C.P.	Reinforced Concrete Pipe
Rect.	Rectangle
Reg.	Regular
Reinf.	Reinforced
Req'd.	Required
Res.	Resistant
Resi.	Residential
Rgh.	Rough
RGS	Rigid Galvanized Steel
R.H.W.	Rubber, Heat & Water Resistant; Residential Hot Water
rms	Root Mean Square
Rnd.	Round
Rodm.	Rodman
Rofc.	Roofer, Composition
Rofp.	Roofer, Precast
Rohe.	Roofer Helpers (Composition)
Rots.	Roofer, Tile & Slate
R.O.W.	Right of Way
RPM	Revolutions per Minute
R.S.	Rapid Start
Rsr	Riser
RT	Round Trip
S.	Suction; Single Entrance; South
SC	Screw Cover
SCFM	Standard Cubic Feet per Minute
Scaf.	Scaffold
Sch.; Sched.	Schedule
S.C.R.	Modular Brick
S.D.	Sound Deadening
S.D.R.	Standard Dimension Ratio
S.E.	Surfaced Edge
Sel.	Select
S.E.R.; S.E.U.	Service Entrance Cable
S.F.	Square Foot
S.F.C.A.	Square Foot Contact Area
S.F. Flr.	Square Foot of Floor
S.F.G.	Square Foot of Ground
S.F. Hor.	Square Foot Horizontal
S.F.R.	Square Feet of Radiation
S.F. Shlf.	Square Foot of Shelf
S4S	Surface 4 Sides
Shee.	Sheet Metal Worker
Sin.	Sine
Skwk.	Skilled Worker
SL	Saran Lined
S.L.	Slimline
Sldr.	Solder
SLH	Super Long Span Bar Joist
S.N.	Solid Neutral
S-O-L	Socketolet
sp	Standpipe
S.P.	Static Pressure; Single Pole; Self-Propelled
Spri.	Sprinkler Installer
spwg	Static Pressure Water Gauge
S.P.D.T.	Single Pole, Double Throw
SPF	Spruce Pine Fir
S.P.S.T.	Single Pole, Single Throw
SPT	Standard Pipe Thread
Sq.	Square; 100 Square Feet
Sq. Hd.	Square Head
Sq. In.	Square Inch
S.S.	Single Strength; Stainless Steel
S.S.B.	Single Strength B Grade
sst	Stainless Steel
Sswk.	Structural Steel Worker
Sswl.	Structural Steel Welder
St.;Stl.	Steel
S.T.C.	Sound Transmission Coefficient
Std.	Standard
STK	Select Tight Knot
STP	Standard Temperature & Pressure
Stpi.	Steamfitter, Pipefitter
Str.	Strength; Starter; Straight
Strd.	Stranded
Struct.	Structural
Sty.	Story
Subj.	Subject
Subs.	Subcontractors
Surf.	Surface
Sw.	Switch
Swbd.	Switchboard
S.Y.	Square Yard
Syn.	Synthetic
S.Y.P.	Southern Yellow Pine
Sys.	System
t.	Thickness
T	Temperature; Ton
Tan	Tangent
T.C.	Terra Cotta
T & C	Threaded and Coupled
T.D.	Temperature Difference
Tdd	Telecommunications Device for the Deaf
T.E.M.	Transmission Electron Microscopy
TFE	Tetrafluoroethylene (Teflon)
T. & G.	Tongue & Groove; Tar & Gravel
Th.;Thk.	Thick
Thn.	Thin
Thrded	Threaded
Tilf.	Tile Layer, Floor
Tilh.	Tile Layer, Helper
THHN	Nylon Jacketed Wire
THW.	Insulated Strand Wire
THWN;	Nylon Jacketed Wire
T.L.	Truckload
T.M.	Track Mounted
Tot.	Total
T-O-L	Threadolet
T.S.	Trigger Start
Tr.	Trade
Transf.	Transformer
Trhv.	Truck Driver, Heavy
Trlr	Trailer
Trlt.	Truck Driver, Light
TTY	Teletypewriter
TV	Television
T.W.	Thermoplastic Water Resistant Wire
UCI	Uniform Construction Index
UF	Underground Feeder
UGND	Underground Feeder
U.H.F.	Ultra High Frequency
U.L.	Underwriters Laboratory
Unfin.	Unfinished
URD	Underground Residential Distribution
US	United States
USP	United States Primed
UTP	Unshielded Twisted Pair
V	Volt
V.A.	Volt Amperes
V.C.T.	Vinyl Composition Tile
VAV	Variable Air Volume
VC	Veneer Core
Vent.	Ventilation
Vert.	Vertical
V.F.	Vinyl Faced
V.G.	Vertical Grain
V.H.F.	Very High Frequency
VHO	Very High Output
Vib.	Vibrating
V.L.F.	Vertical Linear Foot
Vol.	Volume
VRP	Vinyl Reinforced Polyester
W	Wire; Watt; Wide; West
w/	With
W.C.	Water Column; Water Closet
W.F.	Wide Flange
W.G.	Water Gauge
Wldg.	Welding
W. Mile	Wire Mile
W-O-L	Weldolet
W.R.	Water Resistant
Wrck.	Wrecker
W.S.P.	Water, Steam, Petroleum
WT., Wt.	Weight
WWF	Welded Wire Fabric
XFER	Transfer
XFMR	Transformer
XHD	Extra Heavy Duty
XHHW; XLPE	Cross-Linked Polyethylene Wire Insulation
XLP	Cross-linked Polyethylene
Y	Wye
yd	Yard
yr	Year
Δ	Delta
%	Percent
~	Approximately
Ø	Phase
@	At
#	Pound; Number
<	Less Than
>	Greater Than

Index

Index

Notes

Notes

Notes

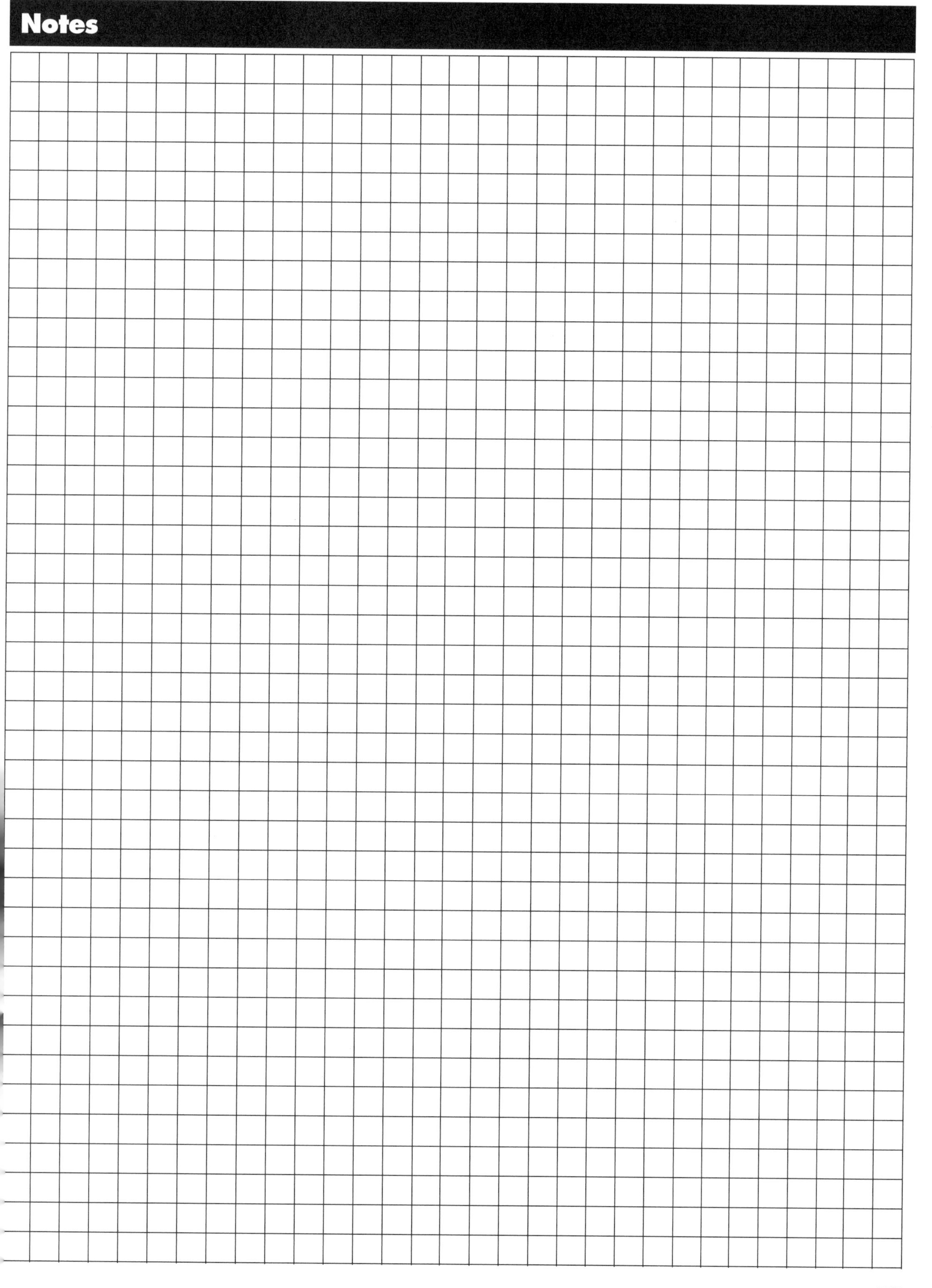

Notes

Notes

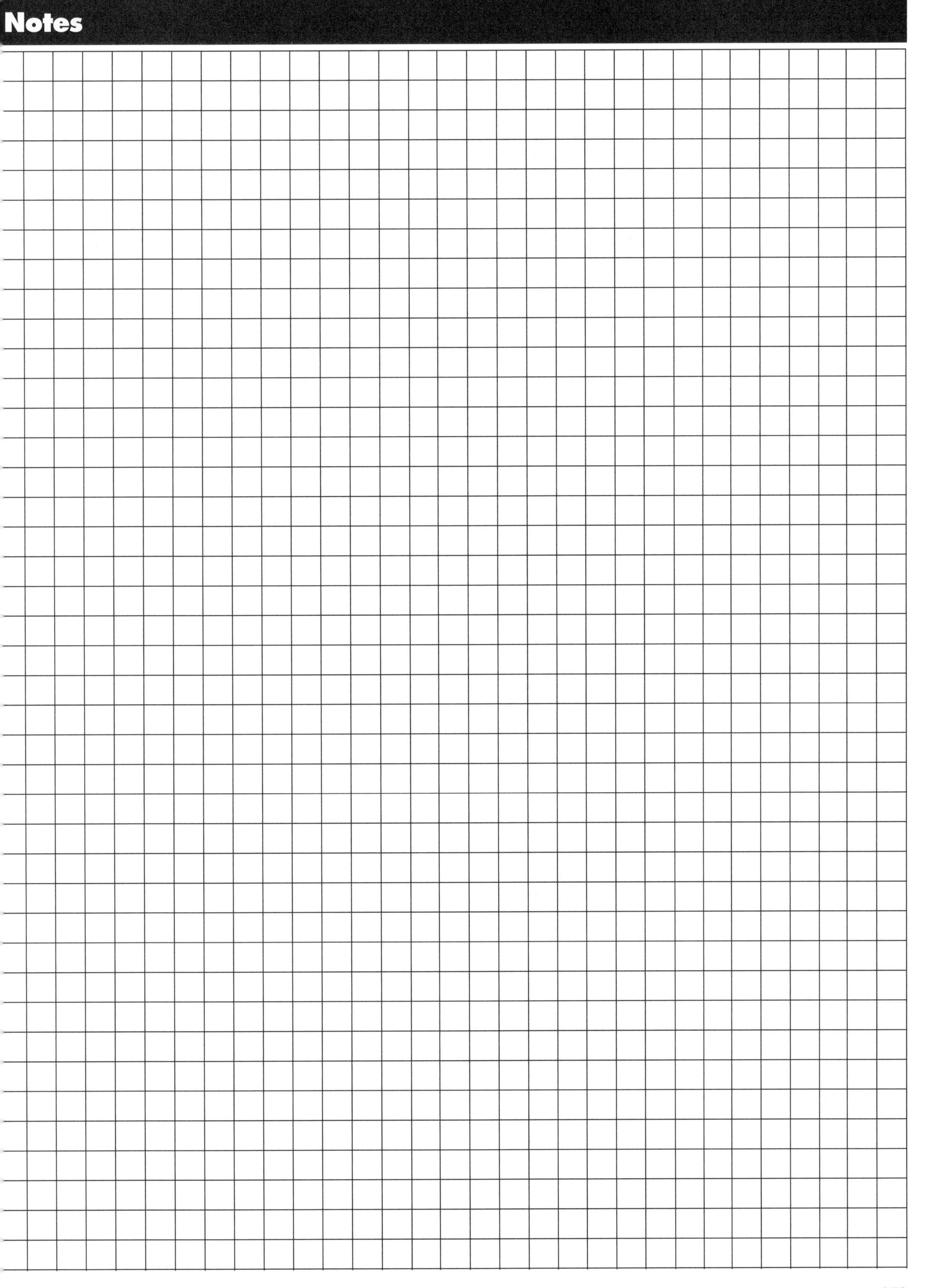

Notes

Notes

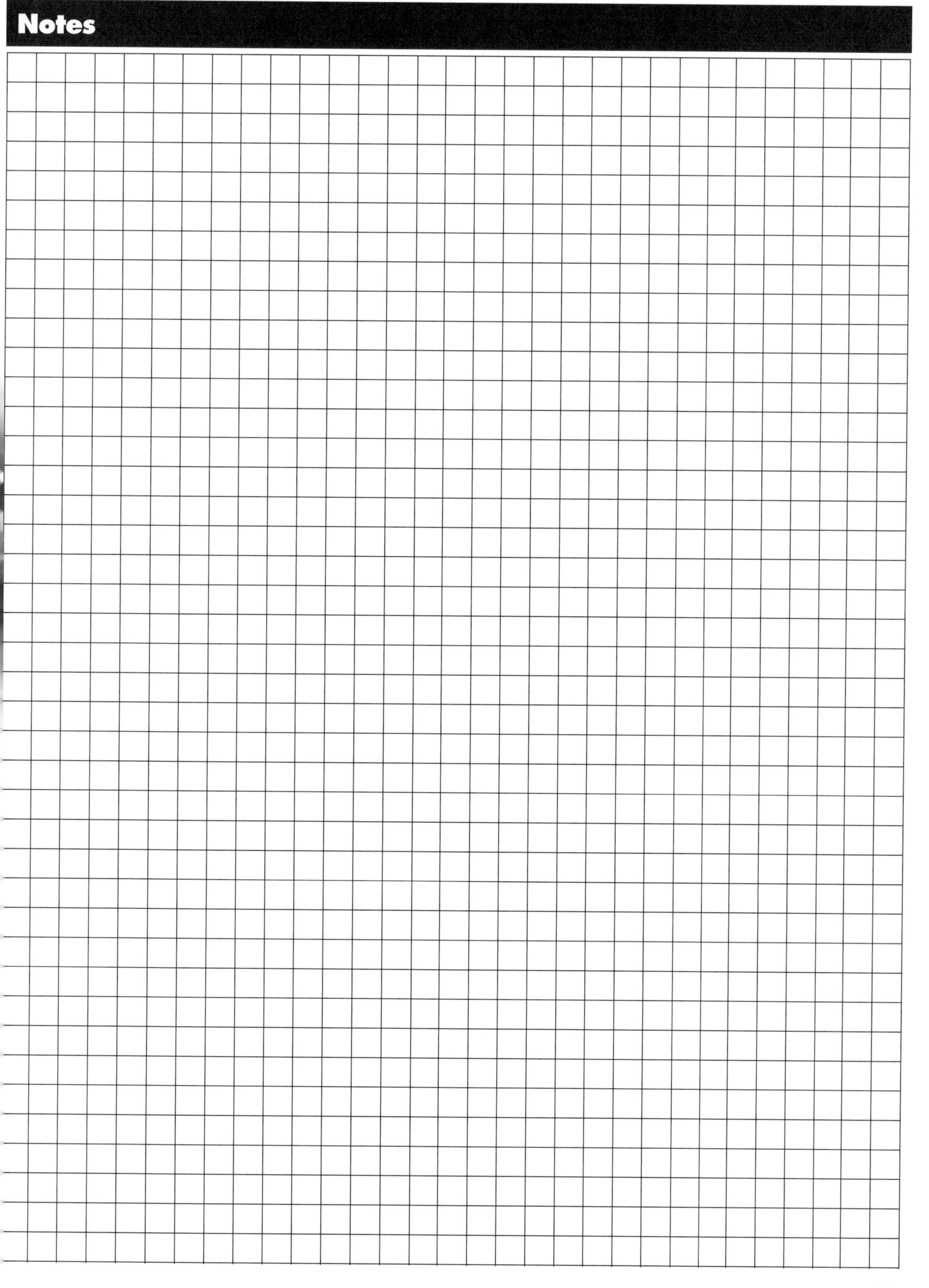

Notes

Notes

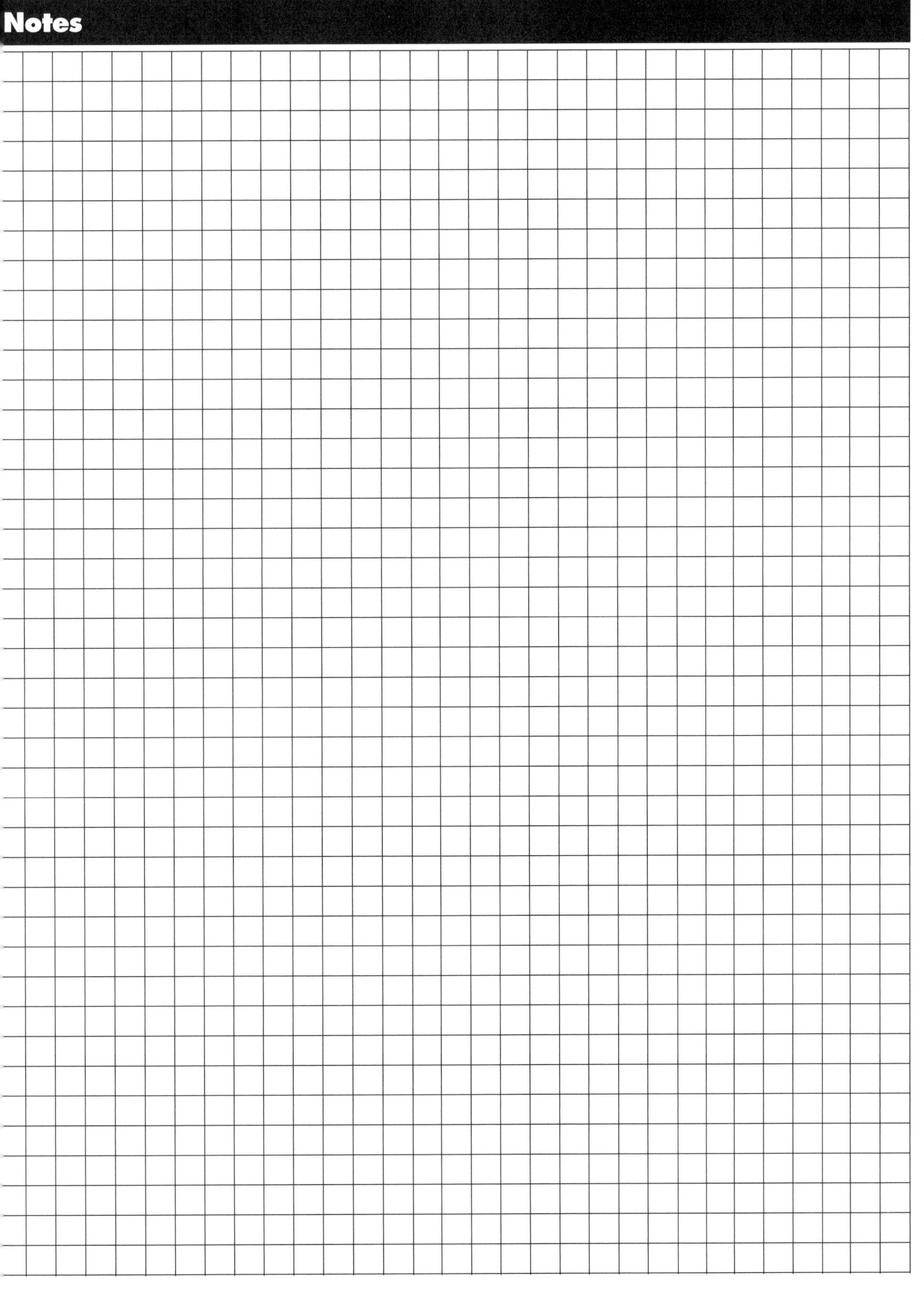

Notes

Notes

Reed Construction Data, Inc.

Reed Construction Data, Inc., a leading worldwide provider of total construction information solutions, is comprised of three main product groups designed specifically to help construction professionals advance their businesses with timely, accurate and actionable project, product and cost data. Reed Construction Data is a division of Reed Business Information, a member of the Reed Elsevier plc group of companies.

The Project, Product and Cost & Estimating divisions offer a variety of innovative products and services designed for the full spectrum of design, construction and manufacturing professionals. Through its International companies, Reed Construction Data's reputation for quality construction market data is growing worldwide.

Cost Information
RSMeans, the undisputed market leader and authority on construction costs, publishes current cost and estimating information in annual cost books and on the *Means CostWorks* CD-ROM. RSMeans furnishes the construction industry with a rich library of corresponding reference books and a series of professional seminars that are designed to sharpen professional skills and maximize the effective use of cost estimating and management tools. RSMeans also provides construction cost consulting for owners, manufacturers, designers, and contractors.

Project Data
Reed Construction Data provides complete, accurate and relevant project information through all stages of construction. Customers are supplied industry data through leads, project reports, contact lists, plans and specifications surveys, market penetration analyses and sales evaluation reports. Any of these products can pinpoint a county, look at a state, or cover the country. Data is delivered via paper, e-mail, CD-ROM or the Internet.

Building Product Information
The First Source suite of products is the only integrated building product information system offered to the commercial construction industry for comparing and specifying building products. These print and online resources include *First Source,* CSI's SPEC-DATA™, CSI's MANU-SPEC™, First Source CAD, and Manufacturer Catalogs. Written by industry professionals and organized using CSI's MasterFormat™, construction professionals use this information to make better design decisions. FirstSourceONL.com combines Reed Construction Data's project, product and cost data with news and information from Reed Business Information's *Building Design & Construction* and *Consulting-Specifying Engineer magazines.* This industry-focused site offers easy and unlimited access to vital information for all construction professionals.

International
Reed Construction Data Canada serves the Canadian construction market with reliable and comprehensive project and product information services that cover all facets of construction. Core services include: *BuildSource, BuildSpec, BuildSelect,* product selection and specification tools available in print and on the Internet; Building Reports, a national construction project lead service; CanaData, statistical and forecasting information; *Daily Commercial News,* a construction newspaper reporting on news and projects in Ontario; and *Journal of Commerce,* reporting news in British Columbia and Alberta.

BIMSA/Mexico provides construction project news, product information, cost data, seminars and consulting services to construction professionals in Mexico. Its subsidiary, PRISMA, provides job costing software.

Byggfakta Scandinavia AB, founded in 1936, is the parent company for the leaders of customized construction market data for Denmark, Estonia, Finland, Norway and Sweden. Each company fully covers the local construction market and provides information across several platforms including via subscription, on an ad-hoc basis, electronically and on paper.

Cordell Building Information Services, with its complete range of project and cost and estimating services, is Australia's specialist in the construction information industry. Cordell provides in-depth and historical information on all aspects of construction projects and estimation, including several customized reports, construction and sales leads and detailed cost information.

For more information, please visit our Web site at www.reedconstructiondata.com.

Reed Construction Data, Inc., Corporate Office
30 Technology Parkway South
Norcross, GA 30092-2912
(800) 322-6996
(800) 895-8661 (fax)
info@reedbusiness.com
www.reedconstructiondata.com

Reed Construction Data®

For more information
visit Means Web site
at www.rsmeans.com

Annual Cost Guides

Means Building Construction Cost Data 2006

Offers you unchallenged unit price reliability in an easy-to-use arrangement. Whether used for complete, finished estimates or for periodic checks, it supplies more cost facts better and faster than any comparable source. Over 23,000 unit prices for 2006. The City Cost Indexes now cover over 930 areas, for indexing to any project location in North America.

$126.95 per copy
Over 700 pages, softcover
Catalog No. 60016 ISBN 0-87629-786-6

Means Open Shop Building Construction Cost Data 2006

The open-shop version of the *Means Building Construction Cost Data.* More than 22,000 reliable unit cost entries based on open shop trade labc rates. Eliminates time-consuming searches for these prices. The first book with open shop labc rates and crews. Labor information is itemized b labor-hours, crew, hourly/daily output, equipment, overhead and profit. For contractors owners and facility managers.

$126.95 per copy
Over 700 pages, softcover
Catalog No. 60156 ISBN 0-87629-797-1

Means Plumbing Cost Data 2006

Comprehensive unit prices and assemblies for plumbing, irrigation systems, commercial and residential fire protection, point-of-use water heaters, and the latest approved materials. This publication and its companion, *Means Mechanical Cost Data,* provide full-range cost estimating coverage for all the mechanical trades.

$126.95 per copy
Over 550 pages, softcover
Catalog No. 60216 ISBN 0-87629-796-3

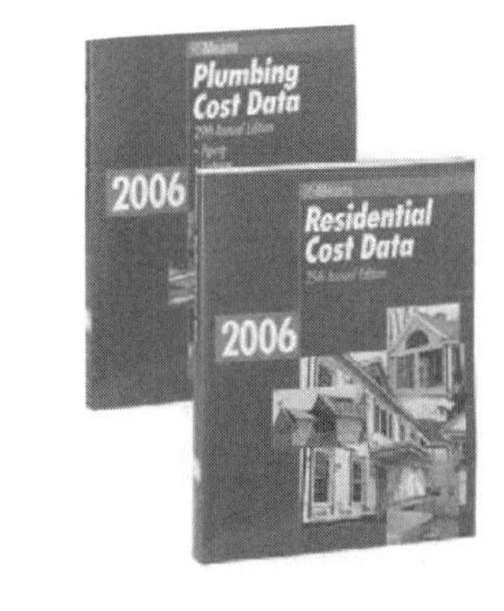

Means Residential Cost Data 2006

Speeds you through residential construction pricing with more than 100 illustrated complete house square-foot costs. Alternate assemblies cost selections are located on adjoining pages, s that you can develop tailor-made estimates in minutes. Complete data for detailed unit cost estimates is also provided.

$107.95 per copy
Over 600 pages, softcover
Catalog No. 60176 ISBN 0-87629-803-x

Means Electrical Cost Data 2006

Pricing information for every part of electrical cost planning: unit and systems costs with design tables; engineering guides and illustrated estimating procedures; complete labor-hour, materials, and equipment costs for better scheduling and procurement. With the latest products and construction methods used in electrical work. More than 15,000 unit and systems costs, clear specifications and drawings.

$126.95 per copy
Over 450 pages, softcover
Catalog No. 60036 ISBN 0-87629-790-4

Means Repair & Remodeling Cost Data 2006

Commercial/Residential

You can use this valuable tool to estimate commercial and residential renovation and remodeling. By using the specialized costs in thi manual, you'll find it's not necessary to force fit prices for new construction into remodeling cost planning. Provides comprehensive unit costs, building systems costs, extensive labor data an estimating assistance for every kind of building improvement.

$107.95 per copy
Over 650 pages, softcover
Catalog No. 60046 ISBN 0-87629-795-5

Means Site Work & Landscape Cost Data 2006

Hard-to-find costs are presented in an easy-to-use format for every type of site work and landscape construction. Costs are organized, described, and laid out for earthwork, utilities, roads and bridges, as well as grading, planting, lawns, trees, irrigation systems, and site improvements.

$126.95 per copy
Over 600 pages, softcover
Catalog No. 60286 ISBN 0-87629-792-0

Means Light Commercial Cost Data 2006

Specifically addresses the light commercial mar which is an increasingly specialized niche in the industry. Aids you, the owner/designer/contracto in preparing all types of estimates, from budget to detailed bids. Includes new advances in methods and materials. Assemblies section allov you to evaluate alternatives in early stages of design/planning.

$107.95 per copy
Over 650 pages, softcover
Catalog No. 60186 ISBN 0-87629-804-8